Recent Advances in GPR Imaging

Recent Advances in GPR Imaging

Special Issue Editors

Mercedes Solla
Susana Lagüela

MDPI • Basel • Beijing • Wuhan • Barcelona • Belgrade

MDPI

Special Issue Editors

Mercedes Solla
Spanish Naval Academy
Spain

Susana Lagüela
University of Salamanca
Spain

Editorial Office
MDPI
St. Alban-Anlage 66
4052 Basel, Switzerland

This is a reprint of articles from the Special Issue published online in the open access journal *Remote Sensing* (ISSN 2072-4292) from 2017 to 2018 (available at: https://www.mdpi.com/journal/remotesensing/special_issues/GPR_imaging)

For citation purposes, cite each article independently as indicated on the article page online and as indicated below:

LastName, A.A.; LastName, B.B.; LastName, C.C. Article Title. *Journal Name* **Year**, *Article Number*, Page Range.

ISBN 978-3-03921-810-3 (Pbk)
ISBN 978-3-03921-811-0 (PDF)

Cover image courtesy of Mercedes Solla.

Contents

About the Special Issue Editors

Mercedes Solla received her Ph.D. degree from the University of Vigo, Spain, in 2010. Her research interests include the use of Ground Penetrating Radar (GPR) for subsurface prospection in archaeology, cultural heritage, civil engineering, GPR signal processing and forward modelling, as well as 3D visualization in Geographic Information System (GIS) environments. In 2009 and 2012, she was a visiting researcher at the Institute for Infrastructure and Environment of the University of Edinburgh, collaborating with the Edinburgh Parallel Computing Center (EPCC) (UK); in 2015 she was a visiting researcher at the Transportation Department of the National Laboratory for Civil Engineering (LNEC) (PT). She has more than 200 publications in peer-refereed journals, book chapters, conference proceedings, and technical reports (with more than 50 scientific papers on prestigious international journals included in the JCR database), and she is co-inventor of a patent. She was the supervisor of 5 Ph.D. theses. She was an active Member of COST Action TU1208 "Civil Engineering Applications of Ground Penetrating Radar" and COST Action SAGA "CA17131—The Soil Science & Archaeo-Geophysics Alliance: going beyond prospection".

Susana Lagüela is the head of the Cátedra Iberdrola VIII Centenary of the University of Salamanca, leading research projects with the international energy company Iberdrola for the inclusion of solar energy in the energy network in Europe, especially in Spain. She is a visiting researcher at UC Berkeley (USA), ETH Zurich (Switzerland), ITC-CNR Padova (Italy), and Delft University of Technology (The Netherlands). As a researcher in European and Spanish projects, she has authored more than 50 scientific papers about energy efficiency and energy 3D modelling of buildings, thermal infrared image processing for the computation of thermal balances, and integration of renewable energy resources for the conversion of the energy network to a network based on distributed resources.

Preface to "Recent Advances in GPR Imaging"

Ground Penetrating Radar (GPR) is a close-range remote sensing and non-invasive geophysical technique that uses electromagnetic pulses to detect buried features and to identify their main geometric and physical properties. The GPR method is commonly employed to provide high-resolution imaging of subsurfaces and buried artifacts and to assess the inner status of structures.

During the last few decades, major advances have been made in the development of GPR array multi-channel systems, airborne platforms, and the use of three-dimensional imaging techniques and processing software. Recent trends also show an increasing interest for the development of new signal processing algorithms and modeling. New approaches focusing on the combined application of GPR with complementary non-destructive techniques are also recommended for high-resolution prospection.

This Special Issue is mainly dedicated to publishing a selection of papers that provide a comprehensive and up-to-date overview of the state-of-the-art research activities dealing with the development of GPR technology and its recent advances on imaging in different fields of application.

Mercedes Solla, Susana Lagüela
Special Issue Editors

remote sensing

MDPI

Editorial

Editorial for Special Issue "Recent Advances in GPR Imaging"

Mercedes Solla [1,2,*] and Susana Lagüela [2,3]

1   Defense University Center, Spanish Naval Academy, Plaza de España s/n, 36920 Marín, Pontevedra, Spain
2   Applied Geotechnologies Research Group, University of Vigo, Rúa Maxwell s/n, Campus
    Lagoas-Marcosende, 36310 Vigo, Spain; sulaguela@usal.es
3   Department of Cartographic and Terrain Engineering, University of Salamanca, Calle Hornos Caleros 50,
    05003 Ávila, Spain
*   Correspondence: merchisolla@cud.uvigo.es

Received: 24 April 2018; Accepted: 25 April 2018; Published: 26 April 2018

The Special Issue (SI) "Recent Advances in GPR Imaging" offers an up-to-date overview of the state of the art of research activities dealing with the development of Ground-Penetrating Radar (GPR) technology and its recent advances on imaging in different fields of application. In fact, the advances experimented during the last decades with regard to the appearance of new GPR systems and of the need to manage large amounts of data have implied an increasing interest in the development of new signal processing algorithms and modeling, as well as in the use of three-dimensional (3D) imaging techniques.

Most of the works present in this SI can be categorized according to their relevant application fields.

The understanding of the GPR data has been a long-term challenge among both the scientific and the non-geophysical community. In this frame, the development of new data processing algorithms and electromagnetic modeling has benefited the interpretation process of field GPR data. The paper from Salinas et al. [1] presents a processing technique to determine the Mean Amplitude of Incoherent Energy (MAEI) for each A-scan, which was applied to the study of the shallow geology in Barcelona and allows zone differentiation (underground streams and paleochannels) depending on the amplitude of the clutter caused by backscattering. Additionally, complementary numerical modeling and passive seismic measurements were applied in order to validate the proposed processing methodology. Prokopovich et al.'s study [2] deals with the development of a time-domain version of the coupled-wave Wentzel–Kramers–Brillouin approximation for the solution of the backscattering problem arising when a pulsed electromagnetic signal impinges on a non-uniform dielectric half-space. The results obtained were compared with those from Finite-Difference Time-Domain (FDTD) modeling, showing very good agreement, which demonstrates the capabilities of the new method to correctly predict the protracted return signals originated by smooth transition layers of the subsurface dielectric medium. Another example of a novel GPR data processing is shown in Fontul et al.'s paper [3], which presents a new approach for the automatic detection of signal variations based mainly on expedite frequency-domain analysis of the GPR signal. Case studies are included with the application of the new approach to railway assessment, with the identification of track events, ballast interventions, and potential locations of malfunctions.

In addition to new algorithms for signal processing and automatic detection, improvements on imaging and interpretation approaches are still needed. In Zhang et al. [4], a new method for GPR imaging based on the Variational Mode Decomposition (VMD) and the Intrinsic Mode Functions (IMFs) is proposed. In this method, the IMFs are generated trace by trace by the VMD, and then these IMFs are sorted and displayed into different profiles (IMF-slices) according to different frequency bands or ranges. Using IMF-slices, some subsurface events could be more clearly identified.

In Dérobert and Pajewski's study [5], a wide dataset of GPR data collected on a full-size geophysical test site in Nantes (France) is presented. The geophysical test site was built to reproduce an urban site (including pipes, cables, stones of various size, and masonry) in a completely controlled environment. A total of 67 profiles were recorded using three different pulsed radar systems equipped with various antenna frequencies from 200 MHz to 900 MHz. An archive containing all the profiles (in raw data) is enclosed to this paper as supplementary material. This dataset is part of the Open Database of Radargrams initiative of COST Action TU1208 "Civil Engineering Applications of Ground-Penetrating Radar" with the aim of providing unified material for the evaluation of processing methodologies and allowing intercomparisons.

Provided the complexity of the interpretation of the measured data (2D-GPR images), the use of 3D imaging techniques advances in the generation of more realistic images of the underground structures. 3D imaging is particularly relevant for archaeological investigations, allowing not only the discovery, but also the 3D reconstruction of buried structures for a more comprehensive archaeological interpretation. Puente et al.'s study [6] deals with the 3D reconstruction of the Roman fort "*Aquis Querquennis*", in Spain, through the combination of three different non-destructive techniques: GPR, Terrestrial Light Detection and Ranging (T-LiDAR), and Infrared Thermography (IRT). Moreover, a novel processing and 3D imaging software "toGPRi" is presented for the creation of the 3D model and the subsequent time-slices at different depths and overlaid imaging. This 3D GPR imaging is georeferenced and then merged with the orthoimages produced by the T-LiDAR, which allowed for a complete interpretation of the Roman site, including its surface geometry.

New approaches focused on the combined application of GPR with complementary non-destructive testing techniques are also recommended for high-resolution prospection. An example of integrated geophysical techniques is shown in Martínez et al. [7], in which GPR and Electrical Resistivity Imaging (ERI) methods were successfully combined for assessing the quality of ornamental rock (marble) in Macael, Spain. In Čeru et al. [8], the GPR was used to select dolines appropriate for further morphometrical and distributive analyses on LiDAR images applied to the study of geomorphological dating of Pleistocene conglomerates in Central Slovenia. The paper by Čeru et al. [9] demonstrates the capability of the GPR method to locate areas of cave sediments at the surface and to determine their spatial extent, which allowed delineating the geometry of unroofed cave systems in Lanski vrh (W Solovenia). Complementary X-ray diffraction (XRD) and X-ray fluorescence (XRF) analyses were performed to analyze the mineral and geochemical compositions of the cave sediments and soils in order to determine which factors might significantly influence the GPR signal propagation.

Acknowledgments: We would like to thank the authors who contributed to this Special Issue and to the reviewers who dedicated their time for providing the authors with very valuable and constructive recommendations.

Conflicts of Interest: The authors declare no conflict of interest.

References

1. Salinas, V.; Santos-Assunçao, S.; Pérez-Gracia, V. GPR Clutter Amplitude Processing to Detect Shallow Geological Targets. *Remote Sens.* **2018**, *10*, 88. [CrossRef]
2. Prokopovich, I.; Popov, A.; Pajewski, L.; Marciniak, M. Application of Coupled-Wave Wentzel-Kramers-Brillouin Approximation to Ground Penetrating Radar. *Remote Sens.* **2018**, *10*, 22. [CrossRef]
3. Fontul, S.; Paixão, A.; Solla, M.; Pajewski, L. Railway Track Condition Assessment at Network Level by Frequency Domain Analysis of GPR Data. *Remote Sens.* **2018**, *10*, 559. [CrossRef]
4. Zhang, X.; Nilot, E.; Feng, X.; Ren, Q.; Zhang, Z. IMF-Slices for GPR Data Processing Using Variational Mode Decomposition Method. *Remote Sens.* **2018**, *10*, 476. [CrossRef]
5. Dérobert, X.; Pajewski, L. TU1208 Open Database of Radargrams: The Dataset of the IFSTTAR Geophysical Test Site. *Remote Sens.* **2018**, *10*, 530. [CrossRef]
6. Puente, I.; Solla, M.; Lagüela, S.; Sanjurjo-Pinto, J. Reconstructing the Roman Site "*Aquis Querquennis*" (Bande, Spain) from GPR, T-LiDAR and IRT Data Fusion. *Remote Sens.* **2018**, *10*, 379. [CrossRef]

7.   Martínez, J.; Montiel, V.; Rey, J.; Cañadas, F.; Vera, P. Utilization of Integrated Geophysical Techniques to Delineate the Extraction of Mining Bench of Ornamental Rocks (Marble). *Remote Sens.* **2017**, *9*, 1322. [CrossRef]

8.   Čeru, T.; Šegina, E.; Gosar, A. Geomorphological Dating of Pleistocene Conglomerates in Central Slovenia Based on Spatial Analyses of Dolines Using LiDAR and Ground Penetrating Radar. *Remote Sens.* **2017**, *9*, 1213. [CrossRef]

9.   Čeru, T.; Dolenec, M.; Gosar, A. Application of Ground Penetrating Radar Supported by Mineralogical-Geochemical Methods for Mapping Unroofed Cave Sediments. *Remote Sens.* **2018**, *10*, 639. [CrossRef]

remote sensing

MDPI

Article

GPR Clutter Amplitude Processing to Detect Shallow Geological Targets

Victor Salinas Naval [1], Sonia Santos-Assunçao [2] and Vega Pérez-Gracia [3],*

[1] Geophysical Technician, World Sensing Barcelona, C/Viriat 47, 08014 Barcelona, Spain; vsalinas@worldsensing.com
[2] Department of Fluid Mechanics, Campus Diagonal Besòs—Edifici A (EEBE), Universitat Politècnica de Catalunya-BarcelonaTech, Av. Eduard Maristany, 16 08019 Barcelona, Spain; sonia.assuncao@upc.edu
[3] Department of Strength of Materials and Structural Engineering, Campus Diagonal Besòs—Edifici A (EEBE), Universitat Politècnica de Catalunya-BarcelonaTech, Av. Eduard Maristany, 16 08019 Barcelona, Spain
* Correspondence: vega.perez@upc.edu

Received: 26 October 2017; Accepted: 7 January 2018; Published: 11 January 2018

Abstract: The analysis of clutter in A-scans produced by energy randomly scattered in some specific geological structures, provides information about changes in the shallow sedimentary geology. The A-scans are composed by the coherent energy received from reflections on electromagnetic discontinuities and the incoherent waves from the scattering in small heterogeneities. The reflected waves are attenuated as consequence of absorption, geometrical spreading and losses due to reflections and scattering. Therefore, the amplitude of those waves diminishes and at certain two-way travel times becomes on the same magnitude as the background noise in the radargram, mainly produced by the scattering. The amplitude of the mean background noise is higher when the dispersion of the energy increases. Then, the mean amplitude measured in a properly selected time window is a measurement of the amount of the scattered energy and, therefore, a measurement of the increase of scatterers in the ground. This paper presents a simple processing that allows determining the Mean Amplitude of Incoherent Energy (MAEI) for each A-scan, which is represented in front of the position of the trace. This procedure is tested in a field study, in a city built on a sedimentary basin. The basin is crossed by a large number of hidden subterranean streams and paleochannels. The sedimentary structures due to alluvial deposits produce an amount of the random backscattering of the energy that is measured in a time window. The results are compared along the entire radar line, allowing the location of streams and paleochannels. Numerical models were also used in order to compare the synthetic traces with the field radargrams and to test the proposed processing methodology. The results underscore the amount of the MAEI over the streams and also the existence of a surrounding zone where the amplitude is increasing from the average value to the maximum obtained over the structure. Simulations show that this zone does not correspond to any particular geological change but is consequence of the path of the antenna that receives the scattered energy before arriving to the alluvial deposits.

Keywords: GPR; clutter; backscattering; scattering modelling

1. Introduction

Ground Penetrating Radar (GPR) has been successfully applied to many different problems in shallow geological surveys as a complement to provide answers or help in resolutions. The theoretical principles and applications of the technique can be found in the literature (see, for example, [1–3]). Usual applications in shallow geology are defining geological structures in the first meters (e.g., [4–6]), mapping groundwater (e.g., [7,8]), detecting cavities (e.g., [9,10]) and defining shallow fractures (e.g., [11,12]). Data acquisition is mainly done with a common fixed offset mode and processing is focused on improving the signal to noise ratio and building 2D and 3D images of the subsurface features.

The main goal of those applications is the detection of reflective targets (normally consisting on cavities, fractures and changes in the ground geology). However, in many cases, the anomalies are not clear enough in the radargrams due to the high noise level caused by the dispersion of the energy and the scattering on the surface of the targets (e.g., [13–15]), leading to misinterpretations. The main efforts are focused on improving the signal to noise ratio to obtain high quality radar images (e.g., [16–18]). In quaternary deposits or in alluvial soils there are two possible causes of those unclear records. One cause of blurred A-scans is related to the small difference in the electromagnetic parameters of the targets and the surrounding medium that may lead to inconclusive and uncertain B-scans. In these cases, the amplitude of the anomalies is small, becoming very difficult of even impossible to distinguish between the signals and the background noise. The second cause is associated to irregular materials. Heterogeneities as granular cluster of material with variable size and shape, or small irregularities on the contact between two media, scatterers the energy, blurring the radar images. This effect introduces undesired clutter in the images superimposed to the anomalies. For that reason, the resultant images are noisy and may lead to misinterpretation. This clutter is commonly observed in the analysis of shallow quaternary geology [19], in the assessment of civil structures where pathologies disintegrate specific zones of the media [20] and in fragmented rocks [21]. Therefore, this clutter might be related to physical characteristics of the media, such as size grain distribution, fissures, small voids and other heterogeneities. The objective of this paper is to present and apply a methodology to extract information from that clutter and to map zones based on changes in random backscattering amplitude. Consequently, the backscattering phenomenon detected in the study of the shallow geology in Barcelona is studied and simulated. Previous laboratory and field tests have been done to validate the use of changes in clutter to detect changes in the ground ([19,22,23]), allowing mapping areas with clusters and scatterers. This phenomenon was also applied by other authors to evaluate ballast contamination [20], differentiating the zones depending on the amplitude of the clutter caused by backscattering.

2. Backscattering of GPR Waves

Scattering of electromagnetic waves can be elastic or non-elastic. In the first case, the frequency of the scattered waves is the same than the frequency of the incident energy while in the second case part of the energy is absorbed producing ionized radiation. The scattering in GPR surveys is always elastic since there is no change in energy and wavelength. Depending on the size of particles and roughness of the targets' surface, the scattering can be anisotropic (Mie scattering) or isotropic (Rayleigh scattering).

The presence of small targets in the medium produces scattering on the radar signal. Dispersion is caused by random reflections of a part of the incident energy on each one of the small targets and on heterogeneities embedded in the medium. Each one of the heterogeneities becomes a transmitter of part of the incident energy. Part of the scattered energy travels back in the direction of the transmitted energy. Backscattering is the diffuse reflection of the waves due to scattering, arriving to the surface of the medium.

As the transmitted energy decreases as a consequence of scattering, this phenomenon is usually considered one of the causes for the loss of transmitted energy ([14,24]). The amplitude of an electromagnetic wave that propagates through a homogeneous medium attenuates because of the geometrical spreading and the absorption and scattering of the energy (e.g., [25]) being the total attenuation of GPR signals caused by intrinsic attenuation and scattering attenuations [26].

The amplitude of the wave, A(r) at a certain distance r from the source, could be expressed as a function of the wave amplitude at the source and the losses as consequence of those attenuating effects:

$$A(r) = A_O \frac{1}{r} e^{-(\alpha+\mu)r} \tag{1}$$

Being A_O the amplitude of reference at the source, r the distance, α the coefficient of attenuation due to the absorption and μ the coefficient of scattering.

The coefficient of scattering depends on the number of scatterers per unit of volume, n and on the cross section of the scatterers, C_S.

$$\mu = \frac{n \, C_S}{2} \qquad (2)$$

Moreover, the effect of scattering is frequency dependent, because the cross section depends on the frequency of the incident field.

In general, the effect of the scatterers is small but increasing the amount of the small targets also increases the dispersed energy. Scattering produces two effects on the signal: the amplitude of the wave decreases because of the dispersed energy and the clutter recorded as background noise in the A-scans augments.

When the incident wave is perpendicular to a flat surface, the reflected wave returns with the same angle of incidence, being described the trajectory by the Law of Snell. However, in the case of a medium with a large number of scatterers which size is smaller than the wavelength, the scattering is described as the reflection of the wave in deviated trajectories in a random and not expected direction (Figure 1). This effect is similar in the case of incident waves on large rough surfaces.

Backscattering is defined as the amount of energy that is scattered back to the source and can be recorded by an antenna placed at the surface of the medium. The backscattered energy strongly depends on the size and the shape of the targets and big size targets can be considered as planar reflectors.

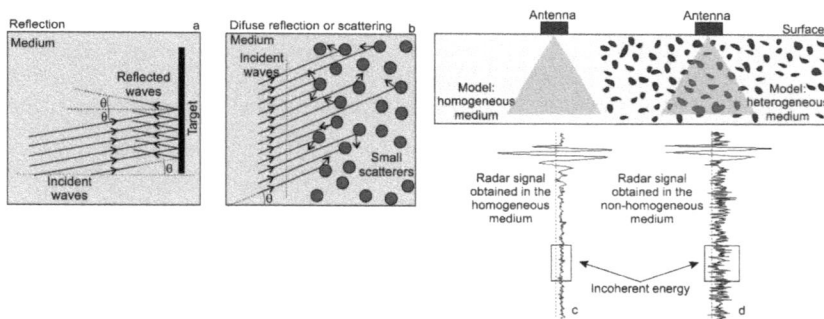

Figure 1. The clutter caused by small and irregular targets embedded in the medium increases the background noise in the A-scans. (**a**) Reflection in a flat target inside the medium; (**b**) Diffuse reflection is small scatterers embedded in the medium; (**c**) GPR A-scan obtained in homogeneous media; (**d**) GPR A-scan from a heterogeneous medium.

3. Heterogeneous Media and Targets with Rough Surfaces

The scattered energy depends on the heterogeneity of the medium or on the roughness of the targets or reflectors embedded in the medium. The Rayleigh criterion allows distinguishing between rough and smooth targets. In the case of a planar wave incident on an irregular and rough target with an angle α, the difference of phase, $\Delta\phi$, between two rays scattered from different points of the surface is:

$$\Delta\phi = 2h \, \frac{2\pi}{\lambda} \cos \beta \qquad (3)$$

Being h the difference of height between the two points of the rough surface of the target embedded in the medium where the scattering is produced, β the incidence angle and λ the wavelength of the electromagnetic wave.

The Rayleigh criterion considers a non-smooth target surface when the phase difference is:

$$\Delta\phi \geq \frac{\pi}{2} \tag{4}$$

In that case, the parameter h could be defined depending mainly on the frequency or wavelength of the incident wave:

$$h = \frac{\lambda}{8 \, \cos\beta} \tag{5}$$

The approach of nearly vertical incident waves allows to approximate the heterogeneity size h as:

$$h = \frac{\lambda}{8} \tag{6}$$

The phenomenon of dispersion and scattering in rough target surfaces has been analysed by several authors ([27–29]). The most usual approaches are the Kirchoff Approximation and the Small Perturbation Model (SPM). The Kirchoff approximation assume that the roughness dimensions of the target surface are large compared to the incident wavelength (Bragg scattering region), while the SPM model undertakes that the roughness is small compared to the wavelength. Therefore, the SPM model is more appropriate for long wavelengths as in the present GPR study, where the radiation is mainly in the I-band (IEEE bands), being the frequency lower than 200 MHz (which represent wavelengths up to 1.5 m). The approach assumes that the parameter h is small compared to the wavelength. In this case, the variation on the height of the targets' surface is small compared to the wavelength and depending on the incident angle, the SPM decomposes the backscattered waves into their Fourier spectrum components.

4. Mean Amplitude of Incoherent Energy

Experimental GPR measurements in areas where high scattering was expected showed an important increase of the clutter in all A-scans [15]. Therefore, in order to relate the clutter with the heterogeneities of the medium, a procedure was developed to enhance the clutter associated to backscattered energy in dispersive targets, applied to the field surveys. The diffuse reflection in these heterogeneities increases the background noise (clutter) in GPR A-scans (see Figure 1). The analysis of the amplitude of this background energy could provide alternative information about changes in the ground, related to the homogeneity of the materials.

In the case of scattering in small particles or in rough surface, it is possible to assume that, for a specific time t, the total electromagnetic field arriving to the receiver can be expressed as the sum of the reflected (E_r) and the scattered (E_s) fields:

$$E_T(t) = E_r(t) + E_s(t) \tag{7}$$

The wave front reflected on high-contrast permittivity surfaces, as well as multiple reflections returns to the receiver as coherent energy. This energy is predominant at short times of the radargram time window due to its great amplitude. At higher times, the amplitude of coherent reflections diminishes due to losses as consequence of previous reflections, geometrical spreading and absorption. Therefore, at a certain time, those amplitudes present similar magnitude than the amplitude of incoherent energy recorded as background noise in the A-scan, as consequence of clutter produced by random backscattering. In addition, that incoherent energy is most likely close to the diffuse regime because the energy arrives from all heterogeneities after being scattered a random number of times following unsystematic paths. This energy becomes predominant in the radar record when the amplitude of the coherent energy decreases in the case of high depth reflectors. A threshold time t_S —in which the predominant field role exchanges—is defined comparing incoherent and coherent energies. This threshold time depends on each particular case studied and must be experimentally determined by comparing amplitudes. Using this parameter, the energy in the B-scans can be split in two zones:

$$E_T(t) \approx \begin{cases} E_r(t) & if \ t < t_S \\ E_s(t) + E_r(t) & if \ t > t_S \end{cases} \tag{8}$$

The changes on the random backscattering must be examined for times $t > t_S$, since analysing data in a properly selected time window (usually $[t_S, 2t_S]$) prevents the consideration of important reflections in targets. Notwithstanding, coherent energy is yet present but can be diminished by applying a subtract average, that reduce it in front of the incoherent energy. The quantitative analysis of the relative density of heterogeneities that perturbs the trace could be a methodology to define different zones in the ground.

The analysis of the B-scans region for times higher than t_s is based on the definition of a single amplitude value that corresponds to the average of the energy recorded in the region. This value was corrected with gain to diminish the effects of geometrical spreading and absorption.

The Mean Amplitude of Incoherent Energy (MAIE) is defined for trace i as:

$$\overline{\overline{A}}_i = \frac{\sum_{j=t_s}^{2t_s} Abs\left(\widetilde{E}_s(t_j)\right)}{\Delta t = t_s} \tag{9}$$

where $\widetilde{E}_s(t_j)$ is the amplitude of the incoherent energy in the selected time window after corrections and filters.

5. Field Data Processing

The processing is divided in three main parts: pre-processing, trace processing and final representation of MAEI versus position (Figure 2). Firstly, the wave velocity in the medium is determined to estimate the dielectric permittivity and consequently, to define the absorption losses. Two methodologies are applied in this part: comparing the GPR images to boreholes and analysing the hyperbolas along each B-scan. In this second case, it is possible to obtain the field of velocities, selecting an average value.

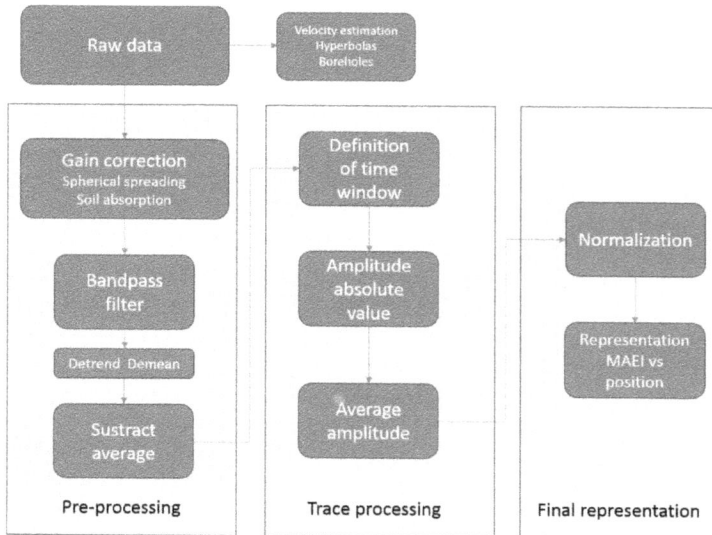

Figure 2. Processing flow diagram.

Secondly, the radargram is pre-processed. This pre-processing consists of a gain correction, a band pass filter with cut-off frequencies 10 MHz and 55 MHz. Data are also processed with demean and detrend before applying a subtract average.

Thirdly, this processing is focused to determine an average value of the amplitude within a 300 ns and 600 ns time window.

The last step of the processing consists of the normalization of the amplitude depending on the geological characteristics, being the normalization depending on a previous geological zonation. Finally, all MAEI values are represented over the position of the traces.

6. Application to Heterogeneous Media

The proposed method was applied in the city of Barcelona (Spain) which is built on a quaternary sedimentary basin. The objective was to determine changes in the ground, associated to variations in the heterogeneity of the medium. The ground of the basin was crossed by many subterranean and seasonal streams and paleochannels. These geological structures are composed by heterogeneous clusters of alluvial sediments in the zones corresponding to the streambeds. As consequence of these heterogeneities, it is expected an increase of the randomly backscattered energy, raising the amplitude of the background noise in the A-scans. Radar data was acquired with a 25 MHz centre frequency antenna. Therefore, the scattering was expected for clusters with sizes of 60 cm or less (Equation (6)). The results obtained from the study of the clutter amplitude caused by backscattering were compared to boreholes, to other geophysical measurements and also to cartographic information from ancient and modern maps.

As mentioned above, the first step was determining an average wave velocity in the medium, comparing the GPR results with known borehole data and with hyperbolas caused by shallow targets. A velocity of about 12 cm/ns was obtained as a mean result along the radar lines [30].

The second step was the application of a gain correction for spherical losses, using the two-way travel time and the velocities. An additional gain was also used to diminish the effects of the energy loss caused by absorption. In this case, the attenuation coefficient was determined by using the approximation of a dielectric and low-losses medium ($\sigma \ll \omega \varepsilon$):

$$\alpha \approx \frac{1}{2}\,\sigma\,\sqrt{\frac{\mu}{\varepsilon}} \tag{10}$$

where σ is the conductivity (an approximate value of 0.3 mS/m was used in the study), ω is the angular frequency, ε is the dielectric permittivity (between 6 and 7) and μ is the magnetic permeability (that was approximated to the unit).

The third step was the application of a band pass filter with cut-off frequencies 10 MHz and 55 MHz. After that processing, the subtracting average was applied to both radargrams (the processed one with the band pass filter and the non-processed one). The results are shown in Figure 3.

The fourth step was selecting the most appropriate time window. Figure 4 shows one of the radar images obtained in one radar line that crosses one of the seasonal subterranean streams of the zone, compared with a map. This image shows that the maximum depth of the anomalies related to anthropogenic constructions is less than 300 ns. This two-way travel time was used to determine the upper limit for the time window, being the lower limit 600 ns. This second value was chosen in order to avoid possible baseline deviations that could increase the level of the background noise amplitude.

The fifth step consists of calculating the absolute value of the amplitude (Figure 5).

The absolute amplitude was normalized based in the maximum amplitude value, comparing all the radar profiles acquired at the same geological unit in the selected time-window. The relative amplitude A_R corresponding to any A-scan is normalized at each geological unit by rescaling the average amplitude determined at the time-window of this A-scan, A. This averge amplitude is

9

determined using the minimum and maximum amplitude values in the same time-window, comparing all A-scans acquired on the same geological zone:

$$A_R = \frac{A - A_{min}}{A_{max} - A_{min}} \tag{11}$$

In this case, the relative amplitude reaches a value 0 when the average amplitude of the time window is equal to the minimum amplitude (assumed as the intrinsic material mean amplitude) and a value 1 when it is equal to the maximum value observed for the clutter' amplitude.

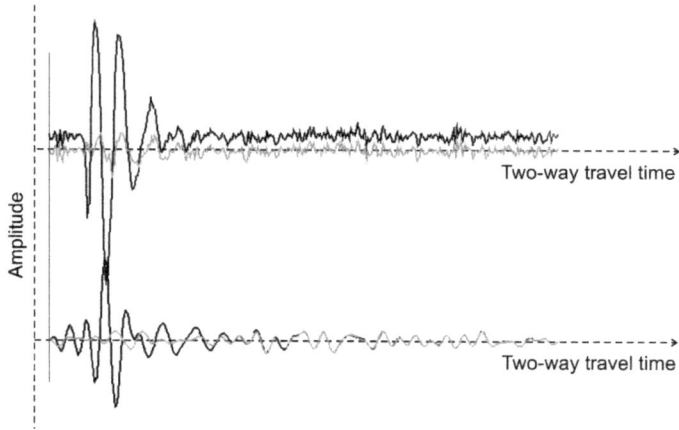

Figure 3. The black A-scan is the original trace. The grey A-scan is the trace obtained after the subtracting average processing. Above, before the application of a band-pass filter. Below, after applying a band pass filter with cut-off frequencies 10 MHz and 55 MHz.

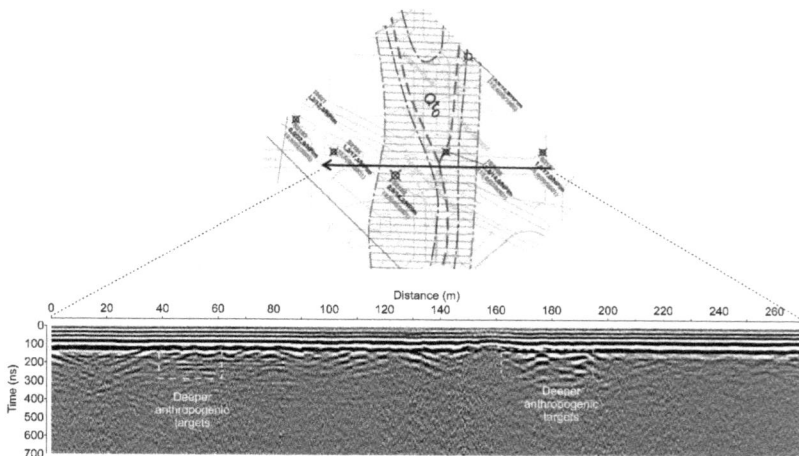

Figure 4. Radar data along 270 m, showing the maximum two-way travel time (TWT) recorded for reflections on anthropogenic targets. The radar line crosses one subterranean stream in the area where the background noise at the B-scan is increased (between 120 m and 160 m).

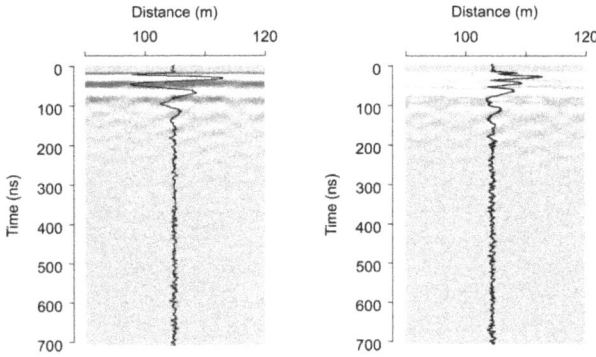

Figure 5. The first radargram presents A-scan and B-scan. The second radargram shows the absolute values of the A-scan and B-scan.

The last step was determining the average normalized amplitude of the clutter in the time-window. Finally, this amplitude was represented versus the distance along the radar lines acquired in the street (Figure 6).

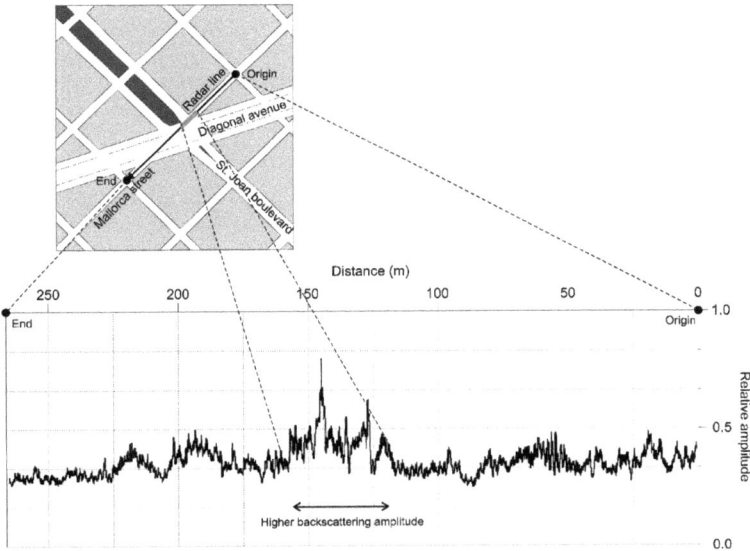

Figure 6. Average relative amplitude in the time-window versus the distance along the radar line and location of the profile at the city map.

The final result of the study includes the analysis of the average amplitudes obtained in all the B-scans that were acquired over the same geological unit.

6.1. Numerical Models

Numerical modelling is a proper method for evaluating the field results, comparing the real images to the numerical ones. The objective of the numerical simulation is to recreate the results of the GPR survey, mainly the backscattering phenomenon under the emission of electromagnetic waves (e.g., [29,31]). The effect takes place mainly under two different conditions: (1) when targets embedded in homogeneous medium present rough surfaces, being the roughness smaller than the wavelength; and (2) when the homogeneous medium is next to a more heterogeneous medium, being the size of the heterogeneities smaller than the wavelength. The presence of scatterers embedded in the homogeneous medium was modelled using the commercial software GPRSim©. This software was developed by the Geophysical Archaeometry Laboratory (Woodland Hills, CA, USA) and it is based on the ray tracing method. The basis and method is widely described in [32]. The software built a synthetic B-scan based on the response of the antenna and the wave type. The modelled medium is digitized into a grid of cells. Each cell contains the parameters defining the electromagnetic properties and also the coefficients of a polynomial function that describes the structural interfaces through the grid cell, allowing an infinite number of possible slope determinations. The synthetic radargram is determined from the rays that are sent into the grid of cells with an initial direction and amplitude, defined from the radiation pattern of the antenna.

Simulation was focused in the evaluation of the second condition, in order to recreate the real field ground. In the GPR field study, the ground was characterized by abrupt changes in the size of particles embedded in the medium as consequence of the existence of subterranean streams and paleochannels. Two models were used in two different computer analyses. The first simulation (called S1) is based in a well-defined band of scatterers and the second simulation (called S2) is based on scatterers in a band embedded in a material with randomly distributed particles (see Figure 7).

The first simulation (S1) of this second condition was studied for three cases, depending on the particles materials. The three cases are based on a simple flat layered structure [Air-Concrete-Clays dry] and the scatterers are randomly distributed in clusters, reproducing the field study conditions in Barcelona. The first stratum is a thin 0.15 cm air layer. The second stratum is a 1.5 m concrete layer (assuming man-made structures), placed over a wider stratum simulating dry clay. Scatterers represent small size targets (0.2 m, 0.4 m and 0.6 m radii) embedded in the third material. Three cases were considered, changing the targets dielectric constants (5, 8 and 30). The values were selected according to the expected values of most common geological materials in the zone. The geometrical distribution of those particles, 10 m to 20 m length restricted bands, intend to replicate the targets (underground seasonal streams and paleochannels) existing in the real medium. In the first case (Figures 7 and 8b), the particles represent dry clay with a slight change in the dielectric permittivity between the matrix and scatterers. In the second case (Figures 7 and 8c) the particles simulate dry sand and in the third case (Figures 7 and 8d) the scatterers represent wet sand. Three different scatterers size are simulated for the same dry sand material. The dielectric constant, electrical conductivity and scatterers radii for the different materials considered in the models are shown in Table 1. Figure 7 presents the scheme of the models for the different computer simulations.

Synthetic traces were obtained every 0.5 m for a set of 5 cases using the same geometrical model in order to carry out a parametric study. In this model, spherical scatterers were randomly distributed into a restricted zone (x = [20, 30] m; y = [10, 30] m). Their position remains fixed for all the cases, which is shown in Figures 7 and 8.

Table 1. Characteristics of the scatterers in the five analysed cases.

Model	Radius (m)	ε_r	S/m	Simulated Material
S1_1	0.2	8	0.0003	Clays-Scatterer
S1_2	0.2	5	0.0001	Dry Sand
S1_3	0.2	30	0.001	Wet Sand
S1_4	0.4	5	0.0001	Dry Sand
S1_5	0.6	5	0.0001	Dry Sand

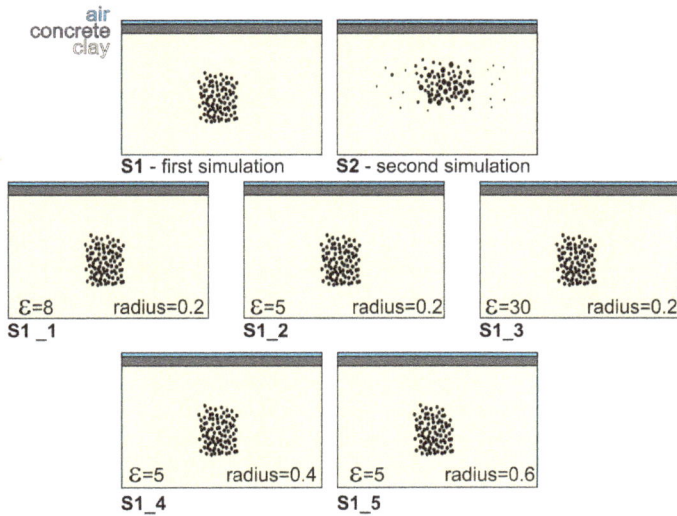

Figure 7. The two simulations (S1 and S2) based on different particles distribution. S1_1, S1_2, S1_3, S1_4 and S1_5 represent the five sub-cases considered in the first simulation (S1).

The results of the first simulation show significant backscattering appearing as consequence of scatterers, even in the case of small difference between the dielectric permittivity of the medium and scatterers. In the case of a greater contrast in the dielectric permittivity (Figure 8d), the synthetic image seems to be produced by a single and irregular target instead of a random distribution of small and individual scatterers. In the case of a larger particle size with the same dielectric permittivity, a slight increase of the backscattering amplitude is noticeable, depending on the radius (Figure 8c,e,f).

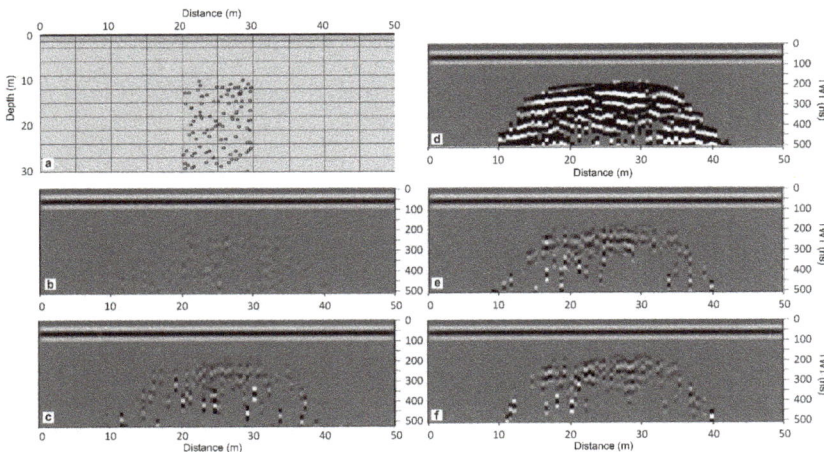

Figure 8. Three layered model with the position of the 100 scatterers (**a**) and synthetic traces for: case S1_1 (**b**); case S1_2 (**c**); case S1_3 (**d**); case S1_4 (**e**) and, case S1_5 (**f**).

Additionally, differences in the radii of the scatterers introduce slight changes on the dispersed amplitudes. Notwithstanding, part of the energy observed in the model with the greater radius (S1_5) at higher times (≈300 ns) should be neglected in the backscattering analysis since it seems to be coherent with boundary reflections.

Other aspect to be considered when comparing to the field B-scans is the width of the area in the synthetic B-scans affected by the backscattering phenomenon. The antenna path produces images in which the wide of the scatterers band does not correspond to the real particle distribution. Shallower particles are detected before the antenna is placed over the band of scatterers.

The second simulation (S2) includes two kinds of scatterers. On one hand, 100 scatterers simulate clay and 0.1 m radius are randomly distributed through the entire clay layer, being the difference in dielectric permittivity very small between particles and matrix. These scatterers represent intrinsic heterogeneities of the layer. On the other hand, for a restricted zone (x = [20, 30] m; y = [5, 25] m) 300 spherical sand scatterers were randomly spread, being their radii randomly selected between 0 and 0.3 m. In this case, the traces were acquired every 0.1 m.

Results (see Figure 9) show that the mean amplitude of the incoherent energy increases when the antenna crosses the band of heterogeneities. In addition, synthetic images present a transition zone of about 5 m at each side of that zone, showing anomalies with an amplitude of about 15% of the maximum value. At higher times, the contribution of the early transmitted phases up to 3 collisions (TRT, TRRT, TTRT, TRRRT, TTRTT...) is smaller compared to the multi-scattered energy.

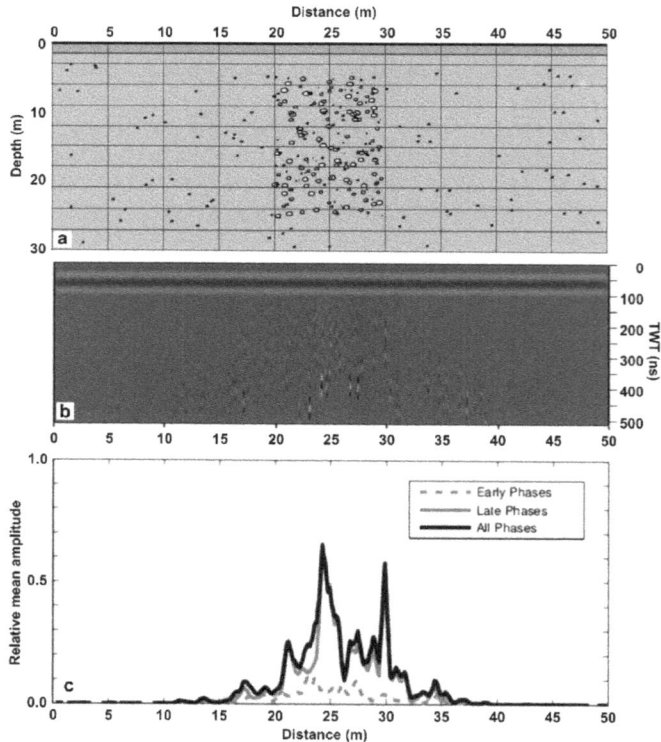

Figure 9. (a) Model used in the second simulation; (b) synthetic traces and (c) mean amplitude of Incoherent Energy for the profile.

6.2. Comparison between GPR Survey and Passive Seismic

GPR was applied to locate those areas with high number of scatterers and field results were compared to computational models. In addition, the survey was also compared to passive seismic measurements and borehole data is specific zones of the profiles. Horizontal to Vertical Spectral Ratio (HVSR) is a passive seismic methodology, applied in many cases to determine the thickness of ground materials (e.g., [33,34]) and to map site effects (e.g., [23,35–37]). A wide description of the methodology and the application can be found in [30,38]. In these maps, each zone is characterized by the mean value of the predominant period. In the surveyed area, there are zones due to underground streams and paleochannels that represent sudden lateral changes in the shallow geology. These geological structures modify the predominant period. Figure 10 exhibits the location of the GPR survey line in Mallorca Street, the passive seismic measurements and the location of three boreholes near the surveyed zone. Boreholes show 15 m of Quaternary materials, approximately, being deeper the anthropogenic fill near the underground stream.

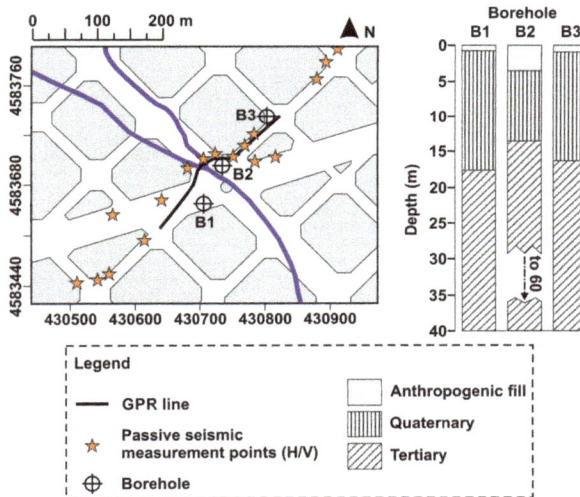

Figure 10. Location of the seasonal underground stream, the GPR line, the passive seismic measurements and the boreholes.

Passive seismic results were tested in different zones previous the measurements in the area of interest. HVSR results obtained near the subterranean streams present in the most cases a double peak in the H/V spectra, increasing also the mean amplitude and the period ([39]). In the particular study of Mallorca Street, a typical double peak denotes the presence of the geological structure associated to the stream (Figure 11). The location of these changes corresponds to the zones of the high backscattering amplitude in GPR B-scans being the results of the survey according to the GPR previous analysis and to the wide anthropogenic fill.

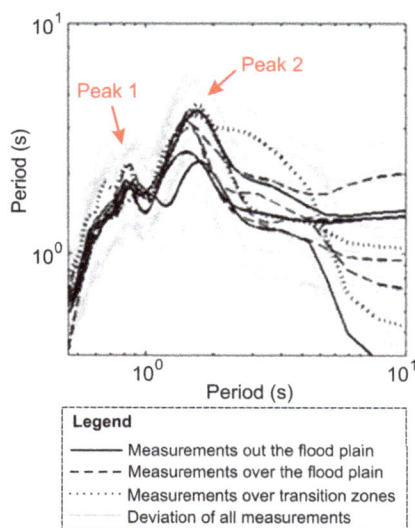

Figure 11. Passive seismic results (horizontal to vertical spectra ratio) in the zone affected by the subterranean stream in Mallorca Street. The arrows indicate the double-peak.

6.3. Discussion and Conclusions

This work described a processing technique based in enhancing the amplitudes associated to random backscattering energy, avoiding the coherent reflections in targets. This method quantified relatively this energy among the different traces, allowing defining zones where this phenomenon is produced.

The application of the procedure to a real scenario shows the correlation between the increase of the clutter in the A-scans at higher times and the existence of subterranean streams. This increase of clutter is most likely consequence of the random scattering of the energy in clusters of small targets corresponding to alluvial deposits.

The analysis of the wave velocity in the case study provides a value near 12 cm/ns. The wavelength in the medium, based in this velocity is about 480 cm. Therefore, assuming the conditions for scattering presented in the paper, the threshold size of the heterogeneities that cause this effect is close 60 cm. Hence, most likely, changes in the clutter amplitude are not caused by changes in the size of sedimentary particles but in the existence of clusters of granular materials, characteristics of fluvial deposits. Most likely, the area historically affected by the subterranean stream is composed by this kind of deposits of materials, being the main cause of the increase of clutter at higher propagation times.

Two simulations have been done in order to compare and validate the results of this procedure in the subsoil evaluations. First, the influence of the grain size and the composition of the scatterers were analysed by means of a parametric study. Comparing the synthetic results obtained in the case of different parameters (grain size and composition), it was observed that amplitudes are strongly dependent on the contrast of impedance between scatterers and surrounding materials. Additionally, the amplitude is higher as the radius of particles grows. However, more analysis could be needed in order to obtain quantitative results.

In the second simulation, there was a higher synthetic trace density and the model included two kinds of scatterers. One of them was distributed randomly in the ground layer, representing the intrinsic layer heterogeneity. The other kind of scatterers represented the clusters, modelling alluvial deposits. The results in this second simulation indicates that the clutter at higher times corresponding to random backscattering increased when the emission and reception of the energy was over the cluster of particles. Moreover, the synthetic radar data showed a transition zone where the backscattering amplitude is increasing from the average value obtained distant from the cluster to the average amplitude obtained over the cluster. Those transition zones were also observed at the case study. Therefore, that result could indicate that the experimental evaluations overestimate the real flood plain that is interpreted in the radargrams as a more extended zone than the real anomalous geological structure. Otherwise, the results of the simulation could indicate that the higher amplitudes obtained through this processing are produced mainly as consequence of multi-scattering.

Concluding, the analysis of the amplitude of clutter caused by random scattering is revealed as a useful tool to distinguish particular sedimentary structures that are not properly identified by anomalies caused by coherent reflections of the energy. The evaluation of the amplitudes at a concrete time window allows the detection of those particular changes in the subsoil.

The results of the proposed methodology indicate the potential of GPR to define sudden changes in the shallow geology in a quick survey. Comparing GPR results to passive seismic measurements, the zones defined by means of the GPR clutter could be most likely defined by the same seismic soil response. Therefore, the GPR method can be applied as a previous evaluation, before the passive seismic measurements. Those preliminary results could help in reducing the number of passive seismic measurements, which are time consuming. In addition, the previous definition of possible zones will increase the accuracy of the final results.

The analysis developed in the current work, could be further extended to other fields such as Civil Engineering, to be used to detect granular accumulation particles in concrete, or non-homogeneous mixture in highways pavements, with the final objective to map pathologies.

Acknowledgments: This research has been partially funded by the Ministry of Economy and Competitiveness (MINECO) of the Spanish Government and by the European Regional Development Fund (FEDER) of the European Union (UE) through projects referenced as: CGL2011-23621 and CGL2015-65913-P (MINECO/FEDER, UE). The study is also a contribution to the EU funded COST Action TU1208, "Civil Engineering Applications of Ground Penetrating Radar".

Author Contributions: All the authors contribute equally in the radar data acquisition, interpretation and writing the paper. Victor Salinas acquired and interpreted the seismic data. Victor Salinas and Vega Perez-Gracia prepared the computer simulation. Sonia SantoS-Assunçao and Vega Perez-Gracia prepared the processing flow. The three authors contribute in the paper preparation and revision.

Conflicts of Interest: The authors declare no conflict of interest.

References

1. Annan, A.P. GPR—History, trends, and future developments. *Subsurf. Sens. Technol. Appl.* **2002**, *3*, 253–270. [CrossRef]
2. Daniels, D.J. *Ground Penetrating Radar*, 2nd ed.; The Institution of Electrical Engineers: London, UK, 2004; 726p.
3. Jol, H.M. (Ed.) *Ground Penetrating Radar Theory and Applications*; Elsevier: Amsterdam, The Netherlands, 2008.
4. Davis, J.L.; Annan, A.P. Ground-penetrating radar for high-resolution mapping of soil and rock stratigraphy. *Geophys. Prospect.* **1989**, *37*, 531–551. [CrossRef]
5. Busby, J.P.; Merritt, J.W. Quaternary deformation mapping with Ground Penetrating Radar. *J. Appl. Geophys.* **1999**, *41*, 75–91. [CrossRef]
6. Grasmueck, M.; Weger, R.; Horstmeyer, H. Three-dimensional ground-penetrating radar imaging of sedimentary structures, fractures, and archaeological features at submeter resolution. *Geology* **2004**, *32*, 933–936. [CrossRef]
7. Benson, A.K. Applications of ground penetrating radar in assessing some geological hazards: Examples of groundwater contamination, faults, cavities. *J. Appl. Geophys.* **1995**, *33*, 177–193. [CrossRef]
8. Arcone, S.A.; Lawson, D.E.; Delaney, A.J.; Strasser, J.C.; Strasser, J.D. Ground-penetratinng radar reflection profiling of groundwater and bedrock in an area of discontinuous permafrost. *Geophysics* **1998**, *63*, 1573–1584. [CrossRef]

9. Ulugergerli, E.U.; Akca, İ. Detection of cavities in gypsum. *J. Balk. Geophys. Soc.* **2006**, *9*, 8–19.
10. Mochales, T.; Casas, A.M.; Pueyo, E.L.; Pueyo, O.; Román, M.T.; Pocoví, A.; Soriano, M.A.; Ansón, D. Detection of underground cavities by combining gravity, magnetic and ground penetrating radar surveys: A case study from the Zaragoza area, NE Spain. *Environ. Geol.* **2008**, *53*, 1067–1077. [CrossRef]
11. Grasmueck, M. 3-D ground-penetrating radar applied to fracture imaging in gneiss. *Geophysics* **1996**, *61*, 1050–1064. [CrossRef]
12. Porsani, J.L.; Sauck, W.A.; Júnior, A.O. GPR for mapping fractures and as a guide for the extraction of ornamental granite from a quarry: A case study from southern Brazil. *J. Appl. Geophys.* **2006**, *58*, 177–187. [CrossRef]
13. Sun, J.; Young, R.A. Recognizing surface scattering in ground-penetrating radar data. *Geophysics* **1995**, *60*, 1378–1385. [CrossRef]
14. Grimm, R.E.; Heggy, E.; Clifford, S.; Dinwiddie, C.; McGinnis, R.; Farrell, D. Absorption and scattering in ground-penetrating radar: Analysis of the Bishop Tuff. *J. Geophys. Res. Planets* **2006**, *111*. [CrossRef]
15. Santos Assunçao, S. Ground Penetrating Radar Applications in Seismic Zonation: Assessment and Evaluation. Ph.D. Thesis, Universitat Politècncia de Catalunya, Barcelona, Spain, 2014.
16. Vuksanovic, B.; Bostanudin, N.J.F. Clutter removal techniques for GPR images in structure inspection tasks. In Proceedings of the Fourth International Conference on Digital Image Processing (ICDIP 2012), Kuala Lumpur, Malaysia, 7–8 April 2012; International Society for Optics and Photonics: Bellingham, WA, USA, 2012; p. 833413.
17. Chen, C.S.; Jeng, Y. A data-driven multidimensional signal-noise decomposition approach for GPR data processing. *Comput. Geosci.* **2015**, *85*, 164–174. [CrossRef]
18. Wang, X.; Liu, S. Noise suppressing and direct wave arrivals removal in GPR data based on Shearlet transform. *Signal Process.* **2017**, *132*, 227–242. [CrossRef]
19. Santos-Assuncao, S.; Perez-Gracia, V.; Salinas, V.; Caselles, O.; Gonzalez-Drigo, R.; Pujades, L.G.; Lantada, N. GPR Backscattering Intensity Analysis Applied to Detect Paleochannels and Infilled Streams for Seismic Nanozonation in Urban Environments. *IEEE J. Sel. Top. Appl. Earth Obs. Remote Sens.* **2015**, *9*, 167–177. [CrossRef]
20. Al-Qadi, I.L.; Xie, W.; Roberts, R. Scattering analysis of ground-penetrating radar data to quantify railroad ballast contamination. *NDT E Int.* **2008**, *41*, 441–447. [CrossRef]
21. Benter, A.; Moore, W.; Antolovich, M. GPR signal attenuation through fragmented rock. *Min. Technol.* **2016**, *125*, 114–120. [CrossRef]
22. Perez-Gracia, V.; Caselles, O.; Salinas, V.; Pujades, L.G.; Clapés, J. GPR applications in dense cities: Detection of paleochannels and infilled torrents in Barcelona GPR applications in dense cities. In Proceedings of the IEEE 13th International Conference on Ground Penetrating Radar (GPR), Lecce, Italy, 21–25 June 2010; pp. 1–5.
23. Salinas, V.; Caselles, J.O.; Pérez-Gracia, V.; Santos-Assunçao, S.; Clapes, J.; Pujades, L.G.; González-Drigo, R.; Canas, J.A.; Martinez-Sanchez, J. Nanozonation in dense cities: Testing a combined methodology in Barcelona City (Spain). *J. Earthq. Eng.* **2014**, *18*, 90–112. [CrossRef]
24. Pettinelli, E.; Burghignoli, P.; Pisani, A.R.; Ticconi, F.; Galli, A.; Vannaroni, G.; Bella, F. Electromagnetic propagation of GPR signals in Martian subsurface scenarios including material losses and scattering. *IEEE Trans. Geosci. Remote Sens.* **2007**, *45*, 1271–1281. [CrossRef]
25. Reichman, J. Determination of absorption and scattering coefficients for nonhomogeneous media. 1: Theory. *Appl. Opt.* **1973**, *12*, 1811–1815. [CrossRef] [PubMed]
26. Harbi, H.; McMechan, G.A. Conductivity and scattering Q in GPR data: Example from the Ellenburger dolomite, central Texas. *Geophysics* **2012**, *77*, H63–H78. [CrossRef]
27. Beckmann, P.; Spizzichino, A. *The Scattering of Electromagnetic Waves from Rough Surfaces*; Artech House, Inc.: Norwood, MA, USA, 1987; 511p.
28. Soldovieri, F.; Hugenschmidt, J.; Persico, R.; Leone, G. A linear inverse scattering algorithm for realistic GPR applications. *Near Surf. Geophys.* **2007**, *5*, 29–41. [CrossRef]
29. Takahashi, K.; Igel, J.; Preetz, H. Modeling of GPR clutter caused by soil heterogeneity. *Int. J. Antennas Propag.* **2012**, *2012*, 643430. [CrossRef]

30. Santos-Assuncao, S.; Perez-Gracia, V.; Gonzalez-Drigo, R. GPR backscattering applied to urban shallow geology: GPR application in seismic microzonation. In Proceedings of the IEEE 8th International Workshop on Advanced Ground Penetrating Radar (IWAGPR), Florence, Italy, 7–10 July 2015; pp. 1–4.

31. Takahashi, K.; Igel, J.; Preetz, H.; Sato, M. Sensitivity analysis of soil heterogeneity for ground—Penetrating radar measurements by means of a simple modeling. *Radio Sci.* **2015**, *50*, 79–86. [CrossRef]

32. Goodman, D. Ground-penetrating radar simulation in engineering and archaeology. *Geophysics* **1994**, *59*, 224–232. [CrossRef]

33. Chandler, V.W.; Lively, R.S. Utility of the horizontal-to-vertical spectral ratio passive seismic method for estimating thickness of Quaternary sediments in Minnesota and adjacent parts of Wisconsin. *Interpretation* **2016**, *4*, SH71–SH90. [CrossRef]

34. Picotti, S.; Francese, R.; Giorgi, M.; Pettenati, F.; Carcione, J.M. Estimation of glacier thicknesses and basal properties using the horizontal-to-vertical component spectral ratio (HVSR) technique from passive seismic data. *J. Glaciol.* **2017**, *63*, 229–248. [CrossRef]

35. Caselles, J.O.; Pérez-Gracia, V.; Franklin, R.; Pujades, L.G.; Navarro, M.; Clapes, J.; Canas, J.A.; García, F. Applying the H/V method to dense cities. A case study of Valencia City. *J. Earthq. Eng.* **2010**, *14*, 192–210. [CrossRef]

36. Panzera, F.; D'Amico, S.; Lombardo, G.; Longo, E. Evaluation of building fundamental periods and effects of local geology on ground motion parameters in the Siracusa area, Italy. *J. Seismol.* **2016**, *20*, 1001–1019. [CrossRef]

37. Qadri, S.T.; Islam, M.A.; Shalaby, M.R.; Khattak, K.R.; Sajjad, S.H. Characterizing site response in the Attock Basin, Pakistan, using microtremor measurement analysis. *Arab. J. Geosci.* **2017**, *10*, 267. [CrossRef]

38. Alfaro, A.; Pujades, L.G.; Goula, X.; Susagna, T.; Navarro, M.; Sanchez, J.; Canas, J.A. Preliminary Map of Soil's Predominant Periods in Barcelona Using Microtremors. *Pure Appl. Geophys.* **2001**, *158*, 2499–2511. [CrossRef]

39. Salinas, V. Detección Fina de Cambios Laterales en la Geología Superficial y en los Suelos y Caracterización de su Influencia en la Respuesta Sísmica Local. Aplicación a Barcelona (Exhaustive Detection of Lateral Changes in Surface Geology and in Soils and Characterization of Their Influence on the Local Seismic Response. Application to Barcelona). Ph.D. Thesis, Universitat Politècncia de Catalunya, Barcelona, Spain, 2015.

remote sensing

MDPI

Article

Application of Coupled-Wave Wentzel-Kramers-Brillouin Approximation to Ground Penetrating Radar

Igor Prokopovich [1], Alexei Popov [1], Lara Pajewski [2] and Marian Marciniak [3,*

[1]   Pushkov Institute of Terrestrial Magnetism, Ionosphere and Radio Wave Propagation (IZMIRAN),
     108840 Moscow, Russia; prokop@izmiran.ru (I.P.); popov@izmiran.ru (A.P.)
[2]   Department of Information Engineering, Electronics and Telecommunications, Sapienza University of Rome,
     00185 Rome, Italy; lara.pajewski@uniroma1.it
[3]   National Institute of Telecommunications, 04-894 Warsaw, Poland
*   Correspondence: m.marciniak@itl.waw.pl; Tel.: +48-225128715

Received: 5 December 2017; Accepted: 20 December 2017; Published: 23 December 2017

Abstract: This paper deals with bistatic subsurface probing of a horizontally layered dielectric half-space by means of ultra-wideband electromagnetic waves. In particular, the main objective of this work is to present a new method for the solution of the two-dimensional back-scattering problem arising when a pulsed electromagnetic signal impinges on a non-uniform dielectric half-space; this scenario is of interest for ground penetrating radar (GPR) applications. For the analytical description of the signal generated by the interaction of the emitted pulse with the environment, we developed and implemented a novel time-domain version of the coupled-wave Wentzel-Kramers-Brillouin approximation. We compared our solution with finite-difference time-domain (FDTD) results, achieving a very good agreement. We then applied the proposed technique to two case studies: in particular, our method was employed for the post-processing of experimental radargrams collected on Lake Chebarkul, in Russia, and for the simulation of GPR probing of the Moon surface, to detect smooth gradients of the dielectric permittivity in lunar regolith. The main conclusions resulting from our study are that our semi-analytical method is accurate, radically accelerates calculations compared to simpler mathematical formulations with a mostly numerical nature (such as the FDTD technique), and can be effectively used to aid the interpretation of GPR data. The method is capable to correctly predict the protracted return signals originated by smooth transition layers of the subsurface dielectric medium. The accuracy and numerical efficiency of our computational approach make promising its further development.

Keywords: ground penetrating radar; electromagnetic propagation in nonhomogeneous media; time-domain analysis

---

1. Introduction

The main goal of subsurface radar probing is the estimation of physical and geometrical properties of a natural or manmade structure, by using electromagnetic waves [1,2]. Ground penetrating radar (GPR) systems emit and receive electromagnetic waves over an ultra-wide frequency range and can work in the time or spectral domain. Time-domain systems are based on the transmission of short electromagnetic pulses; spectral-domain systems transmit a succession of harmonic signals of linearly increasing frequency, in discrete steps. The signal impinging on a GPR receiving antenna results from the interaction of the emitted signal with the structure under test; by processing and interpreting the received signal, physical and geometrical information about the scenario can be deduced. Through exploitation of the inverse Fourier transform (from frequency to time-domain), a spectral-domain GPR provides results equivalent to those of a pulsed GPR. The frequency-domain approach is possible because the environment is regarded as a time-invariant system and the received signal is considered as a linear function of the emitted one.

Laws regulating electromagnetic-pulse radiation and propagation in non-uniform media have to be fully taken into account, in the development of reliable forward and inverse scattering algorithms for the simulation, analysis and interpretation of GPR responses. Closed-form analytical solutions can be found only for very simple scenarios related to canonical scattering problems. Realistic scenarios are complicated and require massive numerical calculations.

The most popular full-wave computational methods combine a relatively simple mathematical formulation with a mostly numerical nature: they are easy to implement and versatile. For example, the finite-difference time-domain (FDTD) technique [3–5] is a full-wave computational method widely used in the GPR community. FDTD is an accurate method and allows to conveniently simulating composite structures; the main drawbacks reside in the approximation limits of the FDTD model itself, in terms of space and time discretization. The calculation time and memory requirements can be prohibitive, for the solution of realistic problems. The criteria for accuracy, stability, and convergence of results are not always straightforward for non-experienced researchers.

Other full-wave formulations have a higher analytical complexity: a deeper physical insight into the considered problem is needed, to develop and implement such techniques [6–11]. Usually, these approaches are less versatile, i.e., they are conceived to solve specific problems rather than to model a wide range of different scenarios. The main advantages of such techniques reside in the possibility to achieve a more comprehensive understanding of the electromagnetic phenomena occurring in the subsurface or structure under test, and a deeper knowledge of how targets get translated into the radargrams. When applicable, these methods turn out to be particularly fast and numerically efficient, hence they are suitable to be embedded into inverse solvers requiring the iterative evaluation of several forward problems.

Electromagnetic scattering problems involving media with one-dimensional (1D) variation of the electromagnetic properties have been widely studied in the literature and still are of high interest [11–16]. Approaches for the solution of such problems find application not only in the GPR field: they are important for the interpretation of data measured with other electromagnetic non-destructive testing methods as well, such as Time Domain Reflectometry (TDR) for moisture evaluation and material analysis [17,18].

One-dimensional problems where the permittivity varies on a wavelength scale are difficult to tackle and only a few permittivity profiles allow for exact analytical solutions [19]. Usually, scenarios involving this kind of inhomogeneous media are modeled by using numerical techniques, such as the already mentioned FDTD method, the finite integration technique (FIT) [20], time-domain integral equation (TDIE) approaches [21], and more. The Green's function method [22] offers some advantages: if different incident waveforms need to be considered, the wave equation does not have to be solved for each of them, and some simplifications can be done analytically [23]; moreover, the wave field does not have to be computed throughout the entire medium but only at the receiver position. Methods specifically conceived for dealing with absorbing inhomogeneous layers and anisotropic inhomogeneous media have been also proposed and tested, with various degrees of success, see for example [24,25].

When the permittivity variation takes place along one direction and in a much larger scale than the wavelength, the propagation of electromagnetic waves can be successfully described by using semi-analytical techniques. Substantially, Maxwell's equations can be solved in a series of homogeneous layers with constant permittivity, and the wave fields can be joined at the interfaces with appropriate continuity conditions. If the thickness of the homogeneous layers tends to zero, such a procedure results in a classical Wentzel-Kramers-Brillouin (WKB) approximation. This approach, originally proposed in quantum mechanics [26], became a powerful tool for the mathematical description of acoustical and electromagnetic wave propagation in natural media with gradually varying dielectric permittivity [27]. Unfortunately, the standard version of the WKB approach cannot deal with backward reflections originated by smooth permittivity gradients, which are of interest in GPR applications. In that respect, the rectification of the WKB technique developed in the frequency domain by Bremmer and Brekhovskikh looks particularly promising [27–29]. Such method, also called "coupled-wave WKB method" or "two-way WKB", consists in an iterative solution of coupled ordinary differential equations of WKB type; it is capable to take into account the backscattered signals and provides a good accuracy over a wide frequency range [27].

The possibility of application of the two-way WKB method to GPR was studied in [30] for the first time: it was demonstrated that the time-domain counterpart of the Bremmer-Brekhovskikh method can accurately describe the waveform of the reflected signal in the presence of permittivity discontinuities or gradual variations. Moreover, it was shown that the method allows to effectively reconstruct the properties of subsurface layers, starting from the signal received by the radar.

The aim of our work is to further develop the promising WKB approach and apply the Bremmer-Brekhovskikh approximation to a more realistic scenario. In particular, we developed, implemented and tested a new semi-analytical method, based on the coupled-wave version of the WKB approximation, to study a two-dimensional (2D) back-scattering problem arising when a pulsed electromagnetic signal impinges on a non-uniform dielectric half-space. Actually, the formulation of our problem is "1.5-dimensional": the subsurface medium is assumed to be horizontally stratified (1D permittivity model) and the source is a line of current stretched along the air-ground interface, which produces a two-dimensional (2D) transient electromagnetic field. We neglect energy losses in the involved media.

The paper is structured as follows. The theoretical approach is presented in Section 2. We consider a simplified 1D-scenario in Section 2.1 to explain the basis of the technique; in Section 2.2, we extend the method to the above-mentioned 1.5-dimensional scenario. In the numerical implementation of our technique, the key point is the solution of a functional equation, to determine the complex poles of an integrand that appears in the explicit representation of the analytical solution. Its physical interpretation in terms of geometrical optics is given in Section 2.3 and a simplification achieved in case of moderate separation between the transmitting and receiving antennas is discussed in Section 2.4. An accurate numerical quadrature algorithm for the arising singular integrals is proposed in Section 3. In Section 4, numerical results are presented. Firstly, the proposed method is compared with the FDTD technique; different soil parameters and configurations are considered. Next, a successful application of our approach to real scenarios is presented. In the first example, the method is employed to aid the interpretation of radargrams collected in 2013 during an IZMIRAN expedition, where GPR was used to search for a large fragment of the Chelyabinsk meteorite in Lake Chebarkul bottom [31,32]. In the second example, the method is used for the simulation of GPR probing aimed to the estimation of the water content in lunar regolith near the poles [33]. Conclusions are drawn in Section 5, where plans for future work are also outlined.

2. Theoretical Method

2.1. One-Dimensional Problem

In this subsection, we resume the simplified 1D-probing scheme proposed in [30] to explain the basis of our approach.

Let us consider the 1D propagation of an electromagnetic pulse, with electric field $E(ct, z)$, in a non-uniform half-space $z > 0$ characterized by a real-valued relative permittivity profile $\varepsilon(z)$ and a vacuum magnetic permeability μ_0 (i.e., the half-space is assumed to be a lossless non-magnetic medium). Here and in the following, t is the time, z is the spatial coordinate and c is the light velocity in vacuum. This phenomenon is governed by the wave equation

$$\partial^2 E(s,z)/\partial z^2 = \varepsilon(z)\partial^2 E(s,z)/\partial s^2 \quad (z > 0,\ s > 0), \tag{1}$$

where $s = ct$ is introduced for convenience, so that $\partial^2/\partial s^2 = c^{-2}\partial^2/\partial t^2$. The source is in $z = 0$. The trivial initial conditions $E = 0$ and $\partial E/\partial t = 0$ in $t = 0$, $\forall z$, and a non-homogeneous boundary condition given by

$$\partial E(s,z)/\partial s\ \big|_{z=0} - \varepsilon_0^{-1/2}\partial E(s,z)/\partial z\ \big|_{z=0} = 2\ df(s)/ds, \tag{2}$$

define a transient field $E(s,z)$ generated by the pulse $f(s)$ entering the non-uniform half-space $z > 0$ with $\varepsilon_0 = \varepsilon(z \to +0)$. The total wave field at $z = 0$, can be written as $E(s,0) = f(s) + g(s)$, where $g(s)$ is the cumulative backscattered signal born on the subsurface permittivity gradients.

To find a unique solution to the boundary-value problem, the radiation condition

$$\partial E(s,z)/\partial s \sim \varepsilon_\infty^{-1/2}\partial E(s,z)/\partial z, \quad z \to \infty \tag{3}$$

has to be imposed, excluding the waves coming from $z = \infty$. In Equation (3), $\varepsilon_\infty = \varepsilon(z \to \infty)$. The application of the Fourier integral transform

$$\widetilde{E}(k,z) = \int_0^\infty E(s,z)\exp(iks)ds \tag{4}$$

reduces Equation (1) to the 1D Helmholtz equation

$$\partial^2 \widetilde{E}(k,z)/\partial z^2 + k^2\varepsilon(z)\widetilde{E}(k,z) = 0 \tag{5}$$

or to an equivalent set of first-order ordinary differential equations (ODE) [29]

$$\frac{\partial A^\pm(k,z)}{\partial z} = \frac{\varepsilon'(z)}{4\varepsilon(z)}\exp\left[\mp 2ik\int_0^z \varepsilon^{1/2}(z)dz\right]A^\mp(k,z), \tag{6}$$

with $\varepsilon'(z) = d\varepsilon/dz$. Equations (6) govern the amplitudes $A^+(k,z)$ and $A^-(k,z)$ of the direct and backward waves in the total field representation

$$\widetilde{E}(k,z) = \left[\frac{\varepsilon_0}{\varepsilon(z)}\right]^{1/4}\left\{A^+(k,z)\exp\left[ik\int_0^z \varepsilon^{1/2}(z)dz\right] + A^-(k,z)\exp\left[-ik\int_0^z \varepsilon^{1/2}(z)dz\right]\right\}, \tag{7}$$

valid for $z > 0$. The equation set (6) can be solved iteratively, starting from $\partial A^\pm(k,z)/\partial z = 0$. The first approximation gives

$$\begin{cases} A^+(z,k) \approx \widetilde{f}(k) \\[2mm] A^-(z,k) \approx -\frac{\widetilde{f}(k)}{4}\int_z^\infty \frac{\varepsilon'(\zeta)}{\varepsilon(\zeta)}\exp\left[-2ik\int_0^\zeta \varepsilon^{1/2}(\xi)d\xi\right]d\zeta. \end{cases} \tag{8}$$

A backward Fourier transform yields an explicit formula relating the initial pulse $f(s)$ with the total signal $E(s,0) = f(s) + g(s)$, that can be measured in $z = 0$. In particular, the half-space response to the input electromagnetic pulse is

$$g(s) = -\frac{1}{4}\int_0^\infty \frac{\varepsilon'(z)}{\varepsilon(z)}f\left[s - 2\int_0^z \varepsilon(\zeta)^{1/2}d\zeta\right]dz. \tag{9}$$

Equation (9), having the evident meaning of a sum of partial reflections due to the permittivity gradients, can be considered as an integral equation for the unknown function $\varepsilon(z)$. As shown in [30], this equation, having a convolution form, can be solved by exploiting the Fourier–Laplace transform, yielding a parametric solution to the 1D inverse problem

$$\begin{cases} \varepsilon(s) = \varepsilon_0 \exp\left[-4 \int_0^s Q(r)dr\right] \\ \\ z(s) = \frac{\varepsilon_0^{-1/2}}{2} \int_0^s \exp\left[2 \int_0^r Q(r')dr'\right]dr \end{cases} \tag{10}$$

where

$$Q(r) = \frac{1}{2\pi} \int_{i\alpha-\infty}^{i\alpha+\infty} \widetilde{g}(k)\, \widetilde{f}^{-1}(k) \exp(-ikr)dk, \tag{11}$$

and $\widetilde{f}(k)$, $\widetilde{g}(k)$ are the Fourier transforms of the initial pulse $f(s)$ and received backscattered signal $g(s)$, calculated accoding to Equation (4).

2.2. 1.5-Dimensional Problem

In this subsection, we deal with a more realistic model, considering a GPR with separated antennas lying at the air-ground interface. We model the transmitting antenna as a line source, and develop an analytical method that allows to describe the electromagnetic field recorded by the receiving antenna, including the surface wave and all partial reflections by the subsurface permittivity discontinuities and gradients. As is well known, the line source is the two-dimensional counterpart of the hertzian dipole, for geometries with one invariant direction.

Although a line source is an idealized electromagnetic representation of an antenna, it allows a more realistic modeling of GPR problems than sources with an infinite extension, such as the plane wave. For example, in the presence of a two-dimensional variation of the subsurface physical properties, the line source allows to account for the position of the antenna in the scenario; furthermore, a suitable combination of line sources can be used to model the field distribution generated by a more complex antenna.

In our method, we exploit the Fourier–Laplace transform and reduce the time-domain boundary value problem to an ordinary differential equation, which is solved approximately by the Bremmer-Brekhovskikh method. A backward integral transform yields an approximate representation of the time-domain Green function, i.e., of the subsurface medium response to an elementary current jump in the GPR transmitting antenna. This result, in combination with the Duhamel principle [34], gives an approximate solution to the forward electromagnetic scattering problem for an arbitrary electromagnetic pulse and permittivity profile.

Let us therefore consider the 1.5-dimensional scenario of short-pulsed radiation emitted by a line source stretched along the surface of a non-uniform dielectric half-space $z > 0$. We assume that the half-space is horizontally layered, with a real-valued relative permittivity. We also assume a uniform current distribution along the thin wire, which is lying at $x = z = 0$, $-\infty < y < \infty$. The wave perturbation is excited by a current pulse $I(t)$. The 2D wave equation governing the y-component of the electric field $E(t; x, z)$ is:

$$\frac{\partial^2 E}{\partial x^2} + \frac{\partial^2 E}{\partial z^2} - \frac{\varepsilon(z)}{c^2} \frac{\partial^2 E}{\partial t^2} = \frac{4\pi}{c^2} \delta(x)\delta(z)I(t), \tag{12}$$

where $\delta(\cdot)$ is the Dirac delta function. By using integral transforms and by imposing the initial conditions $E = 0$ and $\partial E/\partial t = 0\varepsilon(z)$ in $t = 0$, $\forall z$, Equation (12) can be reduced to an ordinary differential equation. In particular, we apply a Fourier transform with respect to the x coordinate:

$$\begin{cases} \widetilde{E}(t; p, z) = \frac{1}{2\pi} \int_{-\infty}^{+\infty} \exp(-ipx)E(t; x, z)dx \\ \\ E(t; x, z) = \int_{-\infty}^{+\infty} \exp(ipx)\widetilde{E}(t; p, z)dp \end{cases} \tag{13}$$

and we obtain the 2D counterpart of Equation (5):

$$\frac{\partial^2 \widetilde{E}}{\partial z^2} - \frac{\varepsilon(z)}{c^2}\frac{\partial^2 \widetilde{E}}{\partial t^2} - p^2\widetilde{E} = \frac{2}{c^2}\delta(z)\dot{I}(t). \tag{14}$$

Then, by using the Laplace transform with respect to the time variable:

$$\begin{cases} \hat{E}(\gamma; p, z) = \int_0^{+\infty} \exp(-\gamma t)\widetilde{E}(t; p, z)dt \\[2mm] \widetilde{E}(t; p, z) = \frac{1}{2\pi i}\int_{\alpha-i\infty}^{\alpha+i\infty} \exp(\gamma t)\hat{E}(\gamma; p, z)d\gamma \end{cases} \tag{15}$$

we obtain the second-order ODE

$$\frac{\hat{E}(\gamma; p, z)}{\partial z^2} - \left[\frac{\gamma^2}{c^2}\varepsilon(z) + p^2\right]\hat{E}(\gamma; p, z) = \frac{2\gamma}{c^2}\delta(z)\hat{I}(\gamma), \tag{16}$$

where $\hat{I}(\gamma)$ is the Laplace transform of the antenna current $I(t)$. Equation (16) can be reduced to a system of first-order ODE similar to Equation (6). Such a system, satisfying the boundary conditions at the air-ground interface, and the radiation condition for $z \to \infty$, can be solved by iterations, starting from zero wave perturbation. The first approximation gives an integral representation of the initial probing wave and its subsurface reflections

$$\begin{aligned}\hat{E}(\gamma; p, z > 0) = \quad & A_0\,(p, \gamma)\frac{\kappa_0^{1/2}}{\kappa^{1/2}(z)}\left\{\exp\left[-\int_0^z \kappa(\zeta)d\zeta\right]\right. \\ & \left. -\frac{1}{2}\,\exp\left[\int_0^z \kappa(\zeta)d\zeta\right]\int_z^{\infty} \frac{\kappa'(\zeta)}{\kappa(\zeta)}\exp\left[-2\int_0^{\zeta}\kappa(\eta)d\eta\right]d\zeta\right\},\end{aligned} \tag{17}$$

as well as the "aerial" wave propagating in the upper half-space:

$$\hat{E}(\gamma; p, z < 0) = A_0(\gamma, p)\exp(\kappa_A z)\left\{1 - \frac{1}{2}\int_0^{\infty}\frac{\kappa'(\zeta)}{\kappa(\zeta)}\exp\left[-2\int_0^{\zeta}\kappa(\eta)d\eta\right]d\zeta\right\}. \tag{18}$$

Here, $\kappa(z) = \left[q^2\varepsilon(z) + p^2\right]^{1/2}$, $\kappa_0 = \kappa(0) = \left[q^2\varepsilon_0 + p^2\right]^{1/2}$, $\kappa_A = \left[q^2 + p^2\right]^{1/2}$, and $q = \gamma/c$. The amplitude A_0 can be found from the excitation condition with a localized source $2\gamma\delta(z)\hat{I}(\gamma)/c^2$. The differentiation of Equations (17) and (18) yields:

$$\frac{\partial \hat{E}}{\partial z}(\gamma; p, z = +0) = -A_0\,(\gamma, p)\kappa_0\left\{1 + \frac{1}{2}\int_0^{\infty}\frac{\kappa'(\zeta)}{\kappa(\zeta)}\exp\left[-2\int_0^{\zeta}\kappa(\eta)d\eta\right]d\zeta\right\}, \tag{19}$$

and

$$\frac{\partial \hat{E}}{\partial z}(\gamma; p, z = -0) = A_0\,(\gamma, p)\kappa_A\left\{1 - \frac{1}{2}\int_0^{\infty}\frac{\kappa'(\zeta)}{\kappa(\zeta)}\exp\left[-2\int_0^{\zeta}\kappa(\eta)d\eta\right]d\zeta\right\}, \tag{20}$$

where it can be noticed that the derivative $\partial\hat{E}/\partial z$ has a jump at the interface, which is approximately equal to $-A_0(\kappa_0 + \kappa_A)$. Taking this into account, we integrate Equation (16) over the small interval $-0 < z < +0$ and relate the wave amplitude A_0 to the Laplace image of the driving current $\hat{I}(\gamma)$:

$$A_0\,(\gamma, p) = \frac{-2\gamma\hat{I}(\gamma)c^{-2}}{\kappa_0 + \kappa_A + \frac{(\kappa_0-\kappa_A)}{2}\int_0^{\infty}\frac{\kappa'(\zeta)}{\kappa(\zeta)}\exp\left[-2\int_0^{\zeta}\kappa(\eta)d\eta\right]d\zeta}. \tag{21}$$

The electromagnetic field amplitude at the interface $z = 0$, where by assumption the receiver antenna is placed, is given by the inverse Fourier–Laplace transform of the spectral distribution Equations (17) and (18):

$$E(t; p, 0) = \frac{1}{2\pi i}\int_{-\infty}^{+\infty}\exp(ipx)dp\int_{\alpha-i\infty}^{\alpha+i\infty}\exp(\gamma t)\hat{E}(\gamma; p, 0)d\gamma, \tag{22}$$

where

$$\hat{E}(\gamma;p,0) = \frac{-2\gamma \hat{I}(\gamma)}{c^2(\kappa_0+\kappa_A)} \frac{1-\frac{1}{2}\int_0^\infty \frac{\kappa'(\zeta)}{\kappa(\zeta)}\exp\left[-2\int_0^\zeta \kappa(\eta)d\eta\right]d\zeta}{1+\frac{1}{2}\frac{\kappa_0-\kappa_A}{\kappa_0+\kappa_A}\int_0^\infty \frac{\kappa'(\zeta)}{\kappa(\zeta)}\exp\left[-2\int_0^\zeta \kappa(\eta)d\eta\right]d\zeta} \approx$$

$$\approx \frac{-2\gamma \hat{I}(\gamma)}{c^2(\kappa_0+\kappa_A)}\left[1-\frac{\kappa_0}{\kappa_0+\kappa_A}\int_0^\infty \frac{\kappa'(\zeta)}{\kappa(\zeta)}\exp\left[-2\int_0^\zeta \kappa(\eta)d\eta\right]d\zeta\right],$$

(23)

In Equation (23), we simplified the expression by exploiting the formula of geometric series.

It is convenient to represent the electromagnetic field excited by an arbitrary current pulse as a convolution of the time-domain Green function with the current pulse $I(t)$:

$$E(t;x,z) = \int_0^t \frac{dI}{dt}(t-t')G(ct';x,z)dt'.$$

(24)

To find the Green function, it is necessary to calculate the radiation produced by a unit current step: $I(t) = 1$ for $t > 0$ and $I(t) = 0$ for $t < 0$, corresponding to $\hat{I}(\gamma) = 1/\gamma = 1/cq$. Having no temporal scale, it is natural to use the uniform space-like variables ($s = ct; x, z$).

From Equation (23), we find the boundary value of the spectral Green function:

$$\hat{G}(\gamma;p,0) = \frac{-2}{c^2(\kappa_0+\kappa_A)} \times$$
$$\left\{1-\frac{1}{2}\frac{\kappa_0}{\kappa_0+\kappa_A}\int_0^\infty \frac{\varepsilon'(z)}{\varepsilon(z)+(p/q)^2}\exp\left[-2q\int_0^z \left[\varepsilon(\zeta)+(p/q)^2\right]^{1/2}d\zeta\right]dz\right\}.$$

(25)

This expression consists of two parts. The first term corresponds to direct pulse propagation along the ground surface (the so-called "direct" wave), the second term represents the cumulative reflection from the subsurface medium gradients.

The "direct" wave $G_d(s;x,z)$, with $s = ct$, can be explicitly found by applying a backward Fourier–Laplace transform to the first term of Equation (25):

$$G_d(s;x,0) = \frac{i}{\pi(\varepsilon_0-1)c}\int_{-\infty}^{+\infty}\exp(ipx)dp\times$$
$$\int_{a-i\infty}^{a+i\infty}\exp(qs)\left[\left(p^2+q^2\varepsilon_0\right)^{1/2}-\left(p^2+q^2\right)^{1/2}\right]\frac{dq}{q^2}.$$

(26)

The inner integral in Equation (26) can be rewritten as two integrals over closed paths circumventing the corresponding branch points. After the substitution $q = ip\eta$ and a change of integration order, the following formula arises, which describes the direct-wave propagation as the sum of two electromagnetic pulses ("aerial" and "ground" waves) moving along both sides of the $z = 0$ interface:

$$G_d(s;x,0) = \frac{4}{(\varepsilon_0-1)c}\int_{-\varepsilon_0^{1/2}}^{\varepsilon_0^{-1/2}}\eta^{-2}(1-\eta^2\varepsilon_0)\delta(x+\eta s)d\eta+$$
$$-\int_{-1}^1 \eta^{-2}(1-\eta^2)\delta(x+\eta s)d\eta = \frac{4}{(\varepsilon_0-1)cx^2}\left[\left(s^2-x^2\varepsilon_0\right)_+^{1/2}-\left(s^2-x^2\right)_+^{1/2}\right].$$

(27)

To find the cumulative signal reflected by the subsurface medium gradients, $G_r(s;x,0)$, we transform into the space-time domain the second part of the spectral function in Equation (25), $G_r(s;x,0) = \int_0^\infty \varepsilon'(z)K(s;x,z)dz$, where:

$$K(s;x,z) = \frac{1}{2\pi ic}\int_{-\infty}^\infty \exp(ipx)dp\times$$

$$\int_{a-i\infty}^{a+i\infty}\frac{q^2\left(p^2+q^2\varepsilon_0\right)^{1/2}\exp\left\{qs-2\int_0^z \left[p^2+q^2\varepsilon(\zeta)\right]^{1/2}d\zeta\right\}}{\left[p^2+q^2\varepsilon(z)\right]\left[\left(p^2+q^2\varepsilon_0\right)^{1/2}+\left(p^2+q^2\right)^{1/2}\right]^2}dq.$$

(28)

In accordance with the problem geometry (absence of scaling parameters), the integrand in Equation (28) is homogeneous with respect to p and q, which allows to simplify calculations by making the substitution $q = |p|w$:

$$K(s; x, z) = \frac{1}{\pi i c} \int_0^\infty \cos(px) dp \times$$

$$\int_{a-i\infty}^{a+i\infty} \frac{w^2 (1+w^2\varepsilon_0)^{1/2} \exp\left\{ pws - 2p \int_0^z [1+w^2\varepsilon(\zeta)]^{1/2} d\zeta \right\}}{[1+w^2\varepsilon(z)]\left[(1+w^2\varepsilon_0)^{1/2} + (1+w^2)^{1/2}\right]^2} dw. \tag{29}$$

We consider the inner Laplace integral in Equation (29) under the two following conditions:

$$s < 2 \int_0^z \varepsilon^{1/2}(\zeta) d\zeta \quad \text{and} \quad s > 2 \int_0^z \varepsilon^{1/2}(\zeta) d\zeta. \tag{30}$$

In the former case, the integration path can be closed in the right half-plane and the integral vanishes due to regularity of the integrand. In the latter case, the integration can be performed along the steepest-descent path Γ where the real part of the exponent is negative (red dashed line in Figure 1). After such path deformation, we can change the integration order and calculate the inner integral:

$$K(s; x, z) = \frac{1}{2\pi i c} \int_\Gamma C(w, z) dw \int_0^\infty \exp[p\Phi(s; w, z)][\exp(ipx) + \exp(-ipx)] dp =$$

$$= \frac{1}{2\pi i c} \int_\Gamma C(w, z) \left[\frac{1}{\Phi(s; w, z) - ix} + \frac{1}{\Phi(s; w, z) + ix} \right] dw. \tag{31}$$

Here, the following notations are introduced:

$$\Phi(s; w, z) = ws - 2 \int_0^z [1 + w^2 \varepsilon(\zeta)]^{1/2} d\zeta$$

$$C(w, z) = \frac{w^2 (1+w^2\varepsilon_0)^{1/2}}{[1+w^2\varepsilon(z)]\left[(1+w^2\varepsilon_0)^{1/2} + (1+w^2)^{1/2}\right]^2}. \tag{32}$$

In the last integral of Equation (31), the integrand vanishes at infinity, so it can be reduced to residues:

$$K(s; x, z) = c^{-1} \sum_j C(w_j, z) / \Phi'_w(s; w_j, z) \tag{33}$$

where $w_j(s; x, z)$ are the roots of the transcendent equation $\Phi(s; w, z) = \pm ix$, lying in the right half-plane; the prime denotes differentiation with respect to w, and

$$\Phi'_w(s; w_j, z) = s - 2w \int_0^z \frac{\varepsilon(\zeta)}{[1 + w^2\varepsilon(\zeta)]^{1/2}} d\zeta \tag{34}$$

The poles of the integrand in Equation (31), lying at the level $\text{Re}[\Phi] = 0$, are schematically marked with crosses in Figure 1. In Figure 2, an example of exact solution to the functional equation $\Phi(s; w, z) = \pm ix$ is presented, for a linear transition layer with $\varepsilon(z) = \varepsilon_0 + (\varepsilon_1 - \varepsilon_0)(z - z_0)/(z_1 - z_0)$. Thus, for a given vertical permittivity distribution $\varepsilon(z)$, the calculation of the essential Green function component, corresponding to the signal due to partial subsurface reflections, requires numerical localization of the poles, summation of the corresponding residues, and substitution of the kernel $K(s; x, z)$ into the integral $G_r(s; x, 0)$.

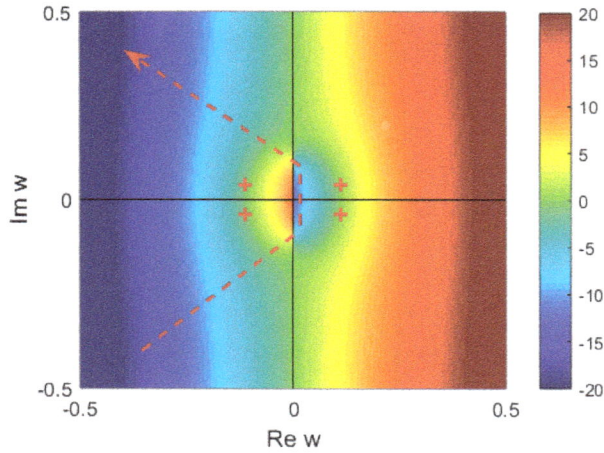

Figure 1. Color map of the exponential in Equation (29), with the steepest descent path, and poles of the integrand.

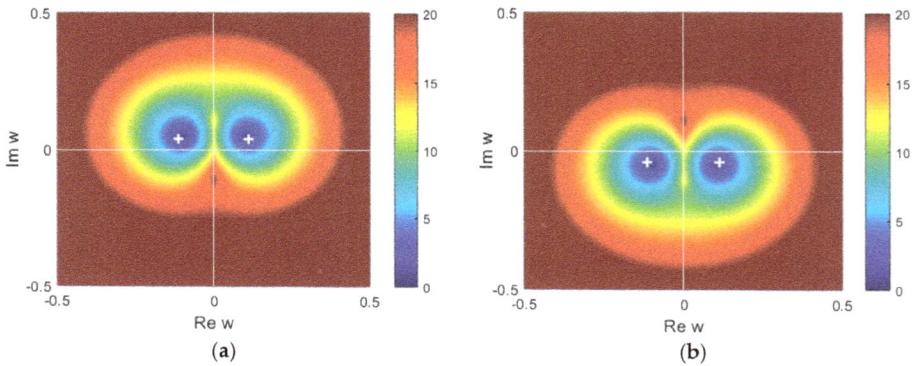

Figure 2. Roots of the functional Equation (35), corresponding to the: upper (**a**); and lower (**b**) signs in the right-hand side, for $x = 4$ m, $\varepsilon_0 = 81$, $\varepsilon_1 = 9$, $z_0 = 2$ m, $z_1 = 6$ m.

Let us finally comment that usually modification of the integration path is performed in order to achieve faster integration (and consequently accelerate forward-modelling calculations)—for example, [23] is a proper reference on this topic. In our case, this procedure allows one to reduce the integral to a sum of residues (33), completely avoiding numerical quadrature.

2.3. Geometrical-Optics Interpretation

Equations (31)–(33) provide an explicit approximate representation of the time-domain Green function for an arbitrary permittivity profile $\varepsilon(z)$, which, in combination with the Duhamel principle [34], solves the electromagnetic forward problem for an arbitrary probing pulse. The key point in the numerical implementation resides in the evaluation of the following functional equation, to determine the poles $w_j(s; x, z)$.

$$\Phi(s; w, z) \equiv ws - 2 \int_0^z \left[1 + w^2 \varepsilon(\zeta) \right]^{1/2} d\zeta = \pm ix \tag{35}$$

By inspecting Equation (35), it can be noted that one of its solutions coincides with the geometro-optical (GO) one, rendering a minimum to the Fermat functional:

$$S(p, \psi, x, z) \equiv \int \varepsilon^{1/2} d\sigma = \\ xp/\cos\psi + 2\int_0^z [\varepsilon(\zeta) - p^2]^{1/2} d\zeta, \quad p = i/w \tag{36}$$

(optical path from an antenna element in $x_0 = z_0 = 0$, $y_0 = x \tan \psi$, to the receiver point in $(x, 0, 0)$, with intermediate specular reflection from $\zeta = z$ plan).

By differentiating Equation (36) with respect to p, ψ and by equating the derivatives $\partial S/\partial p$ and $\partial S/\partial \psi$ to zero, we have:

$$\psi = 0, \quad x = 2p \int_0^z [\varepsilon(\zeta) - p^2]^{-1/2} d\zeta, \\ s = 2\int_0^z \varepsilon(\zeta) [\varepsilon(\zeta) - p^2]^{-1/2} d\zeta \equiv S(x, z). \tag{37}$$

Here, $p = P(x, z)$ is the solution of the second Equation (37), $s = S(x, z)$ being the result of its substitution into the last line of Equation (37), which, apparently, assures the fulfillment of the identity in Equation (35).

As follows from the laws of geometrical optics [19], Equation (37) correspond to a ray trajectory in a horizontally-layered medium, which starts from $(x = 0, y = 0, z = 0)$ at an angle $\theta_0 = \arcsin\left[P(x, z)\varepsilon_0^{-1/2}\right]$ with respect to the z-axis and comes to the observation point $(x = X, y = 0, z = 0)$ after specular reflection from a virtual mirror $\zeta = z$ (see Figure 3). This trajectory lies in the vertical plane $y = 0$ and, evidently, provides the shortest optical path from the line current source to the observation point, among ones touching the given level $\zeta = z$.

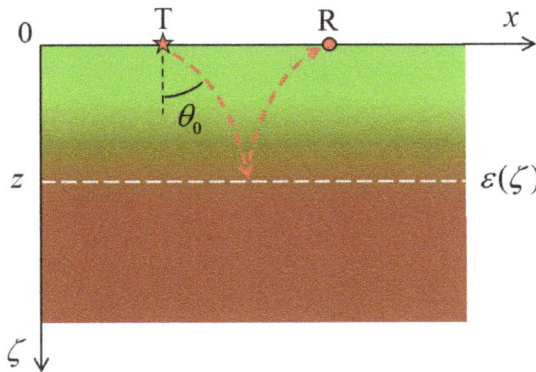

Figure 3. Partial reflection of the probing pulse due to the permittivity gradient. T and R are the transmitter and receiver positions, respectively. The red dashed line represents the GO path, while the white dashed line refers to an effective level of partial reflection.

From physical considerations, one may expect that the main contribution to the time-domain Green function G_r is due to the values of w closest to GO. Ray interpretation suggests an efficient method to solve the functional Equation (36). Let us assume $s = S(x, z) + \mu$, $w = \pm i/(p + v)$, $|\mu| \ll S$,

$|\nu| \ll p$. Substitution of these quantities into Equation (36) gives an approximation, valid for small values of ν:

$$S + \mu = (p + \nu)x + 2p \int_0^z \left[\varepsilon(\zeta) - (p + \nu)^2\right]^{1/2} d\zeta \approx$$

$$\approx (p + \nu)x + 2 \int_0^z \varepsilon(\zeta)\left[\varepsilon(\zeta) - p^2\right]^{-1/2} d\zeta + \tag{38}$$

$$+ 2p(p + \nu) \int_0^z \left[\varepsilon(\zeta) - p^2\right]^{-1/2} d\zeta - 2\nu^2 \int_0^z \left[\varepsilon(\zeta) - p^2\right]^{-3/2} d\zeta.$$

By taking into account the GO Equation (36) and defining

$$T(x,z) = 2 \int_0^z \left[\varepsilon(\zeta) - p^2\right]^{-3/2} d\zeta, \quad p = P(x,z), \tag{39}$$

we get $\mu \approx -T\nu^2/2$. As only the poles $w = \pm i/(p + \nu)$ lying in the right half-plane give a contribution, we define $\nu = \pm i(2\mu/T)^{1/2} = \pm i\{2[s - S(x,z)]/T(x,z)\}^{1/2}$ and obtain their approximate representation:

$$w_\pm(s; x, z) = \frac{1}{\{2[s - S(x,z)]/T(x,z)\}^{1/2} \mp iP(x,z)}. \tag{40}$$

Now, it is easy to calculate the functions in Equations (32) and (34):

$$C(w_\pm, z) \approx \mp \frac{ip(\varepsilon_0 + 1)\left(\varepsilon_0 - p^2\right)^{1/2}}{[\varepsilon(z) - p^2]\left[(\varepsilon_0 - p^2)^{1/2} + (1 - p^2)^{1/2}\right]^2} \tag{41}$$

$$\Phi'_w(s; w_\pm, z) = \mu \mp p(2T\mu)^{1/2}, \quad \mu = s - S(x,z)$$

and the kernel of the time-domain Green function:

$$K(s; x, z) \approx \frac{2ip\left(\varepsilon_0 - p^2\right)^{1/2} \{2T(x,z)[s - S(x,z)]\}^{-1/2}}{c[\varepsilon(z) - p^2]\left[(\varepsilon_0 - p^2)^{1/2} + (1 - p^2)^{1/2}\right]^2} \tag{42}$$

To conclude, in this quasi-optical approximation the search for the poles from Equation (35), which depend on the virtual reflection depth z and normalized time s, is reduced to the calculation of the horizontal GO impulse $P(x,z)$, depending only on z, and to the computation of the integrals $S(x,z)$ and $T(x,z)$ via the explicit formulas given in Equations (37) and (39).

2.4. Quasi-Vertical Sounding

The above analysis reduces our time-domain back-scattering problem to the standard geometrical optics. This provides an efficient modeling tool for the GPR probing of a horizontally layered subsurface media. However, the obtained integral representation Equation (42) is still too heavy for practical applications and for attempts to solve inverse problems. A further simplification can be achieved if the separation between the transmitter and receiver antennas is relatively small. Such a situation is encountered when probing deeper layers of the subsurface medium ($h \geq 10$ m) with a typical antenna offset $x \sim 23$ m. In this case, the angles of arrival are small, we can consider $p/\varepsilon^{1/2} \sim x/(2z)$ as a small parameter and look for the roots of Equation (35) by applying the following approximation:

$$w \approx i/p \to \infty, \quad s = \pm ix/w + 2 \int_0^z \left[\varepsilon(\zeta) + w^{-2}\right]^{1/2} d\zeta \approx \tag{43}$$
$$L(z)w^{-2}/2 \pm ix/w + S_0(z), \quad |w| \to \infty$$

where

$$S_0(z) = 2 \int_0^z \varepsilon(\zeta)^{1/2} d\zeta, \quad L(z) = 2 \int_0^z \varepsilon(\zeta)^{-1/2} d\zeta. \tag{44}$$

In such a way, the equation becomes a quadratic one:

$$(s - S_0)w^2 \mp ixw - L/2 = 0 \tag{45}$$

having two roots in the right half-plane:

$$w_\pm(s; x, z) = \left\{ \pm ix + \left[2L(s - S_0) - x^2\right]^{1/2} \right\} / [2(s - S_0)] \tag{46}$$

The functions introduced above take the form

$$C(w_\pm, z) \approx \varepsilon_0^{1/2} / \left[w_\pm \varepsilon(z) \left(\varepsilon_0^{1/2} + 1 \right)^2 \right], \quad |w| \to \infty$$

$$\Phi'_w(s; w_\pm, z) = 2(s - S_0) \frac{\mp i [2L(s - S_0)/x^2 - 1]^{1/2}}{1 \mp i [2L(s - S_0)/x^2 - 1]^{1/2}} \tag{47}$$

and the kernel of the integral Equation (28) becomes:

$$K(s; x, z) = \varepsilon_0^{1/2} \left(\varepsilon_0^{1/2} + 1 \right)^{-2} \left[2L(z)(s - S_0) - x^2 \right]^{-1/2} / [c\varepsilon(z)] \tag{48}$$

Thus, for a moderate separation between the antennas, $x < 2z$, the essential component of the Green function, responsible for the signal reflected by the permittivity gradients, can be written in a closed form:

$$G_r(s; x, z) = \frac{\varepsilon_0^{1/2}}{c\left(\varepsilon_0^{1/2} + 1\right)^2} \int_0^{Z^+} \frac{\varepsilon'(z)}{\varepsilon(z)} \left[2L(z)(s - S_0) - x^2 \right]^{-1/2} dz. \tag{49}$$

Here, Z^+ is a root of the equation $2L(z)(s - S_0) - x^2 = 0$, corresponding to the depth level from where the partly reflected signal starts towards the receiver, along a geometric-optical path. In virtue of the assumption $p \sim i/w$, our approximation is similar to the method of coupled parabolic equations that was used by Claerbout in the problem of seismic prospecting [35].

3. Numerical Integration

To carry out an accurate numerical quadrature for Equation (49), it is necessary to take into account the algebraic singularity of the kernel $K(s; x, z)$ at the end point Z^+.

Let us introduce the notation $F(z) = \varepsilon'(z)/\varepsilon(z)$, $R(z) = 2L(z)S_0(z) + x^2$ and a uniform discretization grid $z_\mu = [0 : h : z_m]$, where z_m corresponds to $Z^+(s_m)$. By decomposing the integral in Equation (49) into a sum of integrals over the intervals $(z_{\mu-1}, z_\mu)$, we have:

$$G_r(s; 0, z) = \frac{\varepsilon_0^{1/2}}{c\left(\varepsilon_0^{1/2} + 1\right)^2} \sum_{\mu=1}^{m} \int_{z_{\mu-1}}^{z_\mu} \frac{F(z)dz}{[2L(z)(s - S_0) - x^2]^{1/2}}. \tag{50}$$

By expanding the functions $F(z)$, $L(z)$ and $R(z)$ in Taylor series, we find:

$$G_r(s; 0, z) = \frac{\varepsilon_0^{1/2}}{c\left(\varepsilon_0^{1/2} + 1\right)^2} \times$$

$$\sum_{\mu=1}^{m} \int_{z_{\mu-1}}^{z_\mu} \frac{\left[F_{\mu-1} + F'_{\mu-1}(z - z_{\mu-1}) + O(h^2)\right] dz}{\left\{2\left[L_{\mu-1} + L'_{\mu-1}(z - z_{\mu-1})s_m\right] - R_{\mu-1} - R'_{\mu-1}(z - z_{\mu-1}) + O(h^2)\right\}^{1/2}} \tag{51}$$

where $z_\mu = \mu h$, $F'_{\mu-1} = (F_\mu - F_{\mu-1})/h$, etc.

Thus, we have reduced Equation (49) to a sum of standard algebraic integrals that may have singularity of the order $-1/2$:

$$\int_{z_{\mu-1}}^{z_\mu} (A_\mu + B_\mu \zeta)(C_\mu + D_\mu \zeta)^{-1/2} d\zeta. \tag{52}$$

In Equation (52), the following quantities have been introduced:

$$A_\mu = \mu F_{\mu-1} - (\mu - 1)F_\mu, \, A_1 = \varepsilon'(0)/\varepsilon(0) = 0,$$

$$B_\mu = F_\mu - F_{\mu-1}, \, B_1 = F_1,$$

$$C_\mu^m = [\mu L_{\mu-1} - (\mu - 1)L_\mu]R_m/L_m - \mu R_{\mu-1} + (\mu-1)R_\mu, \tag{53}$$

$$C_1^m = R_1 = x^2, \, D_\mu^m = [L_\mu - L_{\mu-1}]R_m/L_m - R_\mu + R_{\mu-1}, \, D_1^m = P_1 R_m / P_m - R_1 + x^2$$

By substituting the well-known analytical expression of integrals Equation (52) into Equation (51), we obtain a numerical quadrature, accurate to $O\left(h^{3/2}\right)$ and suitable to correctly describe weak singularity of the Green function on the reflected wave front:

$$G_r(s;0,z) = \frac{2h\varepsilon_0^{1/2}}{c\left(\varepsilon_0^{1/2}+1\right)^2} \times$$
$$\sum_{\mu=1}^m \left\{ \left(A_\mu - B_\mu C_\mu^m / D_\mu^m\right)\left[\left(C_\mu^m + \mu D_\mu^m\right)^{1/2} - \left(C_\mu^m + (\mu-1)D_\mu^m\right)^{1/2}\right]/D_\mu^m \right. \tag{54}$$
$$\left. + B_\mu \left[\left(C_\mu^m + \mu D_\mu^m\right)^{3/2} - \left(C_\mu^m + (\mu-1)D_\mu^m\right)^{3/2}\right]/3D_\mu^m. \right.$$

4. Results and Discussion

To estimate the accuracy of our approximate analytical solution to the wave Equation (12), we compare our results with those obtained by using the open-source FDTD simulator gprMax [4]. Input data for gprMax are: the geometrical and electromagnetic parameters of uniform fragments of the computation domain, the positions of the transmitter and receiver, and the time-domain waveform of the excitation current. In this paper, we are considering a horizontally layered medium with permittivity gradually varying with depth. In our mathematical formulation of the problem, such medium is defined via the analytical expression of the permittivity distribution $\varepsilon(z)$, to be introduced into the integral representation of the signal received by the radar. As gprMax deals with piecewise-uniform models, to carry out a thorough and accurate comparison between our method and the FDTD technique, we use a uniform discretization grid where the discretization step is the same as in gprMax calculations. For the excitation current waveform, we use the derivative of Gaussian pulse, which in gprMax is referred to as "Ricker waveform":

$$I(t) = -4\pi^2 f_c^2(t - 1/f_c)\exp\left[-2\pi^2 f_c^2(t - 1/f_c)^2\right] \tag{55}$$

Here, f_c is the central frequency of the pulse. In the examples presented below, $f_c = 20$ MHz.

An idealized model of subsurface medium is shown in Figure 4. It consists of a uniform layer with dielectric permittivity ε_0 (for $0 \leq z \leq z_0$) and a half-space with dielectric permittivity ε_1, separated by a transition layer where the dielectric permittivity is $\varepsilon(z)$, $z_0 \leq z \leq z_1$. Note that here we call ε_0 the relative permittivity of the uniform upper layer occupying the region $0 \leq z \leq z_0$ (not the absolute permittivity of a vacuum in SI unit system). The transmitting and receiving antennas, T and R, are placed on the earth surface, at $z = 0$. In the figure, the components of the emitted electromagnetic pulse are shown: **aw** and **gw** indicate the "aerial" and "ground" waves, respectively; **iw** is the incident wave impinging on the transition layer; and **rw** and **tw** are the waves reflected and transmitted by the transition layer, respectively.

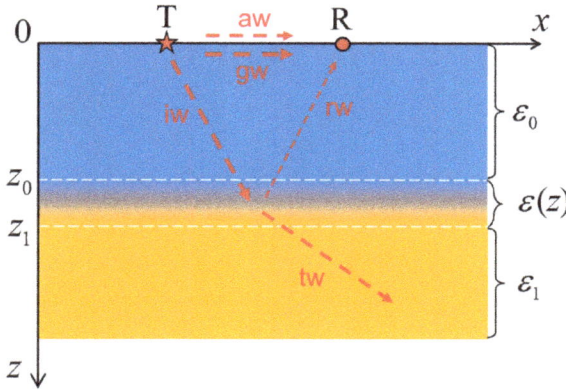

Figure 4. Geometry of the simulated scenario and schematic representation of the radar signal components.

Figure 5a,c model the depth distributions of the dielectric permittivity, corresponding to a gradual transition from pure water ($\varepsilon_0 = 81$) to a hard soil ($\varepsilon_1 = 25$), in a sweet-water pond with silty bottom. The permittivity profile of the transition layer is given by

$$\varepsilon(z) = \frac{\varepsilon_0 + \varepsilon_1}{2} + \frac{\varepsilon_0 - \varepsilon_1}{2} \sin\left[\frac{\pi}{z_1 - z_0}\left(z - \frac{z_0 + z_1}{2}\right)\right] \tag{56}$$

corresponds to a transition layer which is located in $6\text{ m} \leq z \leq 8\text{ m}$ for Figure 5a, and $4\text{ m} \leq z \leq 10\text{ m}$ for Figure 5c. The distance between the transmitter and receiver antennas is $X = 3\text{ m}$. In Figure 5b,d, synthetic radargrams (A-scans) are presented for the scenarios of Figure 5a,c, respectively. Simulations were performed using both our coupled-WKB method (solid line) and gprMax (dashed line). The first double pulse corresponds to the direct surface wave, propagating along both sides of the ground-air interface. A weak signal with longer delay arises due to the cumulative partial reflection from the non-uniform transition layer.

One can note that, notwithstanding the approximate character of WKB method and the additional errors due to the quasi-vertical approximation, the agreement between the two methods is excellent. It is worth pointing out that our semi-analytical approach, implemented in Matlab R2015 (MathWorks, Inc., Natick, MA, USA), provides a computation time about 100 times shorter than gprMax, version 3.0.0b13 [3–5].

A satisfactory qualitative agreement between FDTD and coupled-WKB results persists even for a larger separation between the antennas, when the propagation path is far from the vertical: see Figure 6a,b, where $X = 7\text{ m}$ and 11 m. These plots show an interesting effect: a higher amplitude of the reflected signal when the propagation path is longer. This paradoxical behavior, predicted both by the coupled WKB method and by gprMax, can be explained by considering that, when the separation between the antennas is increased, the propagation path follows a direction which is closer to the total-reflection angle.

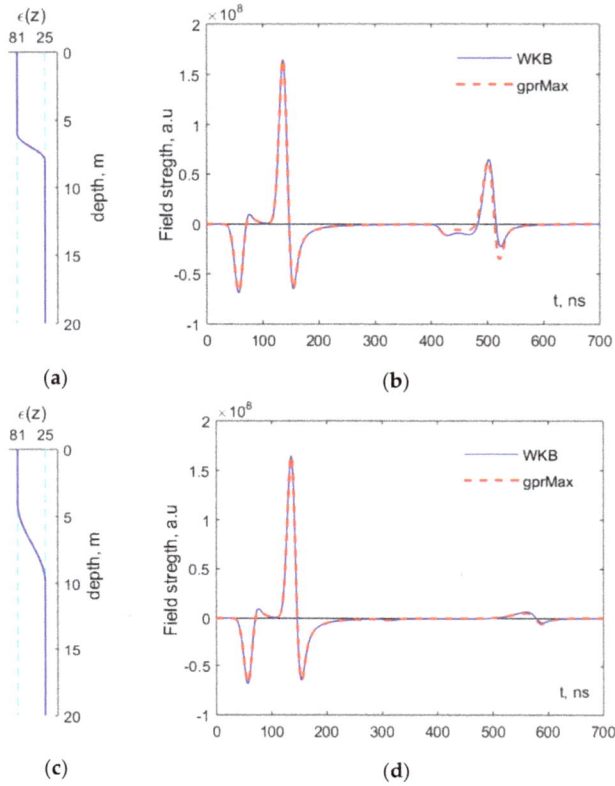

Figure 5. Two vertical profiles of the dielectric permittivity are shown in (**a,c**). The corresponding simulated A-scans (coupled WKB: solid line, gprMax: dashed line) are shown in (**b,d**), respectively.

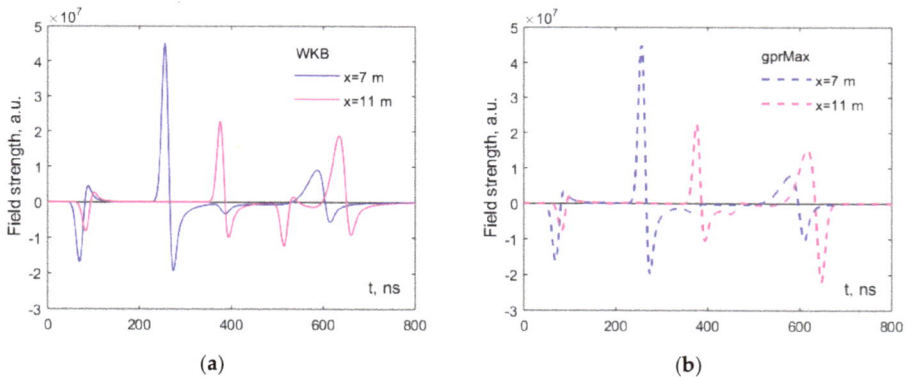

Figure 6. A-scans simulated with: (**a**) our coupled WKB method (solid lines); and (**b**) gprMax (dashed lines), for larger distances between transmitter and receiver.

An application of the developed coupled-WKB simulation technique to a real case study is now presented. In particular, the method is applied to the interpretation of GPR radargrams collected on Lake Chebarkul (Chelyabinsk Region, Russia), on the slopes of the southern Urals, during the IZMIRAN field mission in search of a big fragment of the Chelyabinsk meteorite residing in the silty lake floor [31]. The Chelyabinsk meteor reached the Earth on 15 February 2013, and our data were obtained in March 2013 with a low-frequency "Loza-N" GPR [36].

According to divers' witnesses, the bottom of the lake was covered with a soft silt layer, 2–3 m thick. The experienced "Loza-N" operators assumed that the protracted signals received by the GPR were due to partial reflection from such a loose silt layer. Our numerical simulations with coupled-wave WKB confirm this hypothesis. Indeed, in Figure 7a we present an experimental A-scan showing the aforementioned effect of cumulative partial reflection from a thick layer of bottom sludge; and in Figure 7b we display the numerical results obtained within the framework of our coupled-WKB approximation.

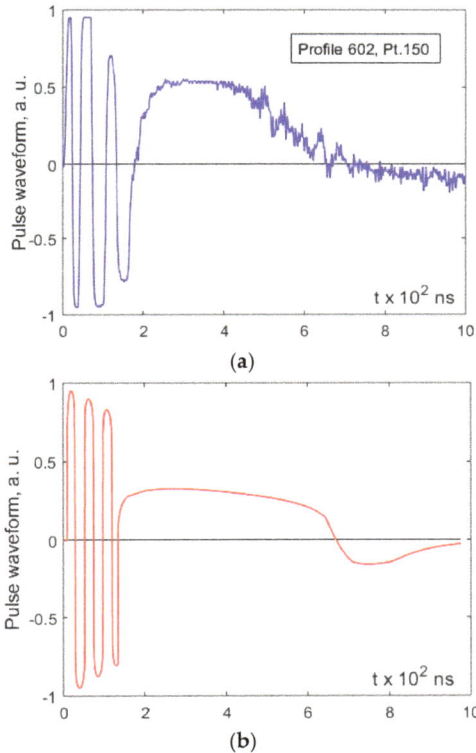

Figure 7. (a) Experimental A-scan with a protracted reflected pulse, recorded on the iced surface of Lake Chebarkul by GPR probing the silty bottom; and (b) synthetic A-scan calculated using our coupled-wave WKB approach.

The following values are employed to carry out the simulation. For the pulse radiated by the line source, a damped sinusoid $I(t) = \sin(\alpha t)\exp(-\beta t)$ is used, with central frequency $f_c = (\alpha^2 - \beta^2)^{1/2} = 20$ MHz. For the ice layer, the relative permittivity is assumed to be $\varepsilon_i = 3$, its thickness is $z_i = 0.8$ m. For the transition silt layer, an approximate permittivity profile deduced from the divers' information and empirically optimized by comparing with the experimental A-scan is $\varepsilon(z) = \varepsilon_0 + (\varepsilon_1 - \varepsilon_0)\tan h^4[(z - z_0)/(z_1 - z_0)]$, with $\varepsilon_0 = 81$, $\varepsilon_1 = 9$, $z_0 = 1$ m and $z_1 = 7$ m.

The pulse received by the radar is calculated by convolving the approximate Green function with the chosen current pulse waveform (Duhamel integral [34]), as follows:

$$E(t) = \frac{1}{c} \int_0^t \frac{dI}{dt}(ct - s')G(s'; x, 0)ds' \tag{57}$$

As can be appreciated by comparing Figure 7a,b, the simulation qualitatively reproduces the aforementioned effect of protracted reflected pulse; the fast oscillating signal in the left part of the plot corresponds to the direct surface wave and its reflection from the lower ice surface. The similarity of the measured and simulated A-scans confirms the applicability of our approach to real scenarios.

We finally present another possible application of the developed method, namely the interpretation of data that could be obtained by GPR probing the lunar regolith during a planned space mission. It is known that a considerable amount of ice is accumulated in lunar regolith near the poles, which may be used in future space missions. To localize and estimate the available volumes of water, mechanical drilling of lunar regolith [37] can be complemented with GPR probing. The example presented in Figure 8 shows that our semi-analytical approach can be successfully used to model and simulate the electromagnetic propagation of a GPR pulse in the upper regolith layer, characterized by smooth gradients of dielectric permittivity due to the changing ice proportion. For this example, we calculate synthetic A-scans and the reference regolith parameters are taken from the literature [33]. A typical permittivity profile is plotted in Figure 8a and the corresponding A-scan is presented in Figure 8b. The main received signal is a bipolar pulse due to the direct wave propagating from the transmitting to the receiving antenna. The backward reflection E_{ref} is too weak to be seen in the scale of the plot, we therefore multiplied it by 10^3 and plotted it as a separate curve. Its waveform reveals the cumulative character of the return signal, which is a superposition of partial reflections from the non-uniform transition layer. Despite the weak power level, the backward reflection can be confidently detected with a deep penetration GPR [36]. Valuable information on the smooth subsurface inhomogeneity can be retrieved by comparing simulation results produced with our method and experimental results.

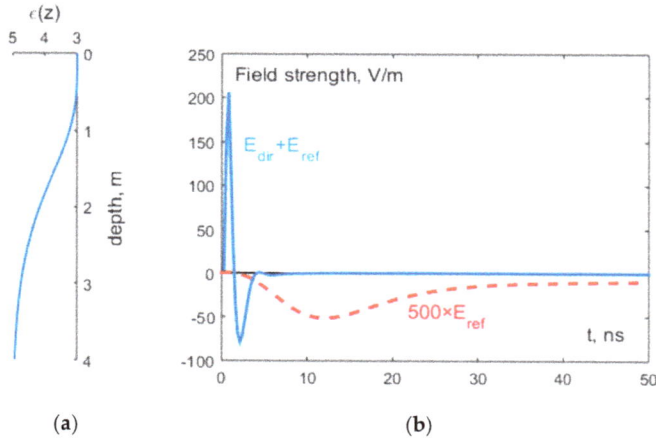

Figure 8. GPR probing of lunar regolith (numerical simulation): (**a**) reference permittivity profile; and (**b**) received pulse, including both the direct wave and weak subsurface reflection (blue), and magnified subsurface reflection (red).

Remote Sens. **2018**, *10*, 22

5. Conclusions

We extended the coupled-wave Wentzel-Kramers-Brillouin method ("two-way WKB" approximation) to the case of Ground-Penetrating Radar (GPR) probing of a horizontally-layered dielectric half-space. In particular, we derived an analytical representation of the electromagnetic field excited by a synchronous ultra wideband current pulse in a thin wire stretched along the ground-air interface. A bistatic sounding scheme, commonly used in GPR surveys, was considered. A physical interpretation of the obtained solution was given in terms of geometrical optics and partial reflections from subsurface permittivity gradients. An efficient numerical algorithm was implemented, including an approximate solution of a complex eikonal equation and a high-precision quadrature of the arising singular integrals. Similarities with the coupled parabolic equation method were pointed out.

While our analytical approach is suitable to model the GPR sensing of shallower and deeper layers, the numerical implementation presented in this paper assumes a relatively small separation between the transmitter and the receiver, which allows to shorten the calculation times and makes our method suitable to be embedded into inversion techniques requiring the fast solution of a high number of forward-scattering problems. Specifically, Formulas (32–34) of Section 2.2 yield a general representation of the time-domain coupled-WKB Green function for arbitrary depths and antenna separation. It holds also for the geometric-optical approximation (42) of Section 2.3, valid for short pulses propagating within a narrow strip ("wave film") behind the GO wave front. Only quasi-vertical approximation discussed in Section 2.4 and implemented in Section 3 requires the assumption of depths exceeding the separation between the transmitter and the receiver. This simplification is valid when modelling monostatic GPR systems (where the same antenna is used for transmitting and receiving signals), as well as bistatic systems if the probed layers are deep enough.

Numerical results of our method were compared with finite-difference time-domain (FDTD) calculations, with very good agreement. Then, two applications to real scenarios were presented. First, our technique was applied to the interpretation of GPR radargrams collected on Lake Chebarkul, in search of a fragment of the Chelyabinsk meteorite. We showed how numerical simulation helps to analyze the protracted return signals originated in smooth transition layers of subsurface dielectric medium. The second example suggests that our method can be used for the estimation of water content in lunar regolith, the upper layer of which contains smooth gradients of permittivity due to gradually increasing fraction of ice.

The good accuracy and numerical efficiency of our semi-analytical computational approach make promising its further development. The approach can be extended to the case of a half-space where the permittivity varies in two directions. Furthermore, we plan to take into account the dissipative and frequency-dispersive behavior of materials by using a complex-valued model of dielectric permittivity in the frequency-domain. The finite length of the antennas and a three-dimensional (3D) gradual variation of the medium parameters will be introduced in a 3D version of the algorithm. We also wish to explore possibilities of hybridization of our approach with FDTD and time-domain integral-equation methods, to capitalize on the strengths of each technique.

Acknowledgments: A considerable part of this work was carried out during two Short-Term Scientific Missions (STSMs) funded by COST (European Cooperation in Science and Technology) Action TU1208 "Civil engineering applications of Ground Penetrating Radar" (www.cost.eu, www.GPRadar.eu). Both missions were kindly hosted by the National Institute of Telecommunications of Poland, in Warsaw. The authors thank COST for funding and supporting the Action TU1208 "Civil engineering applications of Ground Penetrating Radar". The authors thank the National Institute of Telecommunications of Poland for hosting the STSMs and for paying the article processing charge.

Author Contributions: Alexei Popov and Marian Marciniak developed the theoretical approach to the problem. Igor Prokopovich implemented the method and obtained preliminary numerical results during the first STSM, under the supervision of Marian Marciniak. The analysis was completed by Alexei Popov during the second STSM, in collaboration with Marian Marciniak. Lara Pajewski substantively revised the work carried out during the two STSMs. Then, Igor Prokopovich perfected the numerical implementation and calculated the results presented in this paper, under the supervision of Alexei Popov. All authors contributed to the analysis, interpretation and discussion of results. The text of this paper was written by Lara Pajewski, Alexei Popov and Igor Prokopovich, and was checked and revised by all the authors.

Conflicts of Interest: The authors declare no conflict of interest.

References

1. Benedetto, A.; Pajewski, L. (Eds.) *Civil Engineering Applications of Ground Penetrating Radar*; Book Series: "Springer Transactions in Civil and Environmental Engineering"; Springer: New Delhi, India, 2015; p. 385, ISBN 978-3-319-04813-0. [CrossRef]
2. Persico, R. *Introduction to Ground Penetrating Radar: Inverse Scattering and Data Processing*; Wiley-IEEE Press: New York, NY, USA, 2014; p. 392, ISBN 978-1-118-30500-3.
3. Giannopoulos, A. Modelling ground penetrating radar by GprMax. *Constr. Build. Mater.* **2005**, *19*, 755–762. [CrossRef]
4. Warren, C.; Giannopoulos, A.; Giannakis, I. An advanced GPR modelling framework—The next generation of gprMax. In Proceedings of the 8th International Workshop on Advanced Ground Penetrating Radar (IWAGPR 2015), Florence, Italy, 7–10 July 2015; pp. 1–4. [CrossRef]
5. Warren, C.; Giannopoulos, A. Experimental and Modeled Performance of a Ground Penetrating Radar Antenna in Lossy Dielectrics. *IEEE J. Sel. Top. Appl. Earth Obs. Remote Sens.* **2016**, *9*, 29–36. [CrossRef]
6. Frezza, F.; Mangini, F.; Pajewski, L.; Schettini, G.; Tedeschi, N. Spectral domain method for the electromagnetic scattering by a buried sphere. *J. Opt. Soc. Am. A* **2013**, *30*, 783–790. [CrossRef] [PubMed]
7. Frezza, F.; Pajewski, L.; Ponti, C.; Schettini, G.; Tedeschi, N. Cylindrical-Wave Approach for Electromagnetic Scattering by Subsurface Metallic Targets in a Lossy Medium. *J. Appl. Geophys.* **2013**, *97*, 55–59. [CrossRef]
8. Frezza, F.; Pajewski, L.; Ponti, C.; Schettini, G.; Tedeschi, N. Electromagnetic Scattering by a Metallic Cylinder Buried in a Lossy Medium with the Cylindrical-Wave Approach. *IEEE Geosci. Remote Sens. Lett.* **2013**, *10*, 179–183. [CrossRef]
9. Bourlier, C.; Le Bastard, C.; Pinel, N. Full wave PILE method for the electromagnetic scattering from random rough layers. In Proceedings of the 15th International Conference on Ground Penetrating Radar (GPR 2014), Brussels, Belgium, 30 June–4 July 2014; pp. 545–551. [CrossRef]
10. Poljak, D.; Dorić, V. Transmitted field in the lossy ground from ground penetrating radar (GPR) dipole antenna. *WIT Trans. Model. Simul.* **2015**, *59*, 3–11. [CrossRef]
11. Poljak, D.; Dorić, V.; Birkić, M.; El Khamlichi Drissi, K.; Lallechere, S.; Pajewski, L. A simple analysis of dipole antenna radiation above a multilayered medium. In Proceedings of the 9th International Workshop on Advanced Ground Penetrating Radar (IWAGPR 2017), Edinburgh, UK, 28–30 June 2017; pp. 1–6. [CrossRef]
12. Winton, S.C.; Kosmas, P.; Rappaport, C.M. FDTD simulation of TE and TM plane waves at nonzero incidence in arbitrary layered media. *IEEE Trans. Antennas Propag.* **2005**, *53*, 1721–1728. [CrossRef]
13. Diamant, R.; Fernandez-Guasti, M. Light propagation in 1D inhomogeneous deterministic media: The effect of discontinuities. *J. Opt. A Pure Appl. Opt.* **2009**, *11*, 045712. [CrossRef]
14. Zeng, Q.; Delisle, G.Y. Transient analysis of electromagnetic wave reflection from a stratified medium. In Proceedings of the 2010 Asia-Pacific International Symposium on Electromagnetic Compatibility, Beijing, China, 12–16 April 2010; pp. 881–884. [CrossRef]
15. Kaganovsky, Y.; Heyman, E. Pulsed beam propagation in plane stratified media: Asymptotically exact solutions. In Proceedings of the 2010 URSI International Symposium on Electromagnetic Theory, Berlin, Germany, 16–19 August 2010; pp. 829–832. [CrossRef]
16. Diamanti, N.; Annan, A.P.; Redman, J.D. Impact of Gradational Electrical Properties on GPR Detection of Interfaces. In Proceedings of the 15th International Conference on Ground Penetrating Radar (GPR 2014), Brussels, Belgium, 30 June–4 July 2014; pp. 529–534. [CrossRef]
17. Connor, K.M.; Dowding, C.H. *GeoMeasurements by Pulsing TDR Cables and Probes*; CRC Press: Boca Raton, FL, USA, 1999; p. 424, ISBN 9780849305863.
18. Schlaeger, S. Inversion of TDR Measurements to Reconstruct Spatially Distributed Geophysical Ground Parameter. Ph.D. Thesis, University of Karlsruhe, Karlsruhe, Germany, 2002.
19. Epstein, P.S. Reflection of waves in an inhomogeneous absorbing medium. *Proc. Natl. Acad. Sci. USA* **1930**, *16*, 627–637. [CrossRef] [PubMed]

20. Weiland, T. A discretization model for the solution of Maxwell's equations for six-component fields. *Archiv Elektronik Uebertragungstechnik* **1977**, *31*, 116–120.

21. Miller, E.K.; Poggio, A.J.; Burke, G.J. An integro-differential equation technique for the time-domain analysis of thin wire structures. I. the numerical method. *J. Comput. Phys.* **1973**, *12*, 24–48. [CrossRef]

22. Krueger, R.J.; Ochs, R.L., Jr. A Green's function approach to the determination of internal fields. *Appl. Math. Sci.* **1989**, *11*, 525–543. [CrossRef]

23. Lambot, S.; Slob, E.; Vereecken, H. Fast evaluation of zero-offset Green's function for layered media with application to ground-penetrating radar. *Geophys. Res. Lett.* **2007**, *34*, L21405. [CrossRef]

24. Larruquert, J.I. Reflectance optimization of inhomogeneous coatings with continuous variation of the complex refractive index. *J. Opt. Soc. Am. A* **2006**, *23*, 99–107. [CrossRef]

25. Grafstrom, S. Reflectivity of a stratified half-space: The limit of weak inhomogeneity and anisotropy. *J. Opt. A Pure Appl. Opt.* **2006**, *8*, 134–141. [CrossRef]

26. Landau, L.D.; Lifshits, E.M. *Quantum Mechanics: Non-Relativistic Theory*; Butterworth-Heinemann: Oxford, UK, 1977; p. 677, ISBN 0750635398.

27. Brekhovskikh, L.M. *Waves in Layered Media*, 2nd ed.; Academic Press: New York, NY, USA, 1980; 520p, ISBN 9780323161626.

28. Bremmer, H. Propagation of Electromagnetic Waves. In *Handbuch der Physik/Encyclopedia of Physics*; Flugge, S., Ed.; Springer: Berlin/Goettingen/Heidelberg, Germany, 1958; Volume 4/16, pp. 423–639.

29. Jones, A.R. Light scattering for particle characterization. *Prog. Energy Comb. Sci.* **1999**, *25*, 1–53. [CrossRef]

30. Vinogradov, V.A.; Kopeikin, V.V.; Popov, A.V. An approximate solution of 1D inverse problem. In Proceedings of the 10th International Conference on Ground Penetrating Radar (GPR 2004), Delft, The Netherlands, 21–24 June 2004; pp. 95–98.

31. Kopeikin, V.V.; Kuznetsov, V.D.; Morozov, P.A.; Popov, A.V.; Berkut, A.I.; Merkulov, S.V.; Alexeev, V.A. Ground Penetrating Radar Investigation of the Supposed Fall Site of a Fragment of the Chelabinsk Meteorite in Lake Chebarkul. *Geochem. Int.* **2013**, *51*, 636–642. [CrossRef]

32. Buzin, V.; Edemsky, D.; Gudoshnikov, S.; Kopeikin, V.; Morozov, P.; Popov, A.; Prokopovich, I.; Skomarovsky, V.; Melnik, N.; Berkut, A.; et al. Search for Chelyabinsk Meteorite Fragments in Chebarkul Lake Bottom (GPR and Magnetic Data). *J. Telecommun. Inf. Technol.* **2017**, *2017*, 69–78. [CrossRef]

33. Olhoeft, G.R.; Strangway, D.W. Dielectric Properties of the First 100 Meters of the Moon. *Earth Planet. Sci. Lett.* **1975**, *24*, 394–404. [CrossRef]

34. Fritz, J. *Partial Differential Equations*, 4th ed.; Springer: New York, NY, USA, 1982; p. 672, ISBN 978-0-387-90609-6.

35. Claerbout, J.F. *Fundamentals of Geophysical Data Processing*; Pennwell Books: Tulsa, OK, USA, 1985; 274p, ISBN 0865423059.

36. Berkut, A.I.; Edemsky, D.E.; Kopeikin, V.V.; Morozov, P.A.; Prokopovich, I.V.; Popov, A.V. Deep penetration subsurface radar: Hardware, results, interpretation. In Proceedings of the 9th International Workshop on Advanced Ground Penetrating Radar (IWAGPR 2017), Edinburgh, UK, 7–10 July 2017; pp. 1–6. [CrossRef]

37. Matveev, Y.I.; Kostenko, V.I. Problems of Moving Ultrasound Penetrative Devices in a Dispersion Medium during Drilling of the Moon's Regolith. *Acoust. Phys.* **2016**, *62*, 633–641. [CrossRef]

remote sensing

MDPI

Article

Railway Track Condition Assessment at Network Level by Frequency Domain Analysis of GPR Data

Simona Fontul [1,2,*], André Paixão [1,3], Mercedes Solla [4,5] and Lara Pajewski [6]

[1] Department of Transportation, National Laboratory for Civil Engineering—LNEC, 1700-066 Lisbon, Portugal; apaixao@lnec.pt

[2] Department of Civil Engineering, Nova University of Lisbon, 2829-516 Caparica, Portugal

[3] CONSTRUCT—LESE, Faculty of Engineering (FEUP), University of Porto, 4099-002 Porto, Portugal

[4] Defense University Center, Spanish Naval Academy, Plaza de España 2, 36900 Marín, Spain; merchisolla@cud.uvigo.es

[5] Applied Geotechnologies Research Group, University of Vigo, School of Mining Engineering, Campus Lagoas Marcosende, 36310 Vigo, Spain

[6] Department of Information Engineering, Electronics and Telecommunications, Sapienza University of Rome, 00184 Rome, Italy; lara.pajewski@uniroma1.it

* Correspondence: simona@lnec.pt; Tel.: +351-218443640

Received: 16 February 2018; Accepted: 2 April 2018; Published: 5 April 2018

Abstract: The railway track system is a crucial infrastructure for the transportation of people and goods in modern societies. With the increase in railway traffic, the availability of the track for monitoring and maintenance purposes is becoming significantly reduced. Therefore, continuous non-destructive monitoring tools for track diagnoses take on even greater importance. In this context, Ground Penetrating Radar (GPR) technique results yield valuable information on track condition, mainly in the identification of the degradation of its physical and mechanical characteristics caused by subsurface malfunctions. Nevertheless, the application of GPR to assess the ballast condition is a challenging task because the material electromagnetic properties are sensitive to both the ballast grading and water content. This work presents a novel approach, fast and practical for surveying and analysing long sections of transport infrastructure, based mainly on expedite frequency domain analysis of the GPR signal. Examples are presented with the identification of track events, ballast interventions and potential locations of malfunctions. The approach, developed to identify changes in the track infrastructure, allows for a user-friendly visualisation of the track condition, even for GPR non-professionals such as railways engineers, and may further be used to correlate with track geometric parameters. It aims to automatically detect sudden variations in the GPR signals, obtained with successive surveys over long stretches of railway lines, thus providing valuable information in asset management activities of infrastructure managers.

Keywords: Ground Penetrating Radar; railways; signal frequency analysis; track geometry; railway events; spectral domain; network level evaluation

1. Introduction

Up to date information on the real condition of transport infrastructures is crucial for asset management activities, such as efficient maintenance and rehabilitation planning, and also to assess the performance and the service quality of the transport system.

Railway track condition assessment mainly consists of measuring parameters related to the rail wearing and positioning (track geometric quality). The main geometric parameters monitored are rail gauge, cant (cross level), cant gradient (twist), longitudinal level and longitudinal alignment, measured using dedicated inspection vehicles [1,2]. The others parameters, that are assessed to analyse the track condition, are the rail profile, rail roughness and its integrity; this last one is measured using ultrasounds [1]. Generally, during maintenance operations some track components are replaced while others remain the same, in particular the substructure [1,3]. For example, the normal causes of a rail longitudinal level defects (i.e., presence of ballast pockets, fouled ballast, poor drainage, subgrade settlements and transition problems) are therefore not detected by this monitoring procedure [4–6].

An important tool, increasingly used for railway monitoring, is Ground Penetrating Radar (GPR), which is a non-destructive testing (NDT) technique that provides an overall image of the subsurface [7]. This method allows operating with suspended antennas (air-coupled antennas), mounted on railway vehicles, and with high rates of data acquisition without disturbing the transport infrastructure, nor the normal traffic flow [8]. A GPR test can be carried out at different stages of the railway service life, from preliminary prospection to prepare the design of a new line, to quality control during construction [9,10], as well as to assess the causes of track geometry problems and to support maintenance planning decisions [2,11]. Nevertheless, GPR is not yet used in a systematic way for railway track characterisation and rehabilitation planning. The "traditional" way, in which the GPR signal is analysed, is still a more or less detailed interpretation of the signal in the time domain, generally performed in specific locations or for a particular purpose such as, for example, subgrade characterisation or fouling detection. These procedures are time consuming and require significant experience, being mainly undertaken by GPR professionals [2,12–17].

The method presented in this manuscript was developed for practical reasons, as a support tool to be used by the industry, namely to fulfil the Industry 4.0 ambitious objective to do an "almost live" diagnosis of the track [18,19]. The expedite macro-scale processing of GPR data proposed herein, enables continuous and efficient (although qualitative) survey and analysis of track condition. The approach contributes to promoting a wider use of GPR in railway area of application and represents a first step towards the development of quality indexes and alert levels for track network evaluation. GPR antennas can be installed in normal trains and used continuously along the year; it is therefore easy and not excessively expensive to upgrade railways networks by equipping a suitable number of trains with GPR instrumentation. The information acquired by GPR units can then be quickly processed by using the method proposed in this paper (and more advanced versions of it, to be developed in the future). Should be highlighted that the methodology developed in this study is not aimed to be used for detailed electromagnetic analysis of the GPR signal on tracks at limited locations, but rather in a preliminary stage of the GPR results analysis, at network level, to detect areas of the railway track that need a more thorough investigation to be done by using other signal processing techniques.

The approach presented in this study consists in analysing the GPR signals, in both time and frequency domains, to identify changes along significant lengths of railway line that can either correspond to known track singularities or to track pathologies. The processing focuses on specific time intervals and frequency ranges of the signal that are considered to be more representative for identifying changes in infrastructure along the track [15,17]. Four parameters were defined and implemented for track changes identification. A script was developed in MATLAB for GPR data processing that can be further integrated with track geometry data and aerial photography for a user-friendly interpretation.

The method allows three different levels of track assessment:

- diagnosis during inspection survey and detection of events and possible track defects;
- systematic events identification, in consecutive surveys, performed yearly in the same season;
- identification of track changes by comparing surveys performed in distinct climate condition along the year (e.g., summer vs. winter).

The latter differences might be caused by changes in water content due to poor drainage conditions or ballast fouling.

In Section 2, a review on the use of GPR for track assessment is provided, with special emphasis on the advantages of looking at data not only in the time domain but also in the spectral domain.

In Section 3, the data processing method developed in this study is described and some examples are presented. In Section 4, the method is applied to a real-world case study; moreover, recommendations are suggested for combining GPR results with track geometry assessment for condition evaluation. Conclusions are presented in Section 5.

2. Ground Penetrating Radar for Track Assessment

2.1. Overview on the Use of GPR for Railway Monitoring

In recent years, there has been an increased interest in using NDT techniques for track evaluation in order to better characterise its behaviour. Apart from GPR, the main geophysical methods applied for railway monitoring are the Electric Resistivity Tomography (ERT), seismic and gravimetry [20–24]. A review on the geophysical methods generally used on railways is presented in [20]. For example, Falling Weight Deflectometer (FWD), Light FWD and prototype vehicles such as Rolling Stiffness Measurement Vehicle were used, at research level, to determine areas with low density, voids and damp [3,25,26]. In [21], a geophysical study is presented to evaluate the stability of railways combining 3D GPR, ERT and microgravity, which allowed for thicknesses measurements and for the detection of collapse, deformation and other defects such as voids. The main limitation of geophysical methods is the reduced area that is evaluated at a time. Therefore, their use can be combined, being first detect with GPR the areas that require more detailed evaluation and after that, more located measures can be performed, such as ERT, SASW and microgravimetry for a more complete diagnosis of in situ information. Additionally, geotechnical prospections and load tests can be performed in representative locations for structural evaluation of the track [20].

For the purpose of this study, namely continuous assessment of significant lengths of railway without traffic interference, the GPR represents the most appropriate tool among the geophysical methods. The main applications to track evaluation are presented herein.

For maintenance purposes, many GPR studies were focused on the thicknesses measurement of the ballast and sub-ballast layers, as well as the material characterization [5,27–30]. There are also some GPR studies with a particular research interest to assess ballast fouling and moisture content [31]. In a scenario where the infrastructure has been in service for several decades, the breakdown of the ballast over time and the upward migration of the fine soil particles from the foundation, along with capillary water, may affect the track structural performance and eventually lead to its failure, for example, in terms of excessive accumulated plastic deformation of the subgrade layers. The early stage detection of ballast fouling is therefore a crucial factor to extend its life cycle. To that effect, GPR has been successfully used to distinguish clean from fouled ballast [32–34]. This differentiation was possible because clean ballast is associated with the diffraction of the electromagnetic waves in its open voids [14]. Some experimental studies have analysed the influence of different fouling levels in the electromagnetic waves. In [35], four different levels were simulated (from 0% to 76% fouling) and two different air-coupled antennas were used with frequencies of 1 GHz and 2 GHz. Regarding the assessment of the moisture content in the ballast, this parameter was easily evaluated by GPR as the dielectric value of the ballast increases with the presence of water [36]. In [37,38], complementary laboratory tests were performed using different antennas (air-coupled antennas and ground-coupled antennas) with different frequencies between 400 MHz and 2 GHz, in order to evaluate the dielectric constant values for different levels of fouled ballast (from 0% to 55%) and different water contents (from 6% to 14%), which demonstrated that the increase of the dielectric value with water content is particularly more relevant in fouled ballast, as the fine soil particles decrease the drainage capabilities of the material. For fouling evaluation, particular studies were based on the development of new GPR signals processing and interpretation. For a better detection of fouled ballast, more complete interpretations were achieved based on scattering and entropy analysis [39–41], spectral analysis [2,42], as well as wavelet and Fourier Transform analysis [17,43]. With respect to the assessment of the subgrade condition and track defects, the GPR method can be used to detect anomalies in the ballast and sub-ballast layers such as voids, water pockets, or subgrade settlement, which allows also for a deeper inspection into the track structure [5,44].

The GPR antennas more commonly used for railway investigation are air-coupled (horn) antennas operating in a frequency range from 1 GHz to 2 GHz for ballast thicknesses measurements and quality control, but also ground-coupled antennas operating in a wider frequency range, from 100 MHz to 2 GHz, for deeper prospection, such as sub-ballast condition and subgrade defect detection. In [45] a comparison was done between air-coupled and ground-coupled antennas (with frequencies on the order of 1 GHz and 2 GHz) to assess the subgrade condition. The obtained results demonstrated that, in terms of resolution, the air-coupled antennas are more suitable for measuring layer continuity, whereas ground-coupled antennas provide better signal to noise ratio and better penetration, thus helping in the identification of anomalous areas such as cracking of the subgrade.

More recently, the use of multi-frequency or array GPR systems has demonstrated to be an effective tool to detect defects as it combines different resolutions and penetrations at the same time to assess defects of various sizes located at different depths [46,47]. In [48], full-resolution three-dimensional (3D) imaging allowed improving the interpretation of the subsoil and to determine layer discontinuities and damage areas on the track.

While GPR is customarily used at road network level in many countries, GPR inspections at railway network level are still uncommon. In [49], the author discusses a railway assessment practice in the United Kingdom; the approach presented therein includes the use of GPR. In Croatia, the EU-funded DESTination RAIL project [50], which is ending in April 2018, has recently investigated efficient solutions for a number of problems faced by European railway network managers; the proposed approach for the assessment of track conditions includes the use of GPR as primary tool, to be employed in combination with other methods (seismic refraction surveys to assess the embankments, use of unmanned aerial vehicles for visual assessment of the condition of existing railways instead of standard visual inspections, spectral and multichannel analysis of surface waves for the evaluation of the structure and shear modulus profile of trackbed and subgrade).

At network level, the GPR interpretation together with track geometry assessment can provide information on the track quality and its condition for traffic comfort and safety. Generally, the GPR antennas are installed in an inspection vehicle and the measurements are performed simultaneously with the track geometry assessment.

2.2. GPR Signal Processing in the Spectral Domain

Most GPR applications achieve the detection and localization of targets and discontinuities, and/or the estimation of electromagnetic properties of materials, by performing signal processing tasks in the time domain. However, time domain methods do not account for the frequency dispersive properties of media and do not consider the signal phase. Further advancement and wider use of spectral-domain techniques, to be exploited in combination with traditional time-domain procedures, is desirable in the GPR field; an integrated approach allows achieving a more complete and accurate characterization of the inspected structure/subsurface.

For detecting changes of material properties, an analysis of the spectrum of the GPR signal turns out to be especially beneficial. For instance, partially saturated material tends to absorb the higher frequencies, due to the polarization of water caused by the field emitted by the GPR and travelling in the material; therefore, variations of frequency shifts along an acquisition line, or over a grid, provide useful insights into variations of water saturation in the inspected material. Another representative case is the detection of honeycomb or aggregates in concrete, which presence causes a stronger attenuation of the signal at higher frequencies; the scattering phenomena generated by the small voids or particles are negligible at lower frequencies, instead.

Example of studies where the spectrum of the GPR signal was analysed in order to investigate material properties are [51–53]. In [51], the frequency-dependent dispersion of high-frequency GPR waves in concrete was addressed. In [52], a method of signal analysis based on the Short Time

Fourier Transform (STSF) was used, to retrieve information about material properties based on spatial frequency distributions; the method was tested on two scenarios, i.e., a lawn with irrigation pipes and a concrete wall with steel rebar. In [53], a combined time-frequency analysis method was proposed and applied to the study of steel bar corrosion, hydration, and moisture content distribution in concrete.

Some resonance effects related to the size and electromagnetic properties of targets, are more evident in the spectral domain. Of course, the same information is present in the time domain data, but in the frequency domain it is possible to remove the phase, if desired, and this allows revealing better some spatial characteristics. In [54], the GPR signatures generated by buried landmine-like targets were analysed in both the frequency and time domains; the resonances in the spectrum were linked with the target size and dielectric properties. In [55], the Authors investigated the exploitation of frequency-domain spectral features to improve the detection of weak-scattering plastic mines and reduce the number of false alarms resulting from clutter; the motivation for this approach came from the fact that landmine targets and clutter objects often have different shapes and/or composition, yielding different energy density spectra, which can be then exploited for their discrimination. In [56,57], a freeware tool is presented for the detection and localization of dielectric and metallic objects in radargrams, which implements a spectral-domain signal-processing approach.

In the frequency domain, it is also easy to detect and filter out electromagnetic interferences. The effects of electromagnetic interferences and approaches to counteract them are thoroughly addressed in [58]. In [59], on the contrary, interferences were on purpose generated in order to facilitate the detection of buried utilities: in particular, radio-frequency tags were attached to buried pipes in order to generate strong resonances in the GPR spectrum and then, by analysing the signal in the time or frequency domain, it was possible to achieve enhanced detection of the pipes.

As far as the GPR assessment of railways is concerned, spectral-domain methods of analysis have been rarely employed.

Sometimes, data fusion was performed between GPR profiles recorded on a railway section by using antennas working at different central frequencies; but, the fused signal was then analysed in the time domain. For instance, in [46] dual-frequency GPR data were acquired at the same location, along the Qinghai-Tibet railway (as already mentioned in Section 2.1, this is often done, in order to get high resolution in the shallow region and deep penetration); the two spectra were fused together through forward S-transform and expressed as a spectrum with broader bandwidth in the frequency domain; finally, the synthesized spectrum was converted back to the time domain via an inverse S-transform and the fused radar signal was analysed.

A patent was developed to be applied for detection of voids in the subsurface of roads and railways [60]. This device using radar type apparatus is capable, while in motion, to determine the presence or the absence of a cavity in the subsoil of a railway track. This type of equipment is considered important for early diagnose of voids and, in this way can avoid major track settlement.

In [42], an attempt was done to investigate the possible effects of the size of railway ballast particles on the spectrum of the GPR signal. Interesting studies where the analysis of GPR data recorded on railways was performed in the frequency domain are [43,61]. In [43], the Authors started developing an automated real-time procedure for the analysis of GPR data with the main objective of evaluating ballast conditions. Time-frequency techniques were used to produce discriminating features for a neural network classifier, to distinguish between clean, mixed, and spent ballasts; in particular, the STFT was applied to GPR signals representing different ballast conditions, and it was shown that the discrimination could be made based on the change in centre frequency of the distribution. In [61], railroad track substructure conditions were assessed by using GPR and a time–frequency technique was implemented to analyse the signal in both domains. Frequency sub-bands of the signal were analysed separately, to assess ballast fouling and quantify moisture content, measure the thickness of clean ballast and detect trapped water along the track. The study was continued in [17,41].

In spite of several laboratory tests performed with GPR for ballast condition evaluation, at network level its application it is quite challenging due to the huge amount of data gathered during inspection and the variability of test conditions.

In [15] a study was performed in Finland, for evaluation of the track at network level. The frequency domain analysis of a 400 MHz antenna signal was used to detect changes in ballast fouling and pumping by comparing the area of the signal in frequency. In order to detect the influence of the ballast grading on the GPR signal, tests were performed in an experimental section, over clean and fouled ballast, artificially produced. The area of the fouled ballast is more reduced. In the graphs presented it can be observed that the frequencies between 0.7GHz and 2 GHz were more influenced ballast and subgrade condition [15].

In Australia another approach was studied for GPR signal processing in frequency domain at network level [16]. The authors present an automatic classification for ballast condition based on the extraction of local maximum points in the magnitude spectra that correspond to the salient frequencies. An 800 MHz antenna was used for experimental tests. The processing is performed using and support vector machines.

Studies that were undertaken aiming at a more efficient approach for signal processing highlight the need for future research. Also, these techniques are developed for specific ballast material, in terms of type and grading, and for certain antennas frequency, therefore the generalisation of the application has to be performed with care.

3. A New Method for the Processing of GPR Data Recorded Over Railway Lines

A new approach for track condition assessment is proposed in this work. This approach may be used to automatically detect variations in the GPR signals, which may be helpful to Railway Network Administrators when analysing large amounts of GPR data obtained with successive surveys over long stretches of railway lines. This section presents: (i) the equipment to be used, including a description of available commercial GPR antennas dedicated to railway assessment and (ii) the step by step design of the methodology, the decisions adopted at each step, the parameters set-up. Examples of the signal processing used in the methodology are presented at the end of this section. Section 4 presents the application of this method to GPR data obtained in a railway line case study.

3.1. GPR Equipment

Currently, most railway infrastructure managers use dedicated recording cars to automatic perform inspections to evaluate the track condition.

The railway inspection equipment normally used in Portugal is an EM 120 vehicle, originally manufactured by Plasser and Theurer and upgraded recently. Besides measuring the geometric parameters, the rail profile and the rail roughness, among others, the EM 120 performs measurements with GPR. The GPR antenna installed in the vehicle is a 400 MHz antenna, manufactured by IDS Ingegneria dei Sistemi S.p.A. This antenna is a high-speed radar system that can measure at more than 300 km/h. It was developed specifically for railway assessment and to function as air-coupled, suspended at 30 cm above the ground [62]. A 30 cm distance between the antenna and ground of course entails that the electromagnetic coupling with the ground is less efficient than in ground-coupled antenna measurements. However, suspended antennas are the only possible solution for high-speed surveys, to avoid damaging them, and they yield good results. The use of air-coupled antennas mounted on vans and suspended at a similar height is very well-established in pavement inspections. For railways, the main additional challenge is the presence of the metallic rails. Antennas have to be installed over the track longitudinal axis (in order to maximize the distance from the rails), their radiation pattern should be as narrow as possible in the direction transversal to the travel one, and the polarization of the emitted field should of course be orientated orthogonal to the rails, in order to prevent the presence of strong reflections in the data generated by the rails themselves.

The normal GPR settings for this 400 MHz are: a time window of 40 ns and a spatial sampling of 0.1163 m along the line, with 512 samples per scan. Data positioning is aided by a Global Positioning System (GPS) and also done by distance, measured with an encoder. Track events, such as the presence of stations, switches, level crossings, among others, shall be marked by an operator in the track geometry file [2,44]. This GPR equipment allows measurements at a travelling speed of up to 120 km/h, therefore it causes minimal interference in the normal train operation in the network.

Normally, each railway line has a specific minimum time interval between successive track geometry inspections, which are established depending on the traffic and relevance of the line, among other aspects, according to the maintenance and inspection plans of the railway network manager. Regarding GPR inspections, there is no established time interval between two successive inspections. However, in the case of the Portuguese network, because the vehicle that performs the track geometry inspections also performs the GPR inspections, the timings for the GPR inspections depend on the timing of the track geometry inspections. In Portugal the entire national railway network is inspected at least twice per year and the main lines, Northern and Southern Lines, are inspected four times yearly [2,5].

In Figures 1 and 2, three examples of GPR are shown to illustrate the typical output of the equipment described above, namely traces denoting (a) clean ballast, (b) old ballast and (c) a switch (track event); these results were recorded along a 50-m track segment. In each plot of Figure 1, the red curve corresponds to the average signal of all traces (in grey) obtained in the 50-m long track segments; in Figure 2, instead, entire sequences of traces (B-Scans) are shown. In both figures, data are plotted in the time domain (upper panels) and in the spectral domain (lower panels).

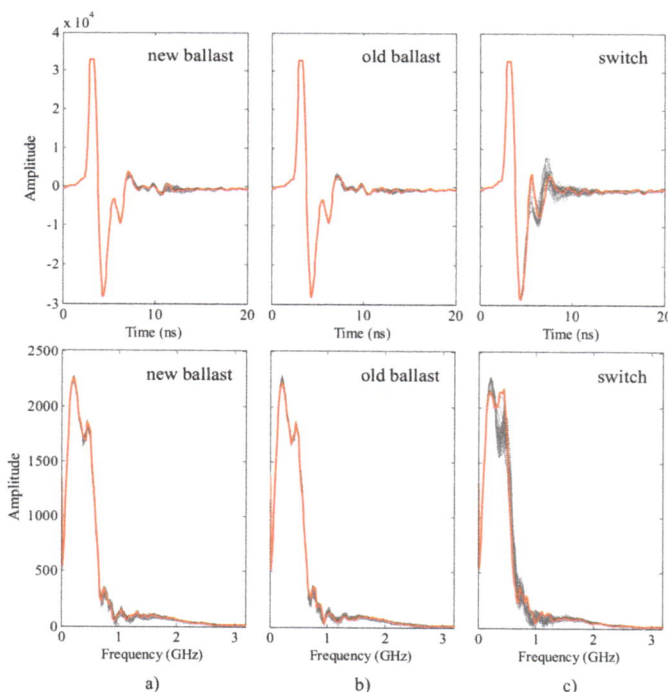

Figure 1. Example of Ground Penetrating Radar (GPR) signal in time and frequency domain for (see Figure 2): (**a**) new ballast, (**b**) old ballast and (**c**) over a switch.

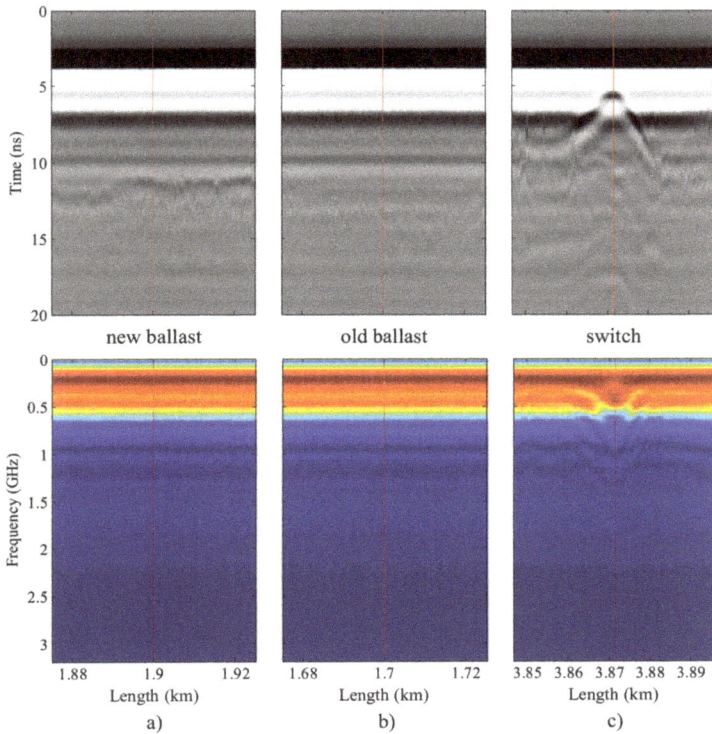

Figure 2. Example of spectrograms in time and frequency domain for (see Figure 1): (a) new ballast; (b) old ballast and (c) over a switch.

By comparing the GPR signals recorded over the three locations, it is apparent that the switch has a strong influence on the overall response; on the other hand, it is noticed that different ballast conditions cause quite subtle differences in the signals, both in the time and frequency domains, as already observed in several studies in literature [2,15,25,29,30,36,61].

It can be observed in Figure 1c that the switch, as other superstructure elements, have a significant influence on frequencies lower than 0.7 GHz. Therefore, the selection of the frequency range to be used in this study does not take this frequencies into consideration as the aim is to detect ballast and subgrade condition changes.

3.2. Methodology Set-Up

The main purpose of this study was to develop an expedite tool for GPR interpretation able to detect areas with distinct condition along the track, at network level. In order to achieve this, the research was focused in selecting GPR signal parameters that reflect changes in track structure, mainly bellow the sleepers, such as ballast and substructure conditions, which are not visible by videography or other direct methods.

To implement the proposed approach, three main steps were established, as described below.

3.2.1. Range Selection of GPR Signal in Time and Frequency Domains

The first step was to establish a range, in the time and frequency domains, that corresponds to segments of each acquired trace of the GPR signals that are considered to better reflect the changes in the track condition at subsurface layers level [15,35,40,63].

- In the time domain, several in situ GPR measurements performed on existing lines, with different characteristics (sleeper's type and material, ballast fouling level, age of the track, with and without sub-ballast layer) were analysed [2]. Based on this analysis, a time window between 7 ns and 16 ns was selected for this study (see, for example Figures 1 and 2), which was considered representative of the conditions of the ballast and subgrade.
- In the frequency domain, numerous in situ GPR data were analysed in order to detect areas of the electromagnetic spectrum affected by changes occurring along the track. A range between 0.7 GHz and 2.0 GHz was selected (see for example Figures 1 and 2). The changes induced by elements of the superstructure such as switches, sleepers and level crossings, are generally registered at lower frequencies (below 0.7 GHz), they were excluded because the main purpose was to detect ballast and substructure pathologies. The selection of the frequency range is in accordance with the information in the literature [15,63].

It should be noted that both time and frequency ranges should be adjusted to the type of antenna used, to its frequency and to its vertical position with respect to the track.

These threshold values depend on many factors, such as the type of rail, material of the sleepers, material and geometry of the ballast layer, material and geometry of the track platform. Therefore, for the implementation of this method at the network level, a set of threshold values need to be established, which will depend of the characteristics of the track, followed by a preliminary validation with field surveys with some pit holes.

3.2.2. Sliding Window for Track Changes Detection

The second step consisted in comparing some characteristics of the GPR signals in a short (S) sliding window with those in a long (L) sliding window.

The lengths of the sliding windows adopted in this study are $S = 10$ m and $L = 200$ m, for the short and long windows, respectively. These parameters were identified as follow:

- Several dimensions for the long and short windows were tested and also the positioning of the small window within the large one was varied;
- The length of the shorter window (10 m) was selected to be representative of the length of the track defects that typically occur at ballast and subgrade levels [2,64,65];
- The length of the longer window (200 m) was selected in order to reflect the length of the track adopted by railway engineers, when analysing track geometric parameters, for track quality classification and for tamping planning [66,67].

The length of the sliding windows can be adjusted depending on the purpose of the study, namely on the length of the defects or anomalies to be detected.

The differences between the signals acquired in the two sliding windows allow identifying changes in track characteristics, both in the time and in the frequency domain (see Section 3.2.3).

The comparison that it is made between the two windows does not filter out any GPR signal. It is meant to only to detect differences in its amplitude along the track. Therefore, it differs from other approaches, such as the horizontal background removal that is used in other applications, but is clearly unwanted for the assessment of layer interfaces in transport infrastructures, as it can delete continuous horizontal information that corresponds to infrastructure layers.

3.2.3. GPR Expedite Parameters Definition

The third step of the proposed approach consists in calculating four new parameters. These parameters aim to analyse GPR data, obtained in railway tracks, in a more user-friendly way for "non-GPR professionals" and to identify locations in the track where changes in the infrastructure occur. The parameters used in this approach are:

- in time domain, z and dz;
- in frequency domain, Z and dZ;

which are defined by the following equations:

$$z = \int_{t_1}^{t_2} |a| dt \tag{1}$$

$$dz = \int_{t_1}^{t_2} |a_S - a_{L,m}| dt \tag{2}$$

$$Z = \int_{f_1}^{f_2} |A| df \tag{3}$$

$$dZ = \int_{f_1}^{f_2} |A_S - A_{L,m}| df \tag{4}$$

where: a and A are the GPR signal amplitudes in the time and frequency domains, respectively; subscripts S and L denote the short and long sliding windows and m denotes the averaged signal in the specified window; t_1 and t_2 are the lower (7 ns) and upper (16 ns) values of the time interval of the signal under analysis; f_1 and f_2 are the lower (0.7 GHz) and upper (2.0 GHz) values of the frequency range of the signal under analysis. Therefore, z and Z correspond to the areas under $|a|$ and $|A|$ plots, respectively in the selected time and frequency ranges, while dz and dZ correspond to the differences in area of the GPR signal amplitude between the short and long sliding windows, also respectively in the selected time and frequency ranges.

3.3. Example of Signal Processing for Methodology Implementation

Figure 3 presents examples of the original signal amplitude in the time (a) and frequency (b) domains, in terms of the mean or averaged signal (denoted by subscript m) over the shorter ($a_{S,m}$ and $A_{S,m}$) or longer ($a_{L,m}$ and $A_{L,m}$) sliding windows; percentiles 5% and 95% of the longer window are also plotted to illustrate the variability of the signal.

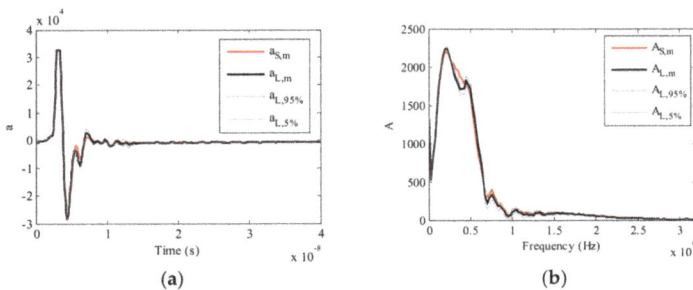

Figure 3. Example of the amplitude of the GPR signal in time domain (a) and in the frequency domain (b).

The examples presented in Figure 4 depict the different steps of GPR signal processing in a normal section of the track: (i) the time (a) and frequency (b) ranges selected for the analysis in order to better reflect the changes in the track condition and (ii) the differences in the amplitudes of the signal (*da* and *dA*) in the time (c) and frequency (d) domains, when comparing the results between the short sliding window and the large sliding window.

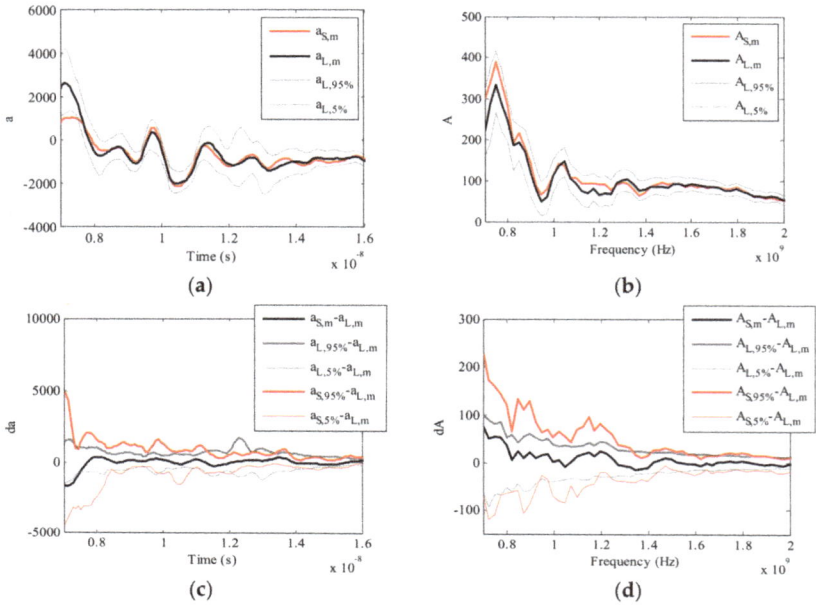

Figure 4. Examples of GPR signal processing steps: (**a**) amplitude in the selected time range; (**b**) amplitude in the selected frequency range; differences in amplitude between the short window, *S* and the larger window, *L*, in time domain, *da* (**c**) and in frequency domain, *dA* (**d**).

Examples of spectrograms of *a* and *A*, over a length of 200 m (the larger sliding window—L), are presented in Figure 5 with the identification of the short sliding window inside the dotted red rectangles.

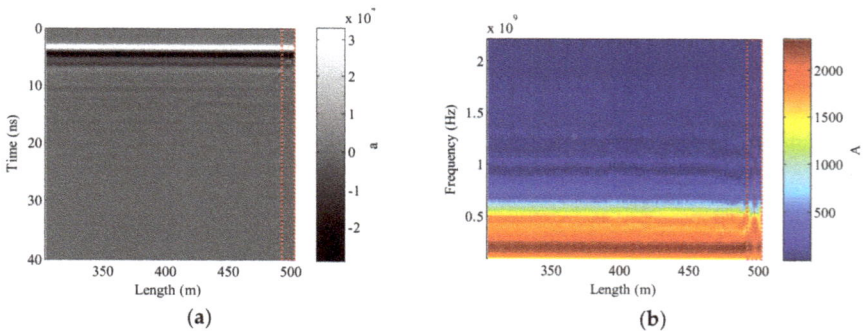

Figure 5. Example of spectrograms regarding the long sliding window of 200 m and the identification of the short sliding window of 10 m (delimited by dotted red lines): (**a**) in the time domain and (**b**) in the frequency domain.

Examples of spectrograms, calculated as the difference between the amplitudes of the short window (a_S or A_S) and the mean signal of the large window ($a_{L,m}$ or $A_{L,m}$) from Figure 5, are presented in Figure 6, in the time (a) and in the frequency (b) domains.

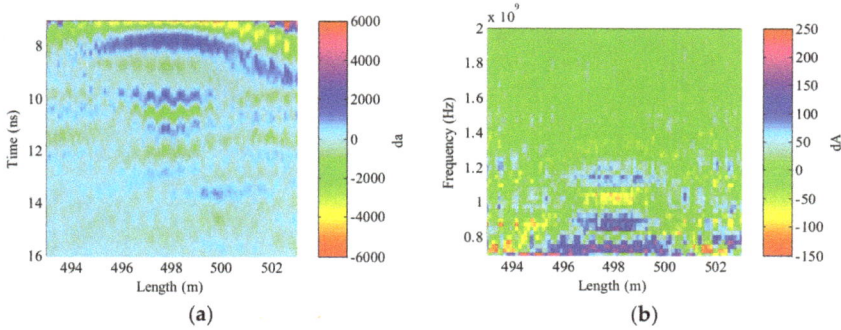

Figure 6. Example of spectrograms calculated as the difference between A_S and $A_{L,m}$: (**a**) in the time, da, and (**b**) in the frequency, dA, domains.

4. Application of the New Method to a Case Study and Discussion of Results

Examples of the proposed methodology application to a track section of an existing line are presented in this section. For confidentiality reasons the GPS location is not revealed and the location reference is not the real one. The GPR data used to illustrate the application of the approach were acquired with the 400 MHz antenna referred in Section 3.1.

Figures 7–10 depict examples of z, dz, Z and dZ values, considering sliding window lengths of L = 200 m and S = 10 m, calculated for a 10 km stretch of an in-service railway line, obtained from GPR signals acquired in the summers of 2015, 2016 and 2017 (in black, blue and red, respectively) and considering $t_1 = 7$ ns, $t_2 = 16$ ns, $f_1 = 0.7$ GHz and $f_2 = 2.0$ GHz, as mentioned above.

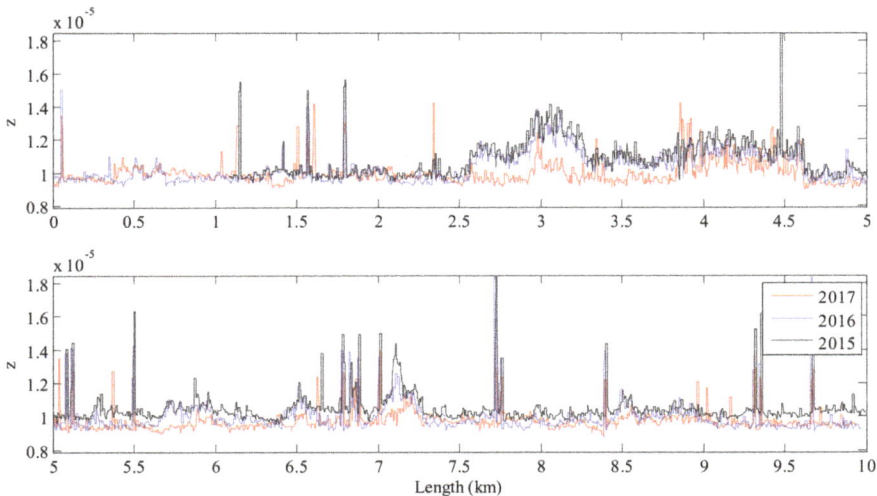

Figure 7. GPR parameter z, in the time domain, regarding data from the summers of 2015, 2016 and 2017.

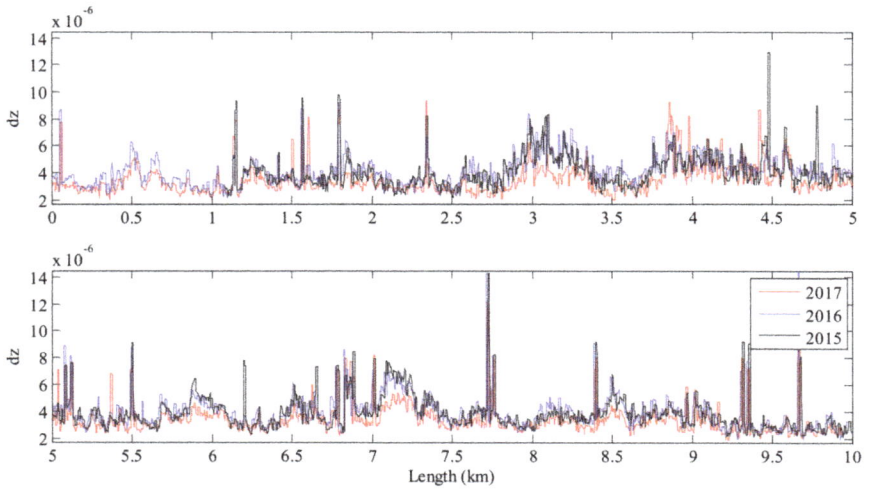

Figure 8. GPR parameter *dz*, in the time domain, regarding data from the summers of 2015, 2016 and 2017.

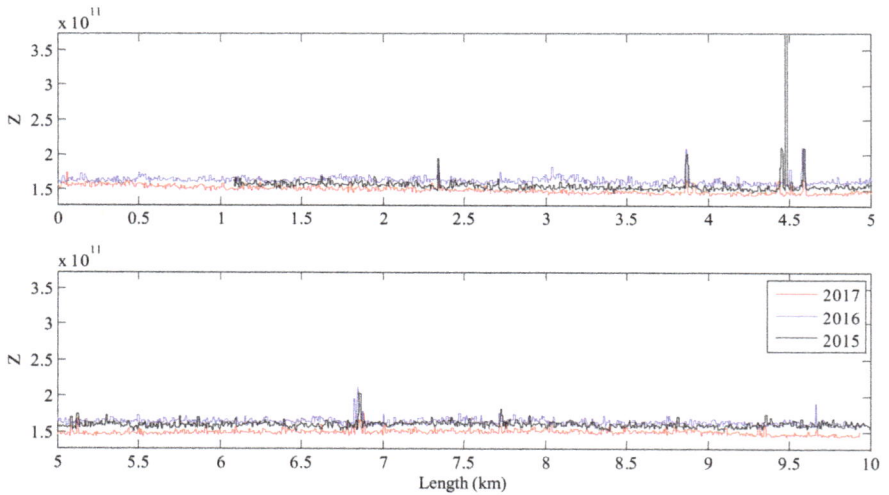

Figure 9. GPR parameter *Z*, in the frequency domain, regarding data from the summers of 2015, 2016 and 2017.

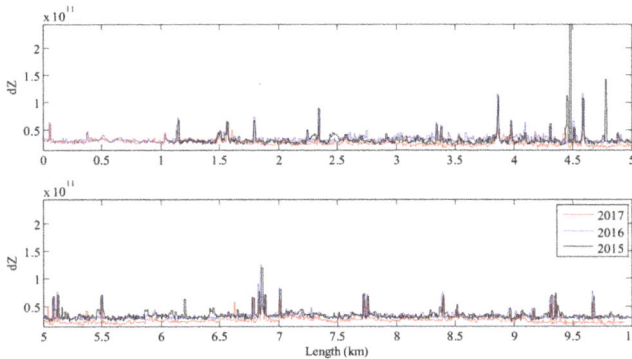

Figure 10. GPR parameter dZ, in the frequency domain, regarding data from the summers of 2015, 2016 and 2017.

It is observed that the variable dZ yields somewhat flat plots interrupted by prominent peaks, which correspond to changes in the GPR signal in the selected frequency range. The authors suggest that this variable might be more appropriate for the proposed approach, though further studies should be performed to test the applicability of this method to other scenarios or other parameters.

4.1. Event Identification

This first application of the methodology it is recommended to be undertaken always in the beginning of the overall analysis of a track, in order to identify track events and to enable a location validation of the GPR results. Consequently, it will provide a good correlation of GPR results with track geometric parameters.

It was verified that the most noticeable peaks in these variables corresponded to the locations of switches and crossings of that line, which significantly disturbed the GPR signal, and were quite clear in all surveys, year after year. Figure 11 presents the identification of the events, cutting zones and sections with new ballast in the dZ parameter of GPR 2017 measurement.

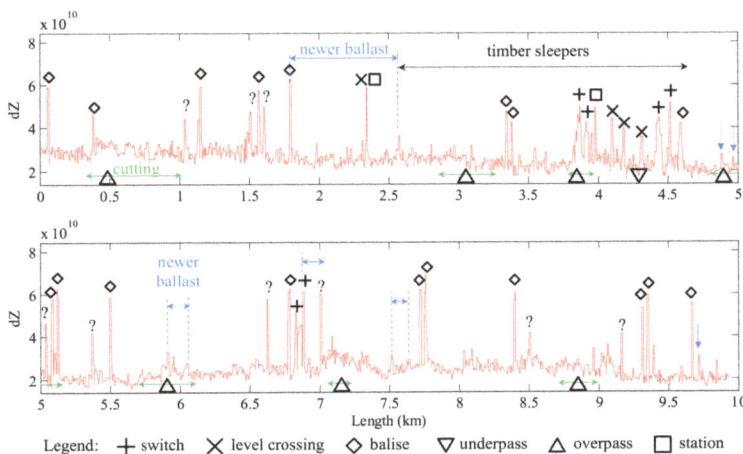

Figure 11. GPR parameter dZ, in frequency domain, regarding data from September 2017 and the identification of track events.

53

In Figure 11, the squares denote the beginning and end of the station platform, the saltires (diagonal crosses) denote the switches, the crosses denote the level crossings, the downward pointing triangles denote underpasses, the upward pointing triangles denote overpasses and the diamonds denote the balises (from the CONVEL system used for the automatic train protection). Many other peaks were identified but were found to be unrelated to any known singularity in the track, which were identified using question marks in Figure 11. It is possible that such remaining peaks could be related to unwanted discontinuities in the track structure or locations where the structure has been altered, with respect to the surrounding sections of the track. Such differences could be related to different track components or to track pathologies such as poor drainage or ballast fouling.

Figure 12 shows different aerial photographs highlighting examples of the identified singularities.

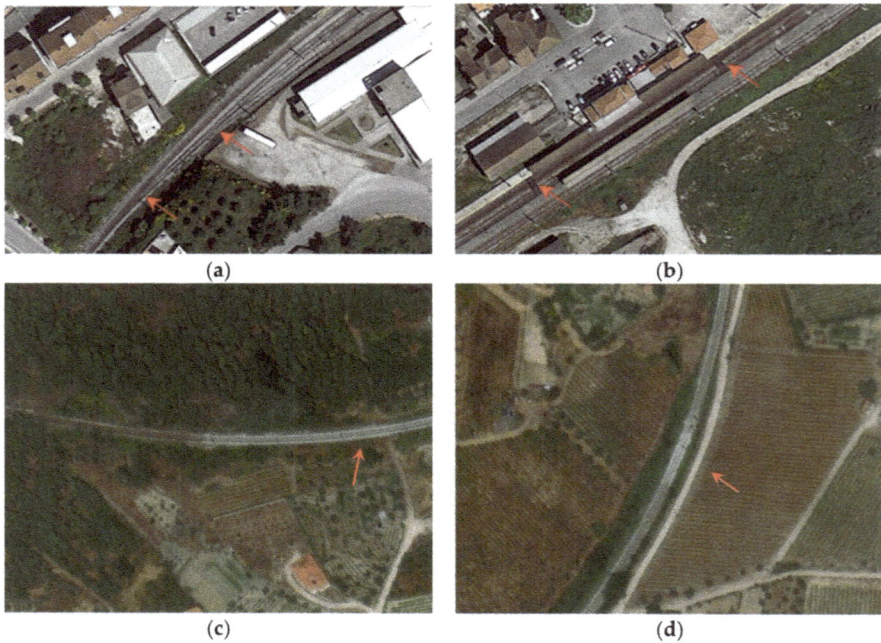

Figure 12. Events of the track: (**a**) two switches at about 3.8 km; (**b**) two level crossings in the station at about 4.1 km; (**c,d**) sections with newer ballast, at about 1.8 km and 7.0 km, respectively (Imagery © 2016 Google).

Figure 13 shows the locations on the track that correspond to the highest peak values identified in the previous plots, considering a stretch of that line with multiple singularities, passing through a train station. It is visible that these singularities are clearly identified using the dZ parameter.

Figure 13. Location of the peaks of dZ that match the location of the track events (Imagery © 2016 Google).

Therefore, it is recommended to use the dZ parameter as pre-analysis of each campaign, in order to correlate the events location and guarantee that the data analysed are in the same locations in different campaigns.

4.2. Sistematic Comparison between Consecutive Campaigns

This second application of the methodology is efficient for track management decisions, as enable the following of the track condition evolution in time. Also, it can be used, when several campaigns are available to forecast the track behaviour.

This can represent an important tool for identification of sections that are deteriorating at a faster rate along the track. To implement the methodology for this purpose it is essentially to have all the information on the maintenance and rehabilitation actions that were undertaken between GPR testing campaigns for cross checking.

To illustrate the potential of the proposed approach to identify changes in the track structure/condition, the authors analysed in more detail the locations with peaks in the variable dZ that were not directly associated with locations of known track events. For this purpose, three different sets of GPR data obtained in the summers of 2015–2017 on the same stretch of line were compared. Firstly, the main purpose was to study the locations identified by the questions marks in Figure 11, and assess their coherence when comparing the three surveys. It was noted that some of those locations (those not related to any track event) were not present in all the surveys.

As an example, two track sections are presented herein. The first one is a section where the ballast was renewed between the 2015 and 2016 GPR surveys. The second one is a section where the dZ values were systematically higher than in the surrounding sections of the line in the three analysed surveys.

The first example corresponds to a location where ballast was renewed in 2 phases: one before the 2015 survey (from 1.80 km to 2.15 km) and the second one after the 2015 survey (from 2.15 km to 2.58 km). Figure 14 shows the sections with renewed ballast in 2015 and in 2017. The graph of dZ obtained in 2015 has generally higher values (black line) in the section that was only rehabilitated later on; on the other hand, the graphs obtained in 2016 and 2017 have similar values, lower than in 2015.

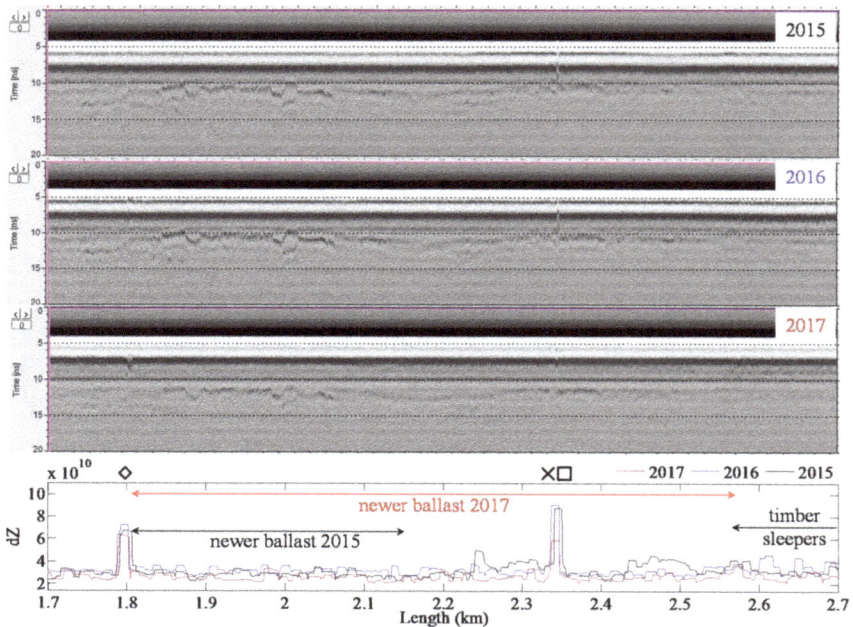

Figure 14. GPR radargrams of the surveys carried out in the summers of 2015, 2016 and 2017 as well as the respective dZ plots.

The results show the potential of the proposed approach to detect distinct areas, when comparing inspection surveys performed in consecutive years. The proposed analysis in the frequency domain allowed observing dissimilar results even if the difference between the GPR raw signals was not clear in the radargrams of the 2015 survey, or in the latter surveys. This dissimilar behaviour can constitute a trigger in future automated GPR data analysis that identifies the need for further processing of the GPR in this section. So, it can be observed that the analysis of the differences in the frequency domain enabled to identify the response of the GPR signal in a section where track rehabilitation was performed.

The second example is a section located between 1.47 km and 1.60 km, which yielded higher dZ values in all the surveys, even in locations that were not associated with any track event. Figure 15 presents the radargrams obtained in this section during the three GPR surveys and the corresponding dZ plots. From the radargrams it can be observed that this section is quite heterogeneous both in terms of ballast thickness as in the "clearness" of the interface between ballast and subgrade layer. The increase in the dZ values may be caused by the change in thickness that occurs at approximately 1.47 km. Again, this result illustrates the potential of the analysis in frequency domain to identify sections that are dissimilar, and therefore require an in-depth interpretation of the GPR data.

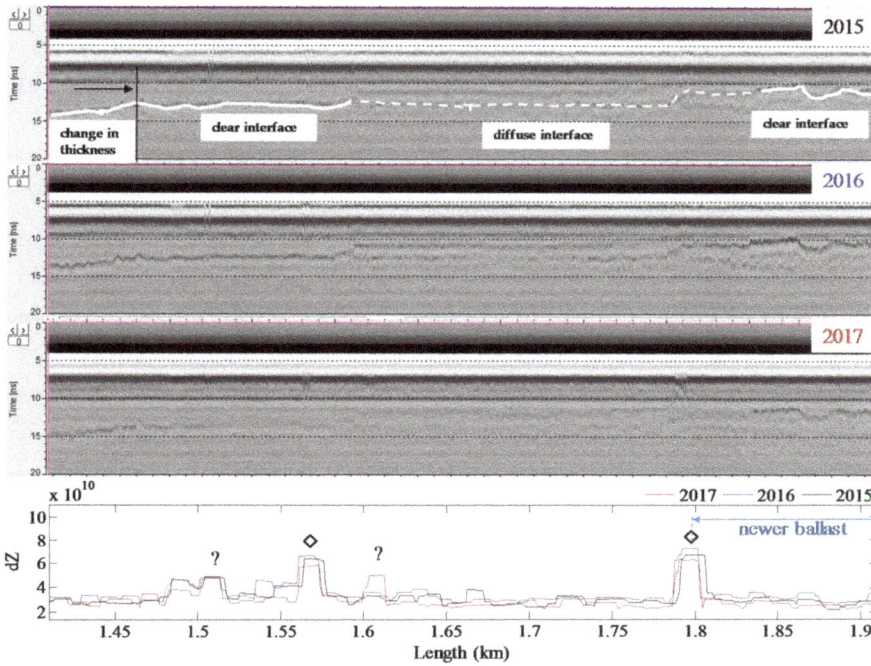

Figure 15. GPR radargrams of the surveys carried out in the summers of 2015, 2016 and 2017 as well as respective dZ plots.

Figure 16 presents an example of a possible visualisation of the GPR signal information together with track geometric parameters, namely the Longitudinal Level (LL). The section presented is the same as the one illustrated in Figure 15. It can be observed that there is an increase in the LL values in the same location where the values of dZ increase, between 1.47 km and 1.60 km. The systematic correlation between those parameters requires further research and validation with extended field data.

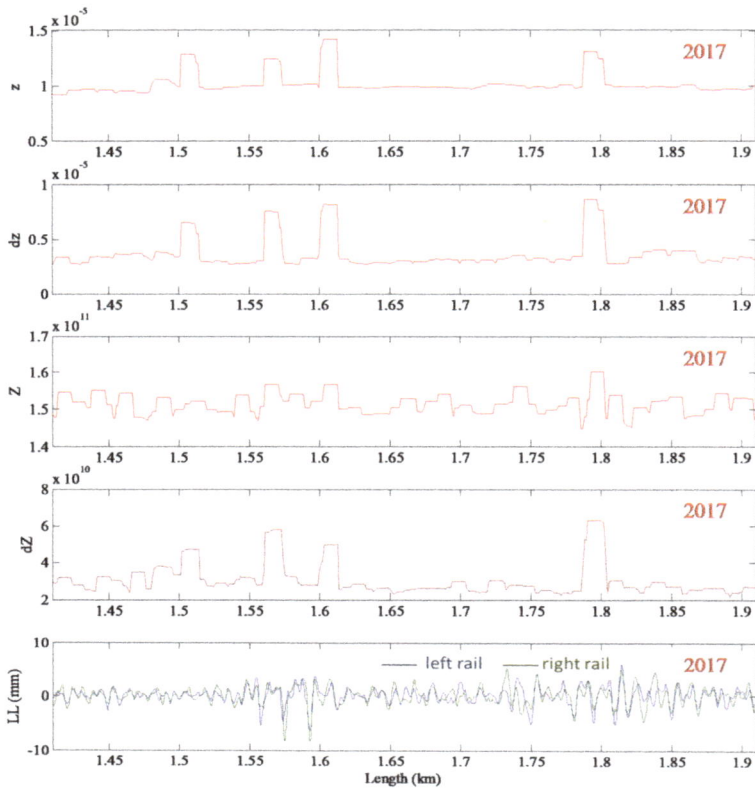

Figure 16. GPR parameters z, dz, Z and dZ and the Longitudinal Level (LL) of the left (blue) and right (green) rails, from the survey in the summer of 2017.

It was demonstrated that the proposed approach can perform an efficient identification of dissimilar sections along the track, which can be considered as requiring a deeper analysis.

After identifying sections with distinct behaviour, more detailed interpretation can be performed and additional NDT can complete the diagnosis of the track condition in these sections. Additionally, information on traffic change along the line is important for a thorough analysis of the track increased rate deterioration.

4.3. Analysis of Seasonal Influence on the GPR Data

The comparing of the GPR campaigns performed in different seasons, with different water contents of the ballast and substructure, represent a tool for identifying sections with ballast fouling and drainage problems. As already referred to in Section 2.1, the fouling of the ballast promotes the water trapping in this layer and, consequently, has a negative impact on track behaviour. This phenomenon is more evident when compare the same section in different hydraulic conditions, namely in dry and wet season. On one hand, if the track structure is healthy and the drainage is proper, no difference is detected between campaigns performed with GPR antennas with the frequency used in this study. On the other hand, if there are drainage problems of ballast fouling, the water content in the track layers will lead to attenuation of GPR signal.

To illustrate this application, two GPR surveys performed in the same year, but in different seasons were analysed. The GPR measurements were undertaken in January, after a few days of rain, and in July, in a dry and warm period. When comparing the values of parameter z (in time domain), it is clear the different behaviour of the GPR signal between winter and summer (see Figure 17). The significantly reduced values in winter may be explained by the attenuation of the signal with higher water content in the assessed media [2]. On the other hand, when comparing the same parameter z for the different surveys performed all during summer, the values are similar (see Figure 7).

The difference represented in the graph (see Figure 17) reflects the GPR signal shape in time domain. The GPR signal is most commonly processed in time domain [2,11,26,38,45] by: (i) calculating the layer thickness using an "default" dielectric value, generally assumed valid for the type of material tested; or (ii) by performing test pits at some locations, measuring the real thickness of the ballast and, based on this, calculating the real dielectric value and adopting this value for processing all the signal along the track. In case of homogeneous media, such as concrete or asphalt, these assumptions are close to the real situation. However, in case of non-homogeneous media such as ballasted railway track, and particularly in old existing lines, there are several factors that affect significantly the dielectric properties of the media along the track. Among those, the two main factors are the level of ballast fouling and the water content. Both affect the GPR signal by reducing the wave speed propagation and attenuating the signal intensity. Nevertheless, they are difficult to dissociate from each other, as their effect is combined [2,25,38], unless in situ samples of material are collected and analysed in laboratory in order to characterise the grading, fouling level and water content. Nevertheless, this only characterises specific locations, as old tracks are quite heterogeneous in terms of materials and drainage conditions. Thus, assuming a constant or "default" dielectric value for the ballast material throughout the line or when comparing between summer and winter GPR surveys, may induce wrong estimates of the thickness of the layers. Therefore, when comparing the two surveys in Figure 17, the differences between them do not correspond to real differences on the track, rather reflecting the influence of the testing conditions, mainly as the different water content of the ballast and the subgrade. In case of GPR processing in time domain, when comparing two campaigns, the information given by z parameter enable the choose of different "default" dielectric constants, to better reflect the influence of the water and turn the processing more realistic.

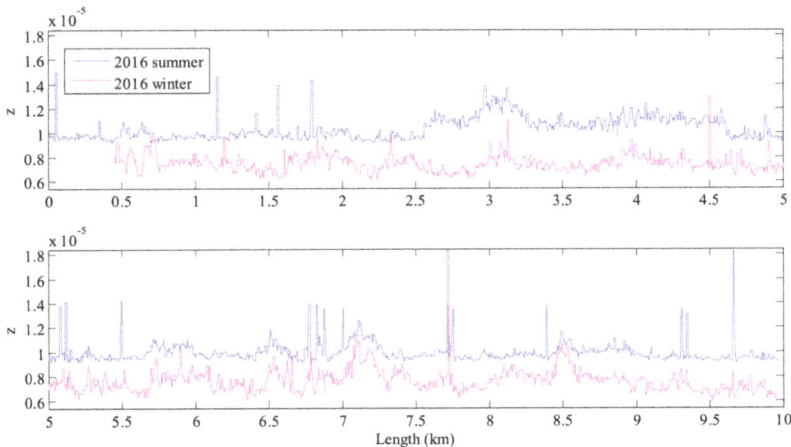

Figure 17. GPR parameter z, in the time domain, measured in January (winter) and in July (summer) of 2016.

Instead of representing the results of the two surveys in the time domain in terms of z, the data can be represented in terms of dz: see Figure 18. In this case, the results become quite similar between the two surveys. It is noted that higher values are observed in the in the section of the track with timber sleepers.

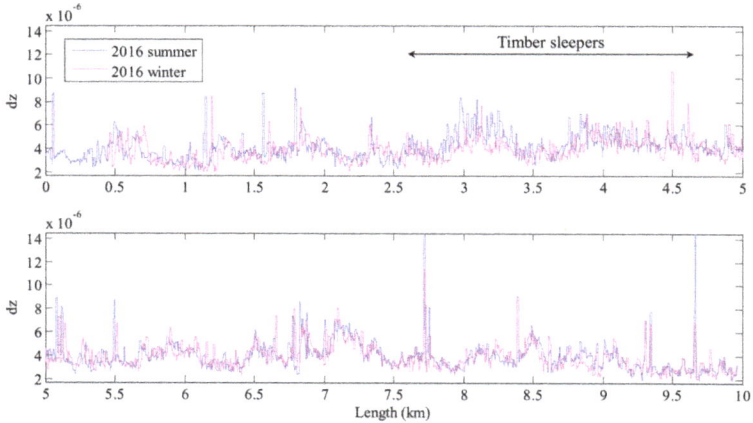

Figure 18. GPR parameter dz, in the time domain, measured in January (winter) and in July (summer) of 2016.

Regarding the analysis in the frequency domain, it enables filtering most of the differences observed between the summer and winter surveys regarding parameters z and dz in the time domain. This is evidenced in Figure 19 and in Figure 20, with plots of Z and dZ values, respectively. Therefore, the authors suggest analysing the GPR results in the frequency domain, in terms of dZ, because it seems less influenced by the water content and, consequently, allows to associate changes in that variable with distinct zones of the track and, eventually, with possible track pathologies. For example, in Figure 17, it is clear the greater variability of dZ values in the zone that corresponds to the track section with timber sleepers.

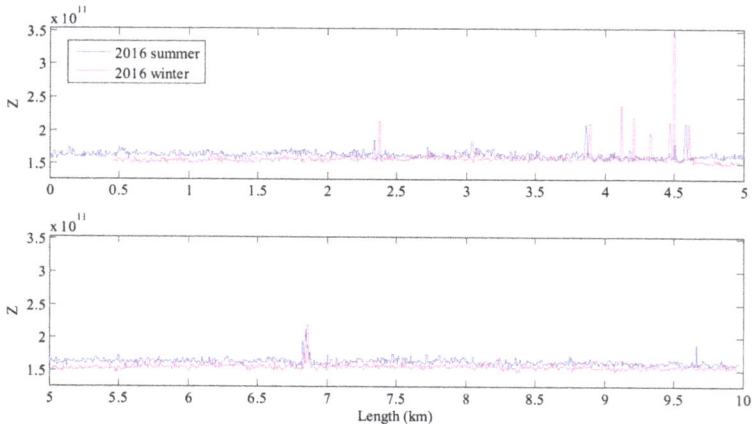

Figure 19. GPR parameter Z, in the frequency domain, measured in January (winter) and in July (summer) of 2016.

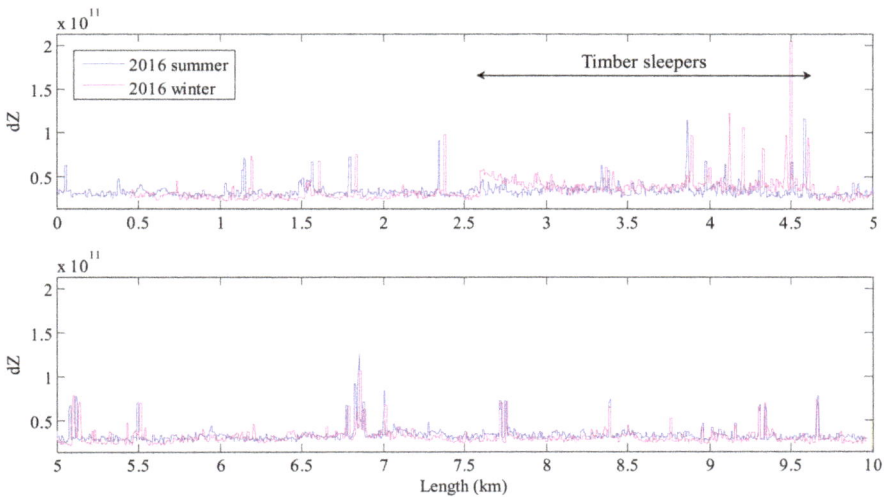

Figure 20. GPR parameter dZ, in the frequency domain, measured in January (winter) and in July (summer) of 2016.

When sections with distinct behaviour are detected by comparing two campaigns performed in different seasons, these can be analysed further on; advanced signal processing techniques have to be used, taking into account scattering and dispersion phenomena as well as the level of ballast fouling, validated with material collected in situ and analysed in laboratory.

4.4. Final Remarks

GPR systems are already installed on inspection vehicles of many railway networks. However, the current and common strategy is to acquire and store the huge amount of data in datacentres, which are accessed punctually in the event of any pathology of the track structure or to estimate the percentage ballast fouling.

Possible future strategies to manage and process the huge amount of data acquired with GPR inspections should include automatic data processing to detect the early development of track pathologies, by using approaches similar to the one presented in this work. In order to improve data interpretation and obtain a more accurate assessment of the track quality, the authors recommend that GPR inspections should be performed at least twice a year: once during the dry season and again in the wet season; and, if possible, inspections should be carried out in the same month in consecutive years for a better comparison between surveys.

The approach presented herein allows the identification of changes in the GPR signal that are related to changes in the track structure. The automatic identification of these locations (after proper validation to discard false positives such as balises, switches and crossings) allows the railway network managers to obtain an overall assessment of the condition of the line and to focus earlier on the eventual development of track pathologies. Thus, the application of this method reduces the data processing effort and allows acting sooner. In this way, maintenance costs decrees and maximum track availability is ensured.

5. Conclusions

The continuous non-destructive monitoring of railway track is an important tool for diagnosing and planning adequate rehabilitation measures. In this context, the use of Ground Penetrating Radar

(GPR) provides valuable information on the track condition, mainly regarding the changes in the infrastructure and pointing to causes of track deterioration located on the subsurface.

In this paper, a novel method is proposed for network level analysis of the GPR signal, which is carried out continuously along the track and performed both in the time and frequency domains. A dedicated script was developed in MATLAB to process, compare and visualize the GPR data. The presented approach allows for the detection of events and possible track defects in a single GPR survey; it is also useful to compare data recorded during different surveys performed during the service life of the track. The GPR results and processing outcomes are presented graphically, enabling a "user friendly" visualisation of the track condition. The method identifies the sections that need more detailed research. Correlations with track geometry and joint analysis are further possible applications of this method.

It is believed that the availability of an expedite processing approach, such as the one proposed herein, will contribute to promoting a wider use of GPR on railways—considering also that GPR antennas can be easily installed on normal trains and be used continuously. Moreover, the method represents a first step in developing quality indexes and alert levels for track network evaluation.

The presented approach was developed for practical purposes, aiming at supporting the railway industry for continuous surveys of track conditions. Indeed, the method allows a time efficient processing of GPR data and is useful to assist the process of decision making regarding effective rehabilitation measures. It can also be used to automatically detect sudden variations in the GPR signal, which may be helpful to Railway Network Administrators when analysing large amounts of GPR obtained with successive surveys over long stretches of railway lines.

Acknowledgments: The authors would like to thank the financial support given through the postdoctoral fellowship SFRH/BPD/107737/2015 that was funded by FCT—Fundação para a Ciência e a Tecnologia, through POCH, being co-funded by ESF and the National Funds of MCTES, Portugal. This study is a contribution to the EU funded COST Action TU1208 "Civil Engineering Applications of Ground Penetrating Radar".

Author Contributions: S.F. and A.P. conceived and designed the procedure, developed the MATLAB code and analysed the data; S.F., A.P., M.S. and L.P. wrote and revised the paper.

Conflicts of Interest: The authors declare no conflict of interest. The founding sponsors had no role in the design of the study; in the collection, analyses, or interpretation of data; in the writing of the manuscript, and in the decision to publish the results.

References

1. Esveld, C. *Modern Railway Track*; MRT-Productions: Zaltbommel, The Netherlands, 2001.
2. De Chiara, F. Improving of Railway Track Diagnosis Using Ground Penetrating Radar. Ph.D. Thesis, University of Rome "Sapienza", Rome, Italy, 2014.
3. Berggren, E. Railway Track Stiffness: Dynamic Measurements and Evaluation for Efficient Maintenance. Ph.D. Thesis, KTH, Stockholm University, Stockholm, Sweden, 2009.
4. Hyslip, J.P.; Chrismer, S.; LaValley, M.; Wnek, J. Track Quality from the Ground Up. In Proceedings of the AREMA Conference, Chicago, IL, USA, 16–19 September 2012.
5. Fontul, S.; Fortunato, E.; De Chiara, F. Non-Destructive Tests for Railway Infrastructure Stiffness Evaluation. In *Proceedings of the 13th International Conference on Civil, Structural and Environmental Engineering Computing*; Tsompanakis, T.Y., Ed.; Civil-Comp Press: Stirlingshire, UK, 2011.
6. Manacorda, G.; Morandi, D.; Sarri, A.; Staccone, G. Customized GPR system for railroad track verification. In Proceedings of the 9th International Conference on Ground Penetrating Radar, Santa Bárbara, CA, USA, 29 April–2 May 2002.
7. Clark, M.; Gordon, M.; Forde, M.C. Issues over high-speed non-invasive monitoring of railway trackbed. *NDT & E Int.* **2004**, *37*, 131–139. [CrossRef]
8. Plati, C.; Loizos, A. Using ground-penetrating radar for assessing the structural needs of asphalt pavements. *Nondestruct. Test. Eval.* **2012**, *27*, 273–284. [CrossRef]
9. Gallagher, G.P.; Leiper, Q.; Williamson, R.; Clark, M.R.; Forde, M.C. The application of time domain ground penetrating radar to evaluate railway track ballast. *NDT & E Int.* **1999**, *32*, 463–468. [CrossRef]

10. Gobel, C.; Hellmann, R.; Petzold, H. Georadar-model and in-situ investigations for inspection of railway tracks. In Proceedings of the 5th International Conference on Ground Penetrating Radar, Kitchener, ON, Canada, 12–16 June 1994.

11. Loizos, A.; Silvast, M.; Dimitrellou, S. Railway trackbed assessment using the GPR technique. *Adv. Charact. Pavement Soil Eng. Mater.* **2007**, *1*, 1817–1826.

12. Vorster, D.J.; Gräbe, P.J. The use of ground-penetrating radar to develop a track substructure characterisation model. *J. S. Afr. Inst. Civ. Eng.* **2013**, *55*, 69–78.

13. Nurmikolu, A. Key aspects on the behaviour of the ballast and substructure of a modern railway track: Research-based practical observations in Finland. *J. Zhejiang Univ. Sci. A* **2012**, *13*, 825–835, ISSN 1673-565X (Print); ISSN 1862-1775 (Online). [CrossRef]

14. Hyslip, J.P.; Smith, S.S.; Olhoeft, G.R.; Selig, E.T. Assessment of railway track substructure condition using ground penetrating radar. In Proceedings of the 2003 Annual Conference of AREMA, Chicago, IL, USA, 5–8 Octorber 2003.

15. Silvast, M.; Levomaki, M.; Nurmikolu, A.; Noukka, J. NDT techniques in railway structure analysis. In Proceedings of the 7th World Congress on Railway Research, Montreal, QC, Canada, 4–8 June 2006.

16. Shao, W.; Bouzerdoum, A.; Phung, S.L.; Su, L.; Indraratna, B.; Rujikiatkamjorn, C. Automatic classification of ground-penetrating-radar signals for railway-ballast assessment. *IEEE Trans. Geosci. Remote Sens.* **2011**, *49*, 3961–3972. [CrossRef]

17. Shangguan, P.; Al-Qadi, I.L.; Leng, Z. Development of Wavelet Technique to Interpret Ground-Penetrating Radar Data for Quantifying Railroad Ballast Conditions. *Transp. Res. Rec. J. Transp. Res. Board* **2012**, *2289*, 95–102. [CrossRef]

18. Hermann, M.; Pentek, T.; Otto, B. Design Principles for Industrie 4.0 Scenarios. In Proceedings of the 49th Hawaii International Conference on System Sciences (HICSS), Koloa, HI, USA, 5–8 January 2016. [CrossRef]

19. Santos, C.; Mehrsai, A.; Barros, A.C.; Araújo, M.; Ares, E. Towards Industry 4.0: An overview of European strategic roadmaps. *Procedia Manuf.* **2017**, *13*, 972–979. [CrossRef]

20. Riveiro, B.; Solla, M. (Eds.) *Non-Destructive Techniques for the Evaluation of Structures and Infrastructure*; CRC Press: London, UK, 2016; ISBN 9781138028104.

21. Sandoval, S.; Mínguez, R.; Nestares, E.; Carbó, A. Multidisciplinary Study of a Ballast Collapse in a High-Speed Railway Track in Spain. In Proceedings of the 6th International Conference on Applied Geophysics for Environmental and Territorial System Engineering AGE, Iglesias, Italy, 28–30 April 2011.

22. Barta, J. A methodology for geophysical investigation of track defects. *Proc. Inst. Mech. Eng. Part F J. Rail Rapid Transit* **2010**, *224*, 237–244. [CrossRef]

23. Fortunato, E.; Bille, J.; Marcelino, J. Application of spectral analysis of surface waves (SASW) in the characterisation of railway platforms. In *Advanced Characterisation of Pavement and Soil Engineering Materials*; Loizos, A., Scarpas, T., Al-Qadi, I.L., Eds.; Taylor & Francis Group: London, UK, 2007.

24. Szwilski, A.B.; Begley, R. *Developing an Integrated Track Stability Assessment and Monitoring System Using Non-Invasive Technologies*; Transportation Research Board: Washington, DC, USA, January 2003.

25. Fortunato, E.; Fontul, S.; Paixão, A.; Asseiceiro, F. Case study on the rehabilitation of old railway lines: Experimental field works. In Proceedings of the 9th International Conference on the Bearing Capacity of Roads, Railways and Airfields, Trondheim, Norway, 25–27 June 2013.

26. Fontul, S.; Fortunato, E.; Paixão, A.; De Chiara, F. Non-destructive tests for evaluation of railway platform. Railways 2012. In Proceedings of the First International Conference on Railways Technology: Research, Development and Maintenance, Las Palmas de Gran Canaria, Spain, 18–20 April 2012.

27. Saarenketo, T. *Electrical Properties of Road Materials and Subgrade Soils and the Use of Ground Penetrating Radar in Traffic Infrastructure Surveys*; Oulu University Press: Oulu, Finland, 2006.

28. Clark, M.R.; Gillespie, R.; Kemp, T.; McCann, D.M.; Forde, M.C. Electromagnetic properties of railway ballast. *NDT & E Int.* **2001**, *34*, 305–311. [CrossRef]

29. Hugenschmidt, J. Railway track inspection using GPR. *J. Appl. Geophys.* **2000**, *43*, 147–155. [CrossRef]

30. Jack, R.; Jackson, P. Imaging attributes of railway track formation and ballast using ground probing radar. *NDT & E Int.* **1999**, *32*, 457–462. [CrossRef]

31. De Bold, R.; O'Connor, G.; Morrisey, J.P.; Forde, M.C. Benchmarking large scale GPR experiments on railway Ballast. *Constr. Build. Mater.* **2015**, *92*, 31–42. [CrossRef]

32. Anbazhagan, P.; Su, L.; Indraratna, B.; Rujikiatkamjorn, C. Model track studies on fouled ballast using ground penetrating radar and multichannel analysis of surface wave. *J. Appl. Geophys.* **2011**, *74*, 175–184. [CrossRef]
33. Fortunato, E.; Pinelo, A.; Matos Fernandes, M. Characterization of the fouled ballast layer in the substructure of a 19th century railway track under renewal. *Soils Found.* **2010**, *50*, 55–62. [CrossRef]
34. Leng, Z.; Al-Qadi, I.L. Railroad Ballast Evaluation Using Ground-Penetrating Radar: Laboratory Investigation and Field Validation. *Transp. Res. Rec. J. Transp. Res. Board* **2010**, *2159*, 110–117. [CrossRef]
35. Benedetto, A.; Tosti, F.; Ciampoli, B.L.; Calvi, A.; Brancadoro, M.G.; Alani, A.M. Railway ballast condition assessment using ground-penetrating radar—An experimental, numerical simulation and modelling development. *Constr. Build. Mater.* **2017**, *140*, 508–520. [CrossRef]
36. Sussmann, T.R.; Maser, K.R.; Kutrubes, D.; Heyns, F.; Selig, E.T. Development of Ground Penetrating Radar for railway infrastructure condition detection. In Proceedings of the Symposium on the Application of Geophysics to Engineering and Environmental Problems, Denver, CO, USA, 4–7 March 2001. RBA-4.
37. Solla, M.; Fontul, S. Non-destructive tests for railway evaluation: Detection of fouling and joint interpretation of GPR data and track geometric parameters. *Ground Penetr. Radar* **2018**, *1*, 75–103. [CrossRef]
38. De Chiara, F.; Fontul, S.; Fortunato, E. GPR Laboratory Tests for Railways Materials Dielectric Properties Assessment. *Remote Sens.* **2014**, *6*, 9712–9728. [CrossRef]
39. Benedetto, F.; Tosti, F.; Alani, A.M. An Entropy-Based Analysis of GPR Data for the Assessment of Railway Ballast Conditions. *IEEE Trans. Geosci. Remote Sens.* **2017**, *55*, 3900–3908. [CrossRef]
40. Forde, M.C.; De Bold, R.; O'Connor, G.; Morrissey, J. New Analysis of Ground Penetrating Radar Testing of a Mixed Railway Trackbed. In Proceedings of the Transportation Research Board 89th Annual Meeting, Washington, DC, USA, 10–14 January 2010.
41. Roberts, R.; Rudy, J.; Al Qadi, I.L.; Tutumluer, E.; Boyle, J. Railroad Ballast Fouling Detection Using Ground Penetrating Radar—A New Approach Based on Scattering from Voids. In Proceedings of the 9th European Conference on NDT, Berlin, Germany, 25–29 September 2006. ECNDT 2006-Th. 4.5.
42. Ciampoli, L.B.; Tosti, F.; Brancadoro, M.G.; D'Amico, F.; Alani, A.M.; Benedetto, A. A spectral analysis of ground-penetrating radar data for the assessment of the railway ballast geometric properties. *NDT & E Int.* **2017**, *90*, 39–47. [CrossRef]
43. Shihab, S.; Zahran, O.; Al-Nuaimy, W. Time-frequency characteristics of ground penetrating radar reflections from railway ballast and plant. In Proceedings of the 7th IEEE on High Frequency Postgraduate Student Colloquium, London, UK, 8–9 September 2002.
44. Fontul, S.; Fortunato, E.; De Chiara, F.; Burrinha, R.; Balderais, M. Railways Track Characterization Using Ground Penetrating Radar. *Procedia Eng.* **2016**, *143*, 1193–1200. [CrossRef]
45. Marecos, V.; Solla, M.; Fontul, S.; Antunes, V. Assessing the pavement subgrade by combining different non-destructive methods. *Constr. Build. Mater.* **2017**, *135*, 76–85. [CrossRef]
46. Xiao, J.; Liu, L. Multi-frequency GPR signal fusion using forward and inverse S-transform for detecting railway subgrade defects. In Proceedings of the International Workshop on Advanced Ground Penetrating Radar, Florence, Italy, 7–10 July 2015.
47. Simi, A.; Manacorda, G.; Miniati, M.; Bracciali, S.; Buonaccorsi, A. Underground asset mapping with dual-frequency dual-polarized GPR massive array. In Proceedings of the 13th International Conference on Ground Penetrating Radar, Lecce, Italy, 19–22 June 2010.
48. Santos-Assunçao, S.; Pedret Rodés, J.; Pérez-Gracia, V. Ground Penetrating Radar Railways Inspection. In Proceedings of the 75th EAGE Conference & Exhibition Incorporating SPE EUROPEC 2013, London, UK, 10–13 June 2013.
49. Musgrave, P. Track bed total route evaluation for track renewals and asset management: A Network Rail perspective. *Constr. Build. Mater.* **2015**, *92*, 2–8. [CrossRef]
50. Kovacevic, M.S.; Gavin, K.; Stipanovic Oslakovic, I.; Bacic, M. A new methodology for assessment of railway infrastructure condition. *Transp. Res. Procedia* **2016**, *14*, 1930–1939. [CrossRef]
51. Lai, W.L.; Kind, T.; Wiggenhauser, H. Frequency-dependentdispersion of high-frequency ground penetrating radar wave in concrete. *NDT & E Int.* **2011**, *44*, 267–273. [CrossRef]

52. Lai, W.L.; Poon, C.S. GPR data analysis in time-frequency domain. In Proceedings of the 14th International Conference on Ground Penetrating Radar (GPR), Shanghai, China, 4–8 June 2012; pp. 362–366. [CrossRef]
53. Lai, W.L.; Kind, T.; Wiggenhauser, H. Using ground penetrating radar and time-frequency analysis to characterize construction materials. *NDT & E Int.* **2011**, *44*, 111–120. [CrossRef]
54. Giovanneschi, F.; González-Huici, M.A.; Uschkerat, U. A parametric analysis of time and frequency domain GPR scattering signatures from buried landmine-like targets. In Proceedings of the SPIE 8709, Detection and Sensing of Mines, Explosive Objects, and Obscured Targets XVIII, Baltimore, MD, USA, 7 June 2013. [CrossRef]
55. Ho, K.C.; Carin, L.; Gader, P.D.; Wilson, J.N. An Investigation of Using the Spectral Characteristics from Ground Penetrating Radar for Landmine/Clutter Discrimination. *IEEE Trans. Geosci. Remote Sens.* **2008**, *46*, 1177–1191. [CrossRef]
56. Meschino, S.; Pajewski, L. SPOT-GPR: A freeware tool for target detection and localization in GPR data developed within the COST Action TU1208. *J. Telecommun. Inf. Technol.* **2017**, *3*, 43–54. [CrossRef]
57. Meschino, S.; Pajewski, L. A practical guide on using SPOT-GPR, a freeware tool implementing a SAP-DoA technique. *Ground Penetr. Radar* **2018**, *1*, 104–122. [CrossRef]
58. Persico, R.; Leucci, G. Interference Mitigation Achieved with a Reconfigurable Stepped Frequency GPR System. *Remote Sens.* **2016**, *8*, 926–937. [CrossRef]
59. Zhang, W.Y.; Hao, T.; Chang, Y.; Zhao, Y.H. Time-frequency analysis of enhanced GPR detection of RF tagged buried plastic pipes. *NDT & E Int.* **2017**, *92*, 88–96. [CrossRef]
60. Foillard, R. EP 1 574 878 B1. Device for Determining the Presence of a Cavity under a Roadway or a Railway, by R. and the Geoscan Company. 2005. Available online: https://patents.google.com/patent/EP1574878B1/en (accessed on 31 March 2018).
61. Al-Qadi, I.L.; Xie, W.; Jones, D.L.; Roberts, R. Development of a time–frequency approach to quantify railroad ballast fouling condition using ultra-wide band ground-penetrating radar data. *Int. J. Pavement Eng.* **2010**, *10*, 260–279. [CrossRef]
62. IDS Georadar. SRS SafeRailSystem. The Fastest Rail Borne System for Railway Ballast Inspection. Available online: http://www.idsgeoradar.com (accessed on 31 March 2018).
63. Pedret Rodés, J. Diseño de un indicador de apoyo a la gestión de Firmes basado en ground penetrating radar. Análisis de la forma del espectro de onda de GPR como indicador de estado de firmes asfálticos. Ph.D. Thesis, Universidade Politècnica de Catalunya, Barcelona, Spain, 2017.
64. Paixão, A.; Fortunato, E.; Calçada, R. A contribution for integrated analysis of railway track performance at transition zones and discontinuities. *Constr. Build. Mater.* **2016**, *111*, 699–709. [CrossRef]
65. Paixão, A. Transition Zones in Railway Tracks: An Experimental and Numerical Study on the Structural Behaviour. Ph.D. Thesis, University of Porto, Porto, Portugal, 2014.
66. EN 13848-5:2008 Track Geometry Quality—Geometric Quality Levels. 2008. European Standard. Available online: https://www.en-standard.eu/csn-13848-5-railway-applications-track-track-geometry-quality-part-5-geometric-quality-levels-plain-line-switches-and-crossings/ (accessed on 1 November 2017).
67. REFER. Tolerâncias dos parâmetros geométricos da via. IT.VIA.018. 2009. Internal Technical Standard. Available online: https://aplicacoes.refer.pt/normas/enormativosWEB/Pesquisa.aspx (accessed on 1 November 2017). (In Portuguese)

remote sensing

MDPI

Article

IMF-Slices for GPR Data Processing Using Variational Mode Decomposition Method

Xuebing Zhang [1], Enhedelihai Nilot [2], Xuan Feng [2,*], Qianci Ren [3] and Zhijia Zhang [1]

[1]   School of Geomatics and Prospecting Engineering, Jilin Jianzhu University, Changchun 130118, China;
     zzzhxb@foxmail.com (X.Z.); zhang804830@foxmail.com (Z.Z.)
[2]   College of Geo-Exploration Science and Technology, Jilin University, Changchun 130026, China;
     ehdlh14@mails.jlu.edu.cn
[3]   College of Earth Sciences, Guilin University of Technology, Guilin 541006, China; renqianci@126.com
*   Correspondence: XuanFeng@jlu.edu.cn; Tel.: +86-134-0435-3645

Received: 14 February 2018; Accepted: 15 March 2018; Published: 19 March 2018

Abstract: Using traditional time-frequency analysis methods, it is possible to delineate the time-frequency structures of ground-penetrating radar (GPR) data. A series of applications based on time-frequency analysis were proposed for the GPR data processing and imaging. With respect to signal processing, GPR data are typically non-stationary, which limits the applications of these methods moving forward. Empirical mode decomposition (EMD) provides alternative solutions with a fresh perspective. With EMD, GPR data are decomposed into a set of sub-components, i.e., the intrinsic mode functions (IMFs). However, the mode-mixing effect may also bring some negatives. To utilize the IMFs' benefits, and avoid the negatives of the EMD, we introduce a new decomposition scheme termed variational mode decomposition (VMD) for GPR data processing for imaging. Based on the decomposition results of the VMD, we propose a new method which we refer as "the IMF-slice". In the proposed method, the IMFs are generated by the VMD trace by trace, and then each IMF is sorted and recorded into different profiles (i.e., the IMF-slices) according to its center frequency. Using IMF-slices, the GPR data can be divided into several IMF-slices, each of which delineates a main vibration mode, and some subsurface layers and geophysical events can be identified more clearly. The effectiveness of the proposed method is tested using synthetic benchmark signals, laboratory data and the field dataset.

Keywords: variational mode decomposition; empirical mode decomposition; IMF-slices; GPR data processing; GPR imaging; time-frequency analysis

---

1. Introduction

Most of the geophysical data are non-stationary. Especially for the ground-penetrating radar (GPR) data, which can be treated as a set of time-series signals, such non-stationarity appears more obviously. Non-stationarity can be defined and measured by certain statistical methods [1], however, geophysicists are prone to using the time-frequency decomposition methods (e.g., wavelet transform [2,3], S-transform [4,5] and matching pursuit [6,7]) to delineate the non-stationary characteristics. This is because these methods provide a visualization and are convenient for further applications.

Recently, empirical mode decomposition (EMD) has been providing alternative solutions with a fresh perspective. With EMD-based methods, the GPR data can be decomposed into a group of stationary sub-components (i.e., intrinsic mode functions, IMFs), and the subsequent processing for subsurface information is thus focused on the analysis of the generated IMFs instead [8]. The decomposed IMFs of GPR data highlight two important features. The first is that the IMFs are close to being stationary. This leads to a direct application of time-frequency analysis of GPR data [9], in which time-frequency structures can be computed steadily within each IMF. The other feature is that the decomposed IMFs correspond to different frequency bands or ranges. This means that any operation performed on the IMFs (e.g., removing or extracting a few IMFs) amounts to filtering in the corresponding frequency domain. In this regard, the derived applications based on IMFs may include denoising [10], subsurface objects identification [11], etc. However, the mode-mixing effect and the lack of a mathematical foundation limit their further applications in GPR data processing and imaging.

To utilize the benefits of the IMFs mentioned above, and avoid the negatives of the EMD, we introduce a new decomposition scheme called variational mode decomposition (VMD) [12] for GPR data processing as well as for imaging. The VMD decomposition is not "empirical" any more, while the mathematical framework is well-established. Specifically, though the decomposition results can still be seen as IMFs, the computation is implemented by solving an optimization problem. Some references have demonstrated that the VMD is able to avoid mode-mixing phenomena [13–15], and therefore the decomposition results match the input signal's intrinsic vibration mode better. In this study, we test VMD decomposition on GPR data, and based on the VMD decomposition results, we propose a new method for GPR imaging which we refer as "the IMF-slice". In the proposed method, the IMFs are generated by the VMD trace by trace, and then each IMF is sorted and recorded into different profiles (i.e., the IMF-slice) according to its center frequency. A GPR profile could be divided into several IMF-slices, each of which delineates a main vibration mode within certain frequency band. Using IMF-slices, some subsurface events are able to be identified more clearly.

As parts of the long-term study of the mathematical transformations of GPR data, this work shows elementary research on the applications of the VMD decomposition for GPR imaging. The objectives of this work are to: (1) introduce the VMD decomposition and test its feasibility on GPR data; (2) compare the decomposition results of GPR data between the VMD and the classic EMD; (3) propose the concept of IMF-slices and discuss their applications in GPR imaging and interpretation. The paper starts with a methods section introducing the VMD decomposition and the IMF-slices for GPR data. We then show some results based on synthetic benchmark tests, laboratory data tests and a field dataset. Finally, the conclusions and discussions are provided in the last section.

2. Methods

2.1. The Variational Mode Decomposition

The VMD algorithm was initially described as solving the following constrained optimization subject to equality constraints by Dragomiretskiy and Zosso [12],

$$\min_{\{u_k\},\{w_k\}} \left\{ \sum_k \left\| \partial_t \left[\left(\delta(t) + \frac{j}{\pi t} \right) * u_k(t) \right] e^{-j\omega_k t} \right\|_2^2 \right\}$$
$$s.t. \sum_k u_k(t) = f(t) \tag{1}$$

In the above form, $\delta(t)$ is the Dirac function, and $\{u_k(t)\}_{k=1,2,...N}$ denotes the decomposed sub-components (i.e., the IMFs) embedded in the input signal $f(t)$, where k identifies the number of IMFs. $\{\omega_k\}_{k=1,2,...N}$ denotes the corresponding center frequencies of each IMF. Within the Lagangian-multiplier framework, the penalty-term and the Lagangian-multiplier λ are introduced. Then the above optimization problem would be turned to solving the defined Lagrange function [16] as follows,

$$\begin{aligned} &L(\{u_k\}, \{\omega_k\}, \lambda) \\ &= \alpha \sum_k \left\| \partial_t \left[\left(\delta(t) + \frac{j}{\pi t} \right) * u_k(t) \right] e^{-j\omega_k t} \right\|_2^2 + \left\| f(t) - \sum_k u_k(t) \right\|_2^2 + \left\langle \lambda(t), f(t) - \sum_k u_k(t) \right\rangle \end{aligned} \tag{2}$$

where $L(\{u_k\}, \{\omega_k\}, \lambda)$ denotes the Lagrange function, and the parameter α is used for balancing the reconstruction error. Its solution can be derived by the well-known alternating direction method of multipliers (ADMM) [17].

Assuming that there are K IMFs decomposed, the frequency-domain expression of the kth IMF derived by the VMD is,

$$\hat{u}_k^{n+1}(\omega) = \frac{\hat{f}(\omega) - \sum\limits_{i=1}^{k-1} \hat{u}_i^{n+1}(\omega) - \sum\limits_{i=k+1}^{K} \hat{u}_i^{n}(\omega) + \frac{\hat{\lambda}^n(\omega)}{2}}{1 + 2\alpha(\omega - \omega_k^n)^2} \tag{3}$$

in which $\hat{f}(\omega)$, $\hat{u}_i(\omega)$, $\hat{\lambda}(\omega)$, and $\hat{u}_k^{n+1}(\omega)$ are the frequency-domain forms of $f(t)$, $u_i(t)$, $\lambda(t)$, $u_k^{n+1}(t)$ using the Fourier transform, and n denotes iteration number. The criterion of iteration termination is

$$\sum_k \|u_k^{n+1}(t) - u_k^{n}(t)\|_2^2 / \|u_k^{n}(t)\|_2^2 \leq \varepsilon, \tag{4}$$

where ε denotes the tolerance of convergence. Based on the decomposed IMFs, the input signal can be represented by the decomposed IMFs and the residual signal $r(t)$,

$$f(t) = \sum_{i=1}^{K} u_i^{n+1}(t) + r(t). \tag{5}$$

It should be noted that, though the final expansion form of the VMD decomposition (i.e., Equation (5)) is similar to that of the EMD, the computation of each IMF in VMD is totally different. Since the detailed deduction and comparisons with the EMD algorithm are not the main purposes of this paper, we will show the decomposition results in the Section 3.

2.2. IMF-Slices of GPR Data

Compared with traditional common-frequency slices, the IMF-slices can be treated as a set of IMFs. In the VMD scheme, the center frequency $\omega_k(t)$ of each decomposed IMF can be obtained simultaneously. Given a GPR profile (i.e., a B-scan), the IMFs can be derived by VMD trace by trace, where the "trace" here means the A-scan. If we pre-define the frequency intervals or ranges for the IMF-slices, the IMFs of each trace can be classified into different IMF-slices. It should be noted that the frequency ranges for each IMF-slice can be estimated empirically using the center frequencies obtained for each IMF of the first five or ten traces. For example, if the frequency range of a IMF-slice is defined as $[\omega_1, \omega_2]$, and combining the median filtering, each decomposed IMF of one trace are evaluated whether it belongs to this range. If yes, it is recorded in the corresponding trace of the IMF-slice, which can be represented as,

$$IMF\text{-}slice(x,t) = \sum_{i=1}^{K} \theta(M(\omega_i(t))) u_i(t) \tag{6}$$

where M denotes the median filtering operator, and the θ is the Heaviside function,

$$\theta(\omega) = \begin{cases} 1, & \text{for } \omega \in [\omega_1, \omega_2], \\ 0, & \text{else.} \end{cases} \tag{7}$$

Equation (6) appears to be a band-pass filter for the IMFs. A schematic diagram based on the VMD decomposition is shown in Figure 1.

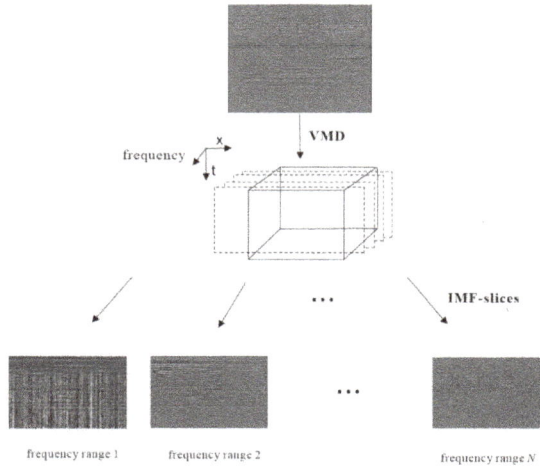

Figure 1. Schematic diagram of IMF-slices.

Similarly, the IMF-slices can be also generated based on EMD. However, since the center frequencies for each IMF, accompanied by the VMD decomposition results, cannot be obtained in the EMD scheme, we select the Taner's instantaneous frequency [18] to replace it,

$$\omega(t) = \frac{1}{2\pi} \frac{f_{IMF}(t)\frac{df_{IMF}^*(t)}{dt} - f_{IMF}^*(t)\frac{df_{IMF}(t)}{dt}}{f_{IMF}^2(t) + f_{IMF}^{*2}(t)}, \tag{8}$$

where $f_{IMF}(t)$ and $f_{IMF}^*(t)$ here denotes the IMF and its conjugate form.

3. Results

3.1. Synthetic Benchmark Tests

In this section, we firstly test a synthetic benchmark signal formulated by Herrera et al. [19,20], which can be characterized as a non-stationary and multi-component signal. The signal (Figure 2a) is comprised of four sub-components, which are relatively stationary in certain time ranges, see Figure 2b–e. Figure 3 shows the corresponding instantaneous frequencies (IFs).

The benchmark signal can be decomposed into a set of IMFs by using both EMD and VMD, as shown in Figures 4 and 5, respectively. It is obvious that the decomposition results using VMD are sparser than those obtained using the EMD method. Additionally, similar vibration modes (with similar time-frequency structures) are always decomposed into different IMFs, which is known as the mode-mixing effect). For example, the third sub-component signal (Figure 2d) is mainly located between 6–10 s and its IF is around 10 Hz. However, the EMD scheme separates them and records them in IMF1 and IMF2, in Figure 4. By contrast, the IMFs derived by the VMD scheme match the benchmark better (see Figure 5). More specifically, in Figure 5, IMF1 refers to Figure 2c, IMF2 refers to Figure 2e, and IMF3 refers to both Figure 2b,d. We remark here that the correspondence can be judged not only by the proximity of frequency ranges, but also by the proximity of amplitude.

Although there still exist some "mode-mixing" phenomena in the decomposition results by VMD, this defect has been extremely reduced compared with EMD. For example in Figure 5, there are some vibrations with low amplitude or energy remaining at 6~10 s of IMF1, but the sub-component signals (Figure 2d) are mainly reflected in the IMF3. From the above benchmark tests, we observe that the

IMFs by VMD are able to reveal the intrinsic properties of the non-stationary signal, whereas EMD may mislead us. Thus, we can draw the conclusion that decomposition by VMD is always more reasonable.

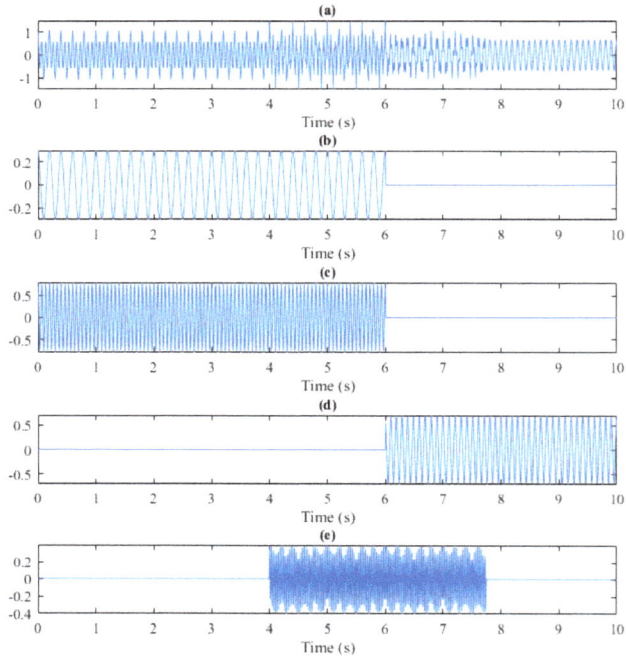

Figure 2. The synthetic data (**a**) composed of four sub-components (**b–e**).

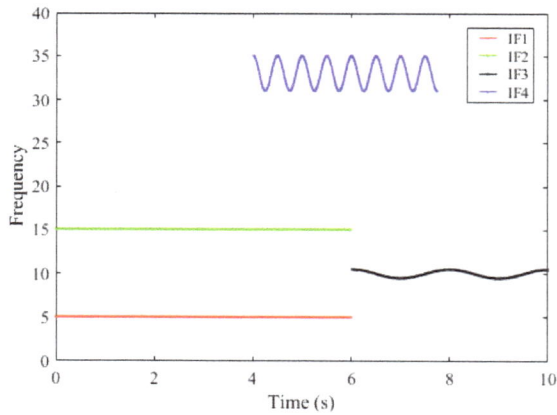

Figure 3. The corresponding instantaneous frequencies of the sub-components in Figure 1.

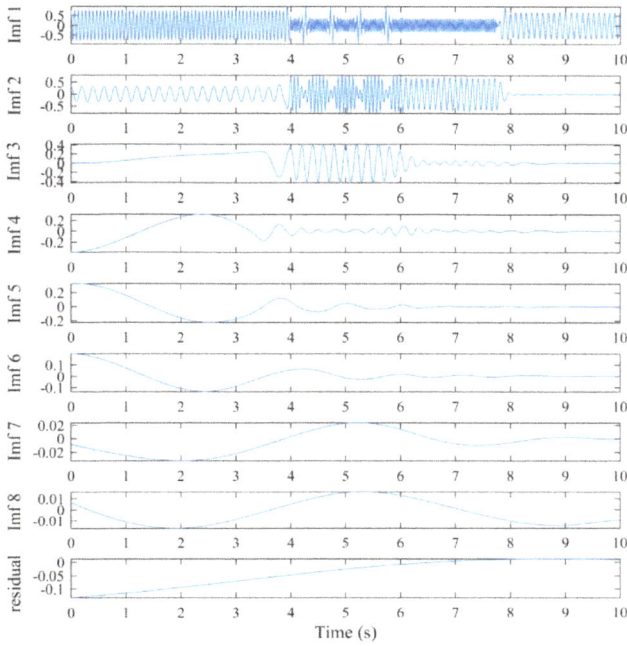

Figure 4. The EMD decomposition results and the residual.

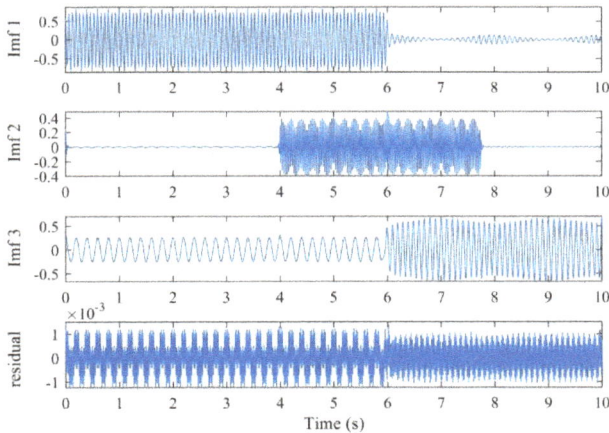

Figure 5. The VMD decomposition results and the residual. IMF1 and IMF2 refer to Figure 2c,e, respectively, and IMF3 refers to both Figure 2b,d.

The corresponding IFs can be computed by Taner's equation using the IMFs decomposed by EMD and VMD. Then the instantaneous spectra can be constructed as shown in Figure 6a,b. Compared with the time-frequency structures in Figure 3, the instantaneous spectrum by VMD is relatively more accurate. This is also because of the fact that VMD decomposition is more reasonable. These synthetic benchmark tests illustrate how the VMD decomposition can be conducted on non-stationary data, avoiding the mode-mixing phenomenon.

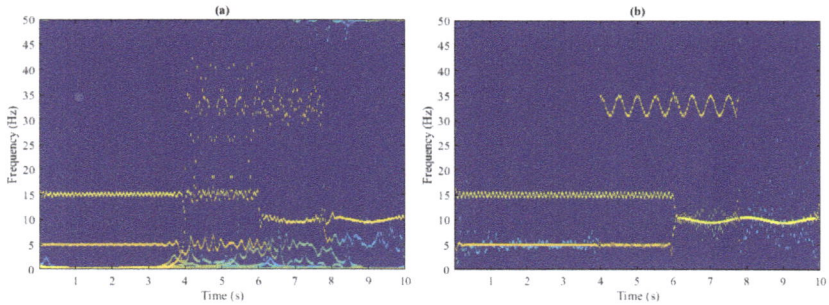

Figure 6. The instantaneous spectrum based on (**a**) EMD and (**b**) VMD.

3.2. Laboratory Data Tests

In order to test the VMD method for GPR data decomposition, we employ the laboratory data (shown in Figure 7) for further analysis. The data was collected in a sand trough as shown in Figure 8a, where the target body (Figure 8b) was set at a certain depth.

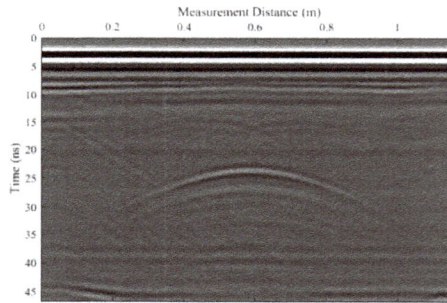

Figure 7. The laboratory data collected.

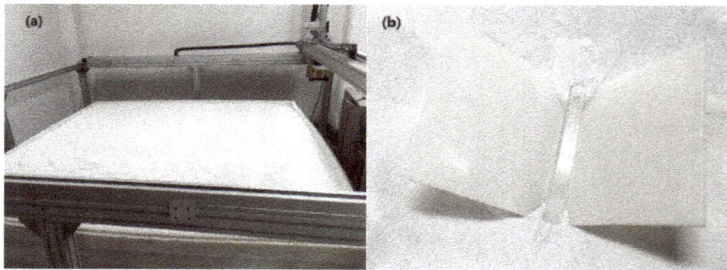

Figure 8. The sand trough (**a**) and the target body (**b**).

Typically, each trace of the GPR data is non-stationary. We apply the VMD method to decompose the GPR data trace by trace. Then we sort the main components and collect them into several profiles (i.e., the IMF-slices), see Figure 9a–d. For comparison, we adopt a similar way of processing the decomposed IMFs using EMD decomposition, and the generated IMF-slices are shown in Figure 10. In Figure 9, the GPR data are divided into four main components, each of which demonstrates a different vibration mode. The reflection of the buried target is identified in the second IMF-slice by the red arrow pointing to it (see Figure 9b). The noise is shown in Figure 9d, without any important information from the subsurface layers being mixed. In Figure 10, the first IMF-slice is meaningless, and the yellow arrows illustrate the so-called mode-mixing effect. Additionally, the reflection indicated by the red arrow in Figure 10b is not as distinct as that achieved using VMD.

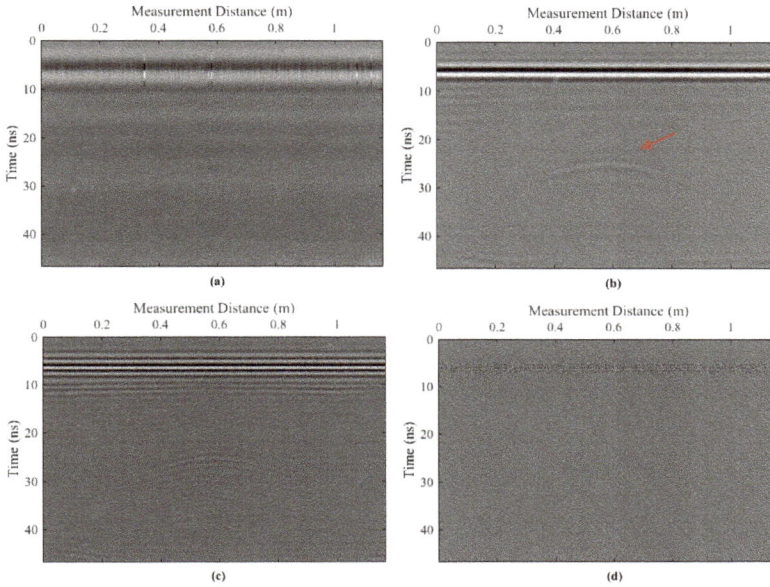

Figure 9. The derived IMF-slices using VMD. (**a**–**d**) show the IMF-slice 1–4 respectively.

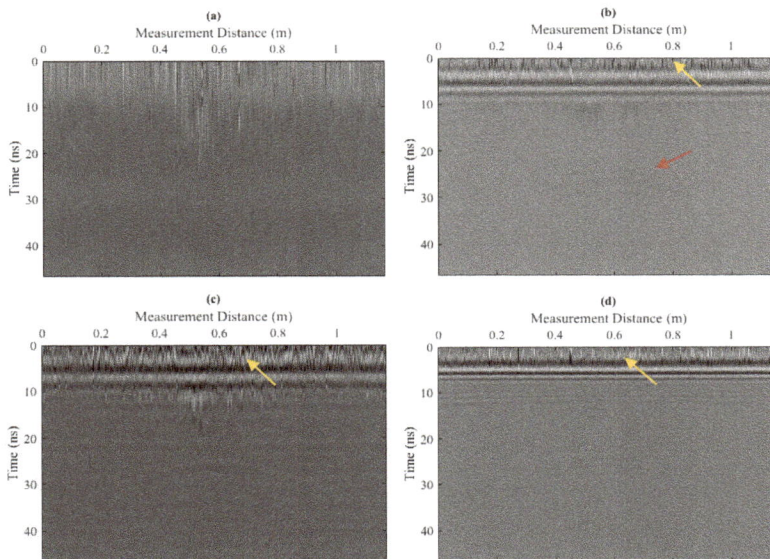

Figure 10. The derived IMF-slices using EMD. (**a**–**d**) show the IMF-slice 1–4 respectively.

4. The Use of VMD with Field Dataset

In this section, we use variational mode decomposition to process a field dataset and show the results of the proposed IMF-slices. Subsequently, the corresponding interpretations are drawn.

The dataset is shown in Figure 11. It was collected in Yan'an, in the Shanxi Province of China. The data had initially been processed; however, the noise still remains. We firstly use VMD and EMD to decompose one trace respectively (Figure 12), and the decomposed results are shown in Figures 13 and 14, respectively. There are four IMFs generated in Figure 13, of which the first three IMFs are the main components. The fourth IMF can be seen as the noise components. In contrast, the EMD generates more IMFs in Figure 14, and the noise is mixed into the first IMF. In this case, the mode-mixing phenomenon that accompanies EMD decomposition can be lessened by the VMD.

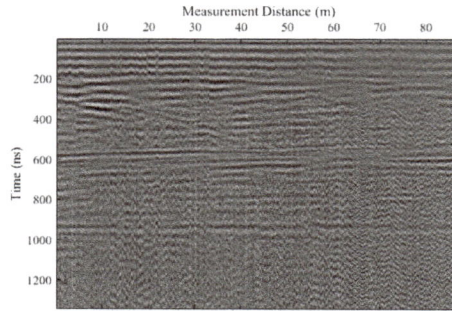

Figure 11. The field GPR dataset.

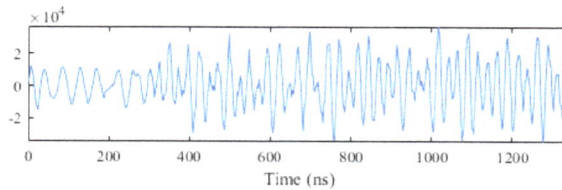

Figure 12. A single trace of the GPR dataset.

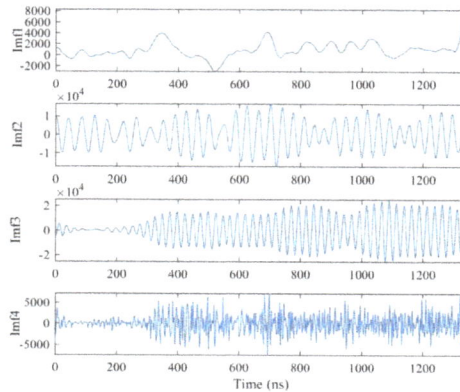

Figure 13. The decomposed IMFs using VMD.

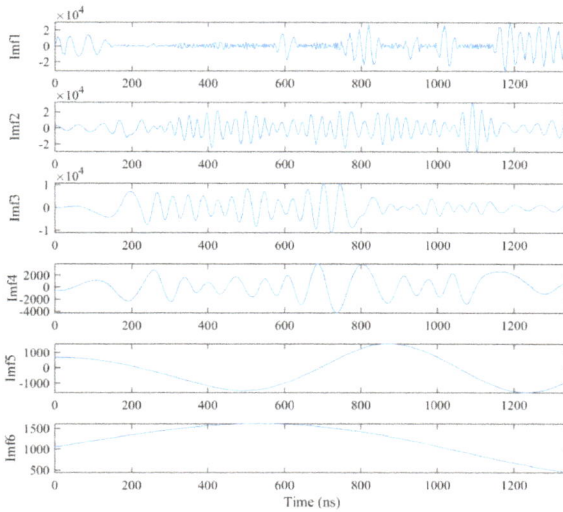

Figure 14. The decomposed IMFs using EMD.

We then compute VMD decomposition for the whole dataset; the derived IMF-slices are shown in Figure 13. The low-frequency components are mainly collected in the first IMF-slice (Figure 15a). We remark here that in this IMF-slice, the DC and the slowly-varying trend term [1] are also collected, so there exists discontinuity of neighboring traces in gray scale. The second and third IMF-slices (Figure 15b–c) demonstrate the subsurface structures at different scales. In the last IMF-slice, the noise of the dataset is extracted, especially in the deep part of the profile.

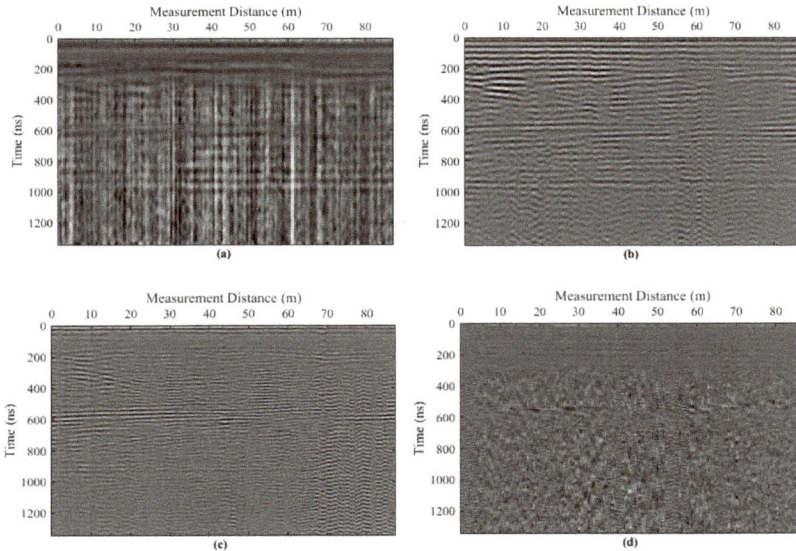

Figure 15. The derived IMF-slices of the field dataset using VMD. (**a–d**) show the IMF-slice 1–4 respectively.

In Figure 16, we also provide comparisons with the IMF-slices based on EMD. In some locations of these IMF-slices, the interpretations are greatly influenced by the mode-mixing of IMFs for different traces. For instance, in Figure 16d the important information in the shallow part and the noise in the deep part are mixed up. It can be observed that, although the IMF-slices by EMD may refer to different frequency ranges, the continuity between neighboring traces is not as good as that achieved with VMD.

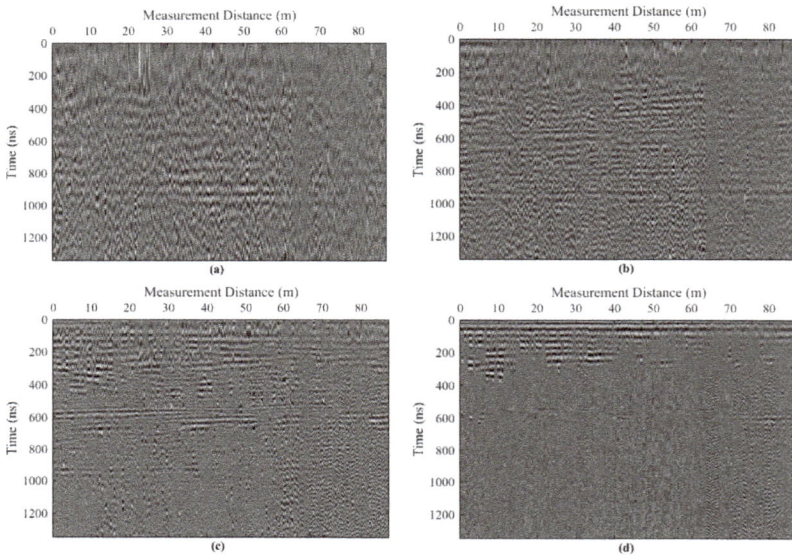

Figure 16. The derived IMF-slices of the field dataset using EMD. (**a–d**) show the IMF-slice 1–4 respectively.

5. Conclusions

EMD-based methods for GPR imaging provide a novel prospect for interpreting GPR data. Using EMD, the GPR data are decomposed, and the relative stationary IMFs are derived and analyzed for subsequent application. In this paper, variational mode decomposition was introduced. Using VMD, the derived IMFs are sparser and match well with the signal's intrinsic properties. Based on the decomposition results, the IMFs are sorted and recorded into the proposed IMF-slices for GPR data imaging. We showed its utility on a group of examples, including synthetic benchmark signals, laboratory data and a field dataset. Using IMF-slices, GPR data could be divided into several IMF-slices, each of which delineates a main vibration mode and some subsurface layer and geophysical events can be identified more clearly.

However, the proposed IMF-slices method still requires empirically pre-defined parameters. If the frequency ranges for each IMF-slice are estimated wrongly, or the time-frequency structures of the traces in the GPR profile vary sharply, it is difficult to find a unified standard to separate the different IMF-slices. In our future work, we need to rethink and seek for more intelligent schemes or more automatic procedures for classifying the derived IMFs for the IMF-slices. Additionally, for the accurate detection of the subsurface targets, some figures of merit could also be employed to assess the generated IMF-slices, such as receiver operating characteristics (ROC) [21]. In this case, VMD decomposition may be extended and applied to more aspects for GPR data processing and imaging.

Acknowledgments: The authors thank Jing Li and the three reviewers for meaningful and inspiring discussions. This work is supported in part by the National Key Research and Development Program of China under grant

2016YFC0600505 and the Scientific Research Program of the Education Department of Jilin Province (the grant number will be issued soon). The language proofreading of this paper during the revision stage is supported by the Provincial Training Program of Innovation and Entrepreneurship for Undergraduates under grant 2017S1027 and 2017S1030.

Author Contributions: "X.Z. and X.F. conceived and designed the experiments; E.N. performed the experiments; X.Z. and Z.Z. analyzed the data and wrote the code; E.N. contributed the field data materials; X.Z. wrote the paper."

Conflicts of Interest: The authors declare no conflict of interest.

Abbreviations

The following abbreviations are used in this paper:

GPR	ground-penetrating radar
EMD	empirical mean decomposition
IMF	intrinsic mode function
VMD	variational mode decomposition
ADMM	alternate direction method of multipliers

References

1. Huang, N.E.; Shen, Z.; Long, S.R.; Wu, M.C.; Shih, H.H.; Zheng, Q.; Yen, N.-C.; Tung, C.C.; Liu, H.H. The empirical mode decomposition and the Hilbert spectrum for nonlinear and non-stationary time series analysis. *Proc. R. Soc. Lond. Seri. A Math. Phys. Eng. Sci.* **1998**, *454*, 903–995. [CrossRef]
2. Rioul, O.; Vetterli, M. Wavelets and signal processing. *IEEE Signal Process. Mag.* **1991**, *8*, 14–38. [CrossRef]
3. Sinha, S.; Routh, P.; Anno, P.; Castagna, J. Spectral Decomposition of Seismic Data with Continuous Wavelet Transform. *Geophysics* **2005**, *70*, P19–P25. [CrossRef]
4. Stockwell, R.G.; Mansinha, L.; Lowe, R.P. Localization of the complex spectrum: The S transform. *IEEE Trans. Signal Process.* **1996**, *44*, 998–1001. [CrossRef]
5. Cheng, Z.; Chen, W.; Chen, Y.; Liu, Y.; Liu, W.; Li, H.; Yang, R. Application of bi-Gaussian S-transform in high-resolution seismic time-frequency analysis. *Interpretation* **2017**, *5*, SC1–SC7. [CrossRef]
6. Feng, X.; Zhang, X.; Liu, C.; Lu, Q. Single-channel and multi-channel orthogonal matching pursuit for seismic trace decomposition. *J. Geophys. Eng.* **2017**, *14*, 90–99. [CrossRef]
7. Zhang, X.; Feng, X.; Liu, C.; Chen, C.; Li, X.; Zhang, Y. Seismic matching pursuit decomposition based on the attenuated Ricker wavelet dictionary. In Proceedings of the 79th EAGE Conference and Exhibition 2017, Paris, France, 12–15 June 2017.
8. Li, J.; Liu, C.; Zeng, Z.; Chen, L. GPR Signal Denoising and Target Extraction With the CEEMD Method. *IEEE Geosci. Remote Sens. Lett.* **2015**, *12*, 1615–1619.
9. Lu, Q.; Liu, C.; Zeng, Z.; Li, J.; Zhang, X. Detection of human's motions through a wall using UWB radar. In Proceedings of the 2016 16th International Conference on Ground Penetrating Radar (GPR), Hong Kong, China, 13–16 June 2016; pp. 1–4.
10. Chen, Y.; Ma, J. Random noise attenuation by f-x empirical-mode decomposition predictive filtering. *Geophysics* **2014**, *79*, V81–V91. [CrossRef]
11. Qin, Y.; Qiao, L.H.; Ren, X.Z.; Wang, Q.F. Using bidimensional empirical mode decomposition method to identification buried objects from GPR B-scan image. In Proceedings of the 2016 16th International Conference on Ground Penetrating Radar (GPR), Hong Kong, China, 13–16 June 2016; pp. 1–5.
12. Dragomiretskiy, K.; Zosso, D. Variational Mode Decomposition. *IEEE Trans. Signal Process.* **2014**, *62*, 531–544. [CrossRef]
13. Liu, W.; Cao, S.; Chen, Y. Applications of variational mode decomposition in seismic time-frequency analysis. *Geophysics* **2016**, *81*, V365–V378. [CrossRef]
14. Liu, W.; Cao, S.; Wang, Z.; Kong, X.; Chen, Y. Spectral Decomposition for Hydrocarbon Detection Based on VMD and Teager-Kaiser Energy. *IEEE Geosci. Remote Sens. Lett.* **2017**, *14*, 539–543. [CrossRef]
15. Liu, W.; Cao, S.; Wang, Z. Application of variational mode decomposition to seismic random noise reduction. *J. Geophys. Eng.* **2017**, *14*, 888–899. [CrossRef]
16. Boyd, S.; Vandenberghe, L. *Convex Optimization*; Cambridge University Press: New York, NY, USA, 2004.
17. Hestenes, M.R. Multiplier and gradient methods. *J. Optim. Theory Appl.* **1969**, *4*, 303–320. [CrossRef]

18. Taner, M.; Koehler, F.; Sheriff, R. Complex seismic trace analysis. *Geophysics* **1979**, *44*, 1041–1063. [CrossRef]

19. Herrera, R.H.; Tary, J.B.; van der Baan, M. Time-frequency representation of microseismic signals using the synchrosqueezing transform. In Proceedings of the GeoConvention 2013, Integration, Calgary, AB, Canada, 6–12 May 2013.

20. Fomel, S. Seismic data decomposition into spectral components using regularized nonstationary autoregression. *Geophysics* **2013**, *78*, O69–O76. [CrossRef]

21. Rodríguez, A.; Salazar, A.; Vergara, L. Analysis of split-spectrum algorithms in an automatic detection framework. *Signal Process.* **2012**, *92*, 2293–2307. [CrossRef]

remote sensing

MDPI

Article

TU1208 Open Database of Radargrams: The Dataset of the IFSTTAR Geophysical Test Site

Xavier Dérobert [1] and Lara Pajewski [2,*]

[1] Institut Francais des Sciences et Technologies des Transports, de l'Amenagement et des Reseaux (IFSTTAR), CS 04–44344 Bouguenais CEDEX, France, xavier.derobert@ifsttar.fr

[2] Department of Information Engineering, Electronics and Telecommunications, Sapienza University of Rome, via Eudossiana 18, 00184 Rome, Italy

* Correspondence: tu1208@gpradar.eu or lara.pajewski@uniroma1.it; Tel.: +39-06-4458-5358

Received: 16 February 2018; Accepted: 26 March 2018; Published: 29 March 2018

Abstract: This paper aims to present a wide dataset of ground penetrating radar (GPR) profiles recorded on a full-size geophysical test site, in Nantes (France). The geophysical test site was conceived to reproduce objects and obstacles commonly met in the urban subsurface, in a completely controlled environment; since the design phase, the site was especially adapted to the context of radar-based techniques. After a detailed description of the test site and its building process, the GPR profiles included in the dataset are presented and commented on. Overall, 67 profiles were recorded along eleven parallel lines crossing the test site in the transverse direction; three pulsed radar systems were used to perform the measurements, manufactured by different producers and equipped with various antennas having central frequencies from 200 MHz to 900 MHz. An archive containing all profiles (raw data) is enclosed to this paper as supplementary material. This dataset is the core part of the Open Database of Radargrams initiative of COST (European Cooperation in Science and Technology) Action TU1208 "Civil engineering applications of Ground Penetrating Radar". The idea beyond such initiative is to share with the scientific community a selection of interesting and reliable GPR responses, to enable an effective benchmark for direct and inverse electromagnetic approaches, imaging methods and signal processing algorithms. We hope that the dataset presented in this paper will be enriched by the contributions of further users in the future, who will visit the test site and acquire new data with their GPR systems. Moreover, we hope that the dataset will be made alive by researchers who will perform advanced analyses of the profiles, measure the electromagnetic characteristics of the host materials, contribute with synthetic radargrams obtained by modeling the site with electromagnetic simulators, and more in general share results achieved by applying their techniques on the available profiles.

Keywords: ground penetrating radar; non-destructive testing; near-surface geophysics; test site

1. Introduction

In this paper, a wide collection of ground penetrating radar (GPR) [1,2] profiles is presented. All data were recorded over the Nantes geophysical test site of the French institute of science and technology for transport, spatial planning, development and networks (Institut français des sciences et technologies des transports, de l'aménagement et des réseaux, IFSTTAR).

The Nantes geophysical test site of IFSTTAR has been in service for twenty years. It was conceived to reproduce a scenario including full-size objects and obstacles commonly found in urban grounds, in a completely controlled environment. Since the design phase, the site was especially adapted to the context of radar-based techniques. The main needs for the urban civil engineering, in which this site intends to provide an experimental field, relate to the non-destructive detection, localization and characterization of hidden structures and not to the assessment of soil properties. The detection of underground structures in urban environments is definitely a crucial problem and geophysical methods, particularly GPR, can perform this task with relatively good efficiency, depending on the physical properties of the host material, the physical and geometrical properties of the sought bodies, and the overall framework. As a tool accessible to professionals, the IFSTTAR test site has been logically used for needs such as tests or comparisons of devices, and validation of measurement or modeling methods.

The test site [3] consists of a pit, 30 m long and 5 m wide, with sloping sides. The useful region of the pit has a variable depth around 4 m. The pit is divided into five sections, filled with different materials such as silt, limestone and gneiss, and separated by vertical interfaces. Several targets are embedded in the test site and accurately geo-referenced. They are representative of objects that can be generally met in trenchless works—such as pipes, cables, stones of various size, and masonry. Precautions were taken to prevent water inflows, to avoid any bias in GPR measurements due to environmental changes. To ensure optimum data acquisition, the site is easily accessible and far from noise sources such as electrical installations and trees.

The dataset subject of this paper includes 67 radargrams recorded by using three commercial pulsed GPR systems equipped with various antennas working on different ranges of the electromagnetic spectrum, and in particular with central frequencies from 200 MHz to 900 MHz. Measurements were performed along 11 parallel lines, crossing the test site in the transverse direction. The dataset is the core part of the Open Database of Radargrams initiative of COST (European Cooperation in Science and Technology) Action TU1208 "Civil engineering applications of Ground Penetrating Radar" [4,5], which consists in realizing and making available an open collection of interesting GPR responses (experimental and synthetic) to enable a very effective benchmark for forward and inverse electromagnetic scattering methods imaging techniques and signal processing algorithms. Actually, numerous studies in the literature present and sometimes compare analytical and/or numerical methods for the analysis and interpretation of GPR data; but, each approach is generally considered individually in a specific context and compared with a limited number of other methods. The availability of a reference set of GPR profiles is a turning point in this field of research, because it facilitates a wider, more effective and rigorous comparison of different techniques. All data of the TU1208 database are being shared with the scientific community via the Action website and/or dedicated open access publications (such as this paper), along with detailed descriptions of the investigated structures and full information about the employed equipment. To the best of our knowledge, TU1208 Open Database of Radargrams is the first open collection of GPR responses: Similar initiatives were never undertaken in the past, in the GPR field.

The paper is structured as follows. In Section 2, COST Action TU1208 is shortly introduced; the objectives of the Open Database of Radargrams are outlined and the context in which the initiative was born is portrayed. In Section 3, the geophysical test site is described in detail: schematic maps and sections of the site, as well as photos shot during the site construction, are presented; moreover, information about the filling materials and buried targets is given. We would like to underline that, although the site was built in 1996, such a comprehensive explanation of its realization method, geometry, and nature of the involved materials, was not published before on the international scientific literature. In Section 4, full information about the GPR equipment used to perform the measurements and relevant data-collection settings is provided. In Section 5, all profiles are plotted in grey-scale maps and several comments are given, which we deem useful for facilitating the comprehension and usage of the data. Conclusions are drawn in Section 6, where plans for future work are traced and ideas for possible investigations based on the presented dataset are suggested.

2. COST Action TU1208 and the Open Database of Radargrams Initiative

COST Action TU1208 was running from April 2013 to October 2017 [4,5]. It involved more than 300 experts from 150 partner institutes in 28 COST countries (Austria, Belgium, Croatia, Czech Republic, Denmark, Estonia, Finland, France, former Yugoslav Republic of Macedonia, Germany, Greece, Ireland,

Italy, Latvia, Malta, the Netherlands, Norway, Poland, Portugal, Romania, Serbia, Slovakia, Slovenia, Spain, Sweden, Switzerland, Turkey, and the United Kingdom), a COST cooperating state (Israel), 6 COST near neighboring countries (Albania, Armenia, Egypt, Jordan, Russia, Ukraine) and 6 COST international partner countries (Australia, Colombia, Hong Kong, The Philippines, Rwanda, the United States). University researchers, software developers, civil and electronic engineers, archaeologists, geophysics experts, non-destructive testing equipment designers and manufacturers, end-users from private companies and public agencies participated in the Action.

The primary objective of COST Action TU108 was to exchange and increase scientific-technical knowledge and experience of GPR technique, whilst promoting a wider and more effective use of this safe and non-destructive method. The scientific structure of the Action included four Working Groups (WGs), which research activities covered all areas of GPR technology, methodology, and applications. The Open Database of Radargrams is a joint outcome of WG 2 and WG 3.

WG 1 focused on the development of novel GPR instrumentation. Within this WG, novel equipment was designed, implemented and tested [6–11]. Moreover, new tests were proposed for checking the performance and stability of GPR systems.

WG 2 focused on the use of GPR in civil engineering. This WG developed guidelines for GPR inspection of flexible pavements, utility detection in urban areas, and evaluation of concrete structures (the afore mentioned system performance compliance tests are included in these guidelines, which are currently being refined before being published). Recommendations for a safe use of GPR were produced [12]. A catalogue of available test sites and laboratories, where GPR equipment, methodology and procedures can be tested, was prepared and is available online [13]. Additionally, WG 2 carried out a wide series of case studies where GPR was successfully employed in civil-engineering works and laboratory tests; some examples are found in [14–20].

WG 3 studied electromagnetic forward [21–24] and inverse [25,26] methods for the solution of near-field scattering problems by buried structures, imaging techniques [27–30] and data processing algorithms [31–33]. The Members of this WG contributed to the Action by developing and releasing free software, including a new and open-source version of the well-known finite-difference time-domain (FDTD) simulator gprMax [34] and further tools for GPR modeling and data analysis [35–38].

WG 4 investigated the joint use of GPR and complementary non-destructive testing methods in civil engineering [39–43]. This WG also dealt with the use of GPR outside from the civil engineering area. The most interesting output of this WG is a wide series of real-field case studies showing how GPR can be effectively employed in well-established and emerging applications; some examples are found in [44–47]. Special attention was paid to the use of GPR for the management of cultural heritage [48–52].

All WGs were active in organizing meetings, workshops, conferences, special sessions, a series of dissemination events (GPR Road Show), and training activities (fifteen Training Schools were offered in four years). The Action also produced an open-access educational package for teaching GPR in the university (TU1208 Education Pack, available on the Action website). As a follow-up of COST Action TU1208, a non-profit association was founded in September 2017 (TU1208 GPR Association), to keep the scientific network alive and further support cooperation between Universities, research centers, private companies and public agencies active in the GPR field.

The Open Database of Radargrams project consists in gathering and organizing a collection of interesting GPR profiles, which are made accessible to scientists willing to test and validate, against reliable data, their electromagnetic modeling, inversion, imaging and signal-processing methods. Descriptions of the inspected structures and employed equipment are provided along with the profiles. It is often difficult, for scientists involved in theoretical and numerical studies, to get access to high-quality usable experimental data, and this is one of the reasons why we believe that the Open Database of Radargrams is a promising initiative. By using analytical and numerical techniques for the solution of forward scattering problems, scientists can try and reproduce the data included in the database, by modeling the scenario at hand. Inversion and imaging methods can be applied to the various radargrams, to try and reconstruct the geometrical and physical properties of the inspected subsurface or structures. The effectiveness of signal processing algorithms and procedures can be verified and tested. An impressive result, which we hope can be achieved by the GPR scientific community in the forthcoming years, will be that of describing the state of the art of the research in the field by applying different techniques to the radargrams included in the database. Let us point out that this open-science initiative does not aim at defining the "best" methods but more properly at indicating the range of reliability and efficiency of each approach in terms of accuracy, potential, advantages, and drawbacks.

The idea to start the database came out during a TU1208 Short-Term Scientific Mission [53]. It was lately discussed during a WG 3 meeting of the Action [54] and presented at the 2015 edition of the European Geosciences Union General Assembly [55]. It takes inspiration by successful past ventures carried out in different areas. For example, the Ipswich database [56–58] in the field of free-space electromagnetic scattering is a collection of experimental X-band data measured on metallic and dielectric scatterers, in anechoic chamber, at the USAF (United States Air Force) Ipswich Measurement Facility. The subsequent Fresnel database [59–62] aims at extending the scope of applications and includes data measured over a wider frequency range, on a series of homogeneous scatterers, in the anechoic chamber of the Centre Commun de Ressourses Microondes at Marseille, France. The Marmousi database [63] in seismic science is a two-dimensional synthetic dataset generated at the Institute Francais du Pétrole at Rueil-Malmaison Cedex, France. The geometry of the model is inspired by a real profile through the North Quenguela in the Cuanza basin, in Angola; the model was used to create complex data, which require advanced processing techniques to obtain a correct image of the earth. The Marmousi database was used for a workshop on practical aspects of seismic data inversion, during the 52nd EAEG (European Association of Exploration Geophysicists) meeting held in Copenhagen in 1990, and was subsequently studied by hundreds of researchers through the years. The Musumeci database [64], of interest for scientists who analyze seismic events, is a dataset of 151 events leading the 17 July–9 August 2001 lateral eruption at Mt. Etna volcano, in Italy.

As said in the Introduction, the dataset presented in this paper is the core of the Open Database of Radargrams, both because it is the most complete (it includes a high number of profiles, measured with pulsed radar systems produced by three different manufactures and equipped with several antennas working in various frequency ranges) and it was obtained in a controlled environment (complete information concerning the geometry of the test site is available). Another interesting dataset included in the database comes from GPR measurements performed over the historical masonry bridge of Traba, in Spain [65]. Data were recorded by using two commercial pulsed GPR systems, both equipped with 250 MHz and 500 MHz antennas; two-dimensional gprMax models of the bridge and relevant simulation results are available, as well. The database comprises also GPR responses recorded on a masonry column of the Hospital de Sant Pau i la Santa Creu in Barcelona, Spain [40]. The column has a complex internal structure and data were recorded with a commercial pulsed system equipped with an antenna having a central frequency of 1600 MHz. In [53], a series of synthetic profiles was produced by implementing and executing two dimensional gprMax models of concrete cells hosting various metallic reinforcing elements, as well as dielectric and metallic pipes. This dataset was subsequently enriched with several new synthetic profiles, obtained by adopting a more accurate representation of concrete (which takes into account its frequency-dispersion properties); the distance between the targets embedded in concrete and their size were also varied. A further interesting synthetic dataset was obtained by implementing and executing a realistic three-dimensional gprMax model of a fictional but realistic landmine detection environment, with several targets. This dataset was prepared for the GPR Imaging Challenge of the 9th International Workshop on Advanced Ground Penetrating Radar (IWAGPR 2017); researchers attending the conference were invited to submit their processing, imaging and inversion results, to be presented and discussed during the event [66].

3. Description of the IFSTTAR Geophysical Test Site

The IFSTTAR geophysical test site was constructed in 1996, in the IFSTTAR Nantes Centre, in France (see Figure 1). The site is located in the northern sector of the research center area, close to the canteen and to the edge of the property. The test site can be used to test, compare and validate various geophysical equipment and methods, including GPR [67]. Further possible uses include training and demonstration activities.

A schematic plan view of the test site is presented in Figure 2, along with a sketch of its longitudinal section; on the plan view, eleven red arrows represent the acquisition lines of GPR profiles. A pit was excavated in the ground, in a region where the soil has a natural slope, which guarantees that the site is easily drainable. The ground support of the pit, visible in the photo reported in Figure 3a, consists of very altered mica schists, with an almost vertical dip and a cleavage perpendicular to the slope, certainly very absorbent to electromagnetic waves. The photo in Figure 3b shows that the pit was dug with the longitudinal axis parallel to the direction of the land slope, to facilitate the evacuation of water. The excavation background is 5 m wide, with a slope of 4%; the pit is 30 m long and its useful region has a variable depth, ranging from 3.30 m to 4.70 m. The pit sides have a slope of 2/1, which means that the pit surface width is between 19 m and 24.60 m, the surface slope being reduced to 1%. The materials filling the five transversal trenches of the site are described in Section 3.1. To avoid any water ingress, whatever incoming (rain, water flow, capillary ingress), the test site is protected with different elements described in detail in Section 3.2. The targets embedded in the site are described in Section 3.3. Finally, Section 3.4 is concerned with the methods employed to geo-locate the site targets.

For interested users, several photos and videos of the construction phases are available on request, as well as a complete topographical file describing the geometry of the pit, the position of each buried object, and the topography of the finished surface. Moreover, photos of each buried object and details of its composition are available, too, and samples of each type of soil.

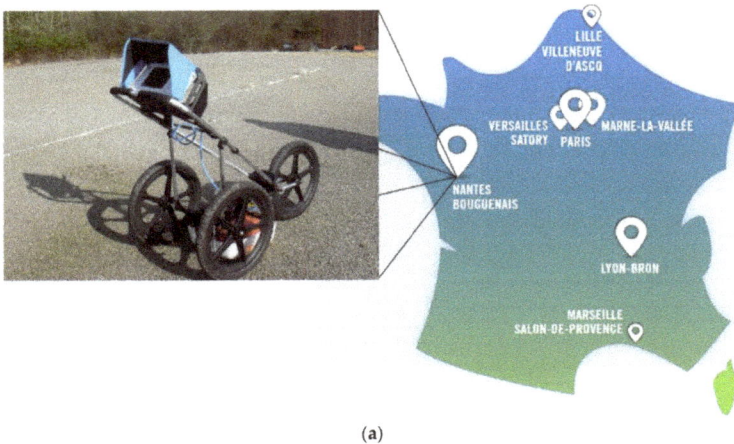

(a)

Figure 1. *Cont.*

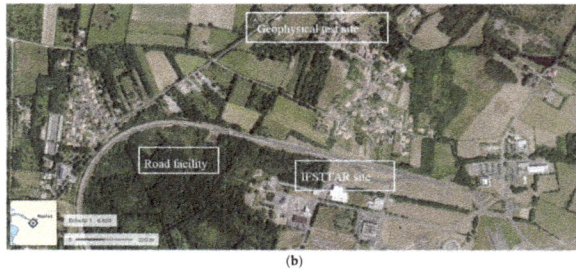

(b)

Figure 1. Geographical position of the IFSTTAR geophysical test site. (**a**) Map of France, where the the test site location is indicated; (**b**) Aerial photo (taken from the website of the French Ministry 'Ministère français de la Transition Écologique et Solidaire', www.geoportail.gouv.fr/carte) showing the test site area.

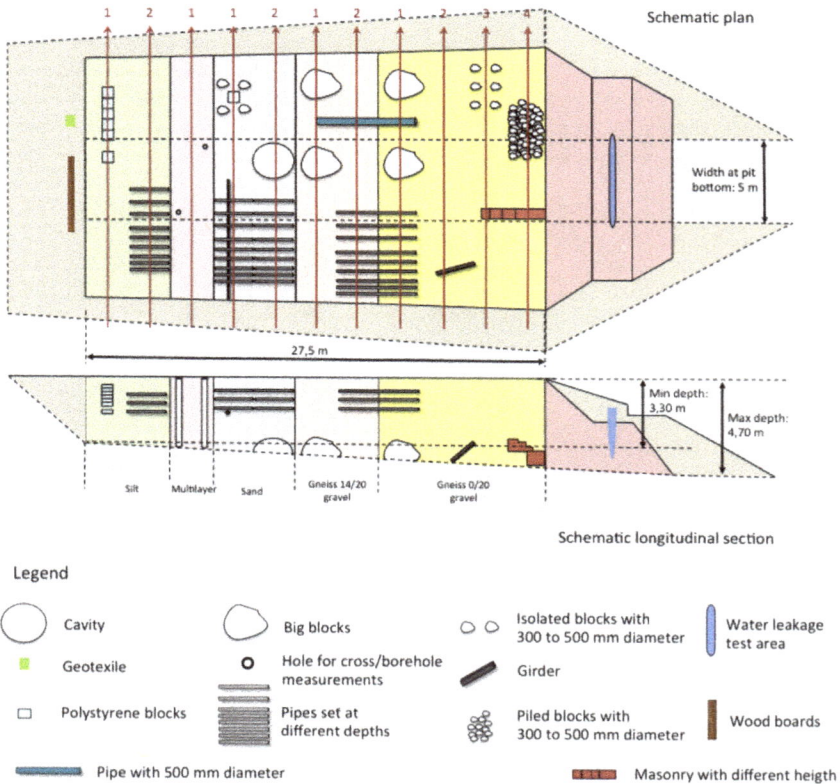

Figure 2. Schematic plan view of the test site and longitudinal section. Red arrows represent the acquisition lines of GPR profiles.

Figure 3. (**a**) Excavation of the mica schist ground support; (**b**) shape of the test site between embankments.

3.1. Filling Materials

The pit is divided into five transversal trenches, corresponding to eleven 2.5-m long sections (transverse slices) for GPR surveys, filled with different materials and separated by vertical interfaces. In particular, there are (Figure 2):

- Two adjacent sections filled with silt (hence, the silt region is 5 m long, altogether);
- A multilayered section, consisting of a stack of layers of different materials which thicknesses from 0.60 m to 1.30 m;
- Two adjacent sections of limestone (hence, the limestone region is 5 m long);
- Two sections filled with Gneiss 14/20 gravel (therefore, the overall length of the low-density gravel region is 5 m; the density is approximately 1.8 t/m^3); and
- Four sections filled with Gneiss 0/20 gravel (all in, the high-density gravel region is therefore 10 m long; and the density is around 2.2 t/m^3).

Gneiss 14/20 and Gneiss 0/20 are crushed gneiss with a grain size from 14 mm to 20 mm, and from 0.1 mm to 20 mm, respectively. Above all sections, a layer of washed 0/2 sand was placed, having a thickness of 10 cm. In each region, it is possible to find places without any buried target, in order to calibrate measurements either over the lateral slope or over the pit bottom. The various filling materials were chosen to be representative of urban environments (e.g., presence of silt, limestone basins), while being relatively "pure", thus allowing a reasonably easy numerical modeling of the test site. The sloping sides were mainly realized with excavated materials.

Materials are characterized by their density, rate of fines, the Methylen Blue Value (MBV, which globally expresses the quantity of clay contained in a material), the maximum diameter of the aggregates and the optimal Proctor value (conceived to determine the optimal moisture content at which a given soil becomes most dense and achieves its maximum dry density). For practical reasons, the MBV value, which follows a French standard, is correlated with two international values [68]: The Cation Exchange Capacity (CEC), which quantifies how many cations can be retained on soil particle surfaces, and the activity Ac, defined by Skempton [69] as the ratio of the plasticity index of clay I_p to its content of clay particles (noted C_2, content of particles smaller than 2 µm). The physical properties of the test site filling materials are resumed in Table 1.

The silt comes from a construction site at "Le Loup du Lac", along the National Route No. RN12 connecting Rennes and St Brieuc, in the Ille-et-Vilaine department of France that is located in the region of Brittany, in the northwest of the country; the classification of the silt in the G.T.R. 92 French standard is A1. The calcareous (limestone) sand comes from an area close to Arthon, about 15 km from Nantes, next to the south bank of the Loire, where a deposit of consolidated (but very friable) sand is present, dated from the Tertiary and more specifically from the Lutetian; the G.T.R. 92 classification is R4; the humidity level in the sand sections of the test site was controlled during the construction of the test site and it was approximately equal to the optimal Proctor value. The Gneiss gravel comes from Chassé, a French village located in the Sarthe department, in the region of the Pays de la Loire. The sand used for a 10-cm layer covering the entire test site comes from a quarry along the "Carrière de Petit Mars" road going from Chasse to Chemin des Masses, close to St Mars du Désert, about 20 km north of Nantes; the aggregate range is 0–5mm, with 15% of fines.

Table 1. Physical characteristics of the filling materials.

Physical Characteristic	Silt	Limestone	Gneiss 14/20	Gneiss 0/20
Dry density (t/m^3)	-	1.74–1.90	1.8–1.9	2.2
Fines < 80 mm (%)	98	13–19	-	< 1
D max aggr. (mm)	≤2	0.8–2	20	20
MBV	0.73–2	0.16–0.56	-	-
CEC (cmol$^+$/kg)	3.11–7.50	1.15–2.53	-	-
Ac = Ip/C$_2$ (mm)	0.35–0.42	0.32–0.34	-	-
Optimum water content (Proctor Opt. %)	10	12–15	-	-

In order to obtain a good compaction, the pit was filled in small layers (slices), having a thickness of 20 cm; beginning at the pit far end, each slice was filled while being separated from the following one by a wood board. A first compaction was performed with hand driven mechanical tools, on each side of the board and all around the buried objects; boards were then pulled out and a 13 T compactor was used over the total layer (Figure 3). The main difficulties were due to: the differences in behavior of the various soils under compaction, so that the limits between the slices tended to bend (due to this, one should consider that the vertical interfaces between different region are not abrupt and planar, and that a melted region about 0.50 m wide is present); the compaction near the lateral slope was more difficult because of the presence of the geosynthetic drain.

3.2. Protection against Water

The protection against water is a key issue in the realization of a test site, where controlled and known conditions are sought, to guarantee reproducibility of the results. A possible solution for making a site completely waterproof is to place a geomembrane below it and a tarp above; else, a building can be constructed, to contain and protect the whole test site. However, a waterproof geomembrane is very expensive, especially in a rocky region as the one where the IFSTTAR geophysical test site is realized. As for the construction of a building, this is obviously even more expensive. Those optimal solutions were excluded when the IFSTTAR geophysical test site was designed, due to financial constraints; several alternatives were considered and studied, and finally, it was decided to realize an underground structure against water inflow and a surface structure against rain and snow, as described in the following.

The underground structure of protection against water was realized before filling the pit and consists of:

- A Gneiss 14/20 gravel layer, 20 cm thick, coated above and below by a geotextile. This drainage layer is present throughout the entire bottom of the pit and it is open at the end of its lowest side; it collects water from the surrounding soil. In the drainage layer, two PVC tubes with a 10 cm diameter are present, which can serve to pass drill-type probes (see Figure 4a).
- A Gneiss 14/20 gravel mask at the beginning of the pit, before the silt region, to protect the site against surrounding subsoil water and to achieve a vertical limit of the silt region.
- A three-layer geotextile for the drainage of the sides of the entire site, composed by: A layer of geotextile (BIDIM 300, which is a standard anti-contaminant layer), a plastic grid made of two crossed wire networks (Tenax grid) to ensure water transmissivity, and another layer of geotextile (BIDIM 300) to ensure permeability. This multilayer drives directly water from the lateral embankments to the bottom drain (see Figure 4b).

The surface protection against rain and snow is achieved by means of:

- A coating layer, realized by spreading a bituminous emulsion with 69% density (having a weight of 1.5 kg/m^2), by lying over it a geotextile with 200 g/m^2 weight, and by finally spreading again a bituminous emulsion with 69% density (having a weight of 2.5 kg/m^2). Some Gneiss 2/4 gravels were embedded in the upper bituminous emulsion of the coating layer, for circulation and protection from the sun (Figure 5).
- Side trenches, located outside the test site platform, at the foot of the embankments.
- An asphalt flange at the downstream end of the site, to evacuate laterally the water that would otherwise trickle longitudinally and erode the slope.

(a)   (b)

Figure 4. (**a**) Bottom and; (**b**) lateral watertight during the placement.

Figure 5. Placement of the 3-layer asphalt wearing.

3.3. Targets

Several targets are present in the test site; they are representative of objects that can be commonly found in trenchless works—such as pipes, cables, stones of various sizes, masonry, and more. Table 2 offers an overview on target types, possible objectives of non-destructive investigations in their presence, and general information about targets actually embedded in the test site. In the following of this sub-section, more detailed information is provided for all the test site targets.

Table 2. Target types, possible objectives of a non-destructive investigation, general properties of the targets embedded in the test-site.

Targets	Aims of a Non-Destructive Investigation	General Information about Targets in the Test Site
Pipes	Detecting the presence of pipes and tracking their location in plan and depth. Identifying pipe crossings, which may cause perturbation. Detecting and tracking particularly large pipes.	Groups of 3 pipes at the same depth; 3 different depths; 4 different enclosing materials. One area with electric cables crossing the pipes. Presence of pipes with Ø500.
Voids	Detecting the presence of voids and evaluating their depth.	Staircase of 1 m^2 polystyrene blocks, simulating voids, in the section filled with the most absorbing material for the electromagnetic waves (silt).

Table 2. *Cont.*

Targets	Aims of a Non-Destructive Investigation	General Information about Targets in the Test Site
Laminate soils	Estimation of thicknesses and material properties.	Multilayered section with holes for crosshole/borehole/tomography measurements.
Rocky blocks	Detecting the presence of rocky blocks, estimating their position and differentiating them according to their size.	Blocks with 3 different sizes (Ø300 mm, Ø500 mm, and 4 m^3), in different host materials. The Ø300 mm and Ø500 mm stones are at 3 different depths; the 4 m^3 blocks are at 2 different depths.
Reproducibility of results—host materials with different density	Checking the GPR performance in the presence of host materials with different density.	Gravel with expected density of 1.8 and 2.2.
Masonry	Detection and localization.	Masonry blocks at different depths.
Metallic objects	Detection, localization, shape estimation.	Girder obliquely buried.

3.3.1. Silt Region

The transversal section of the silt region is schematized in Figure 6. It has to be mentioned that the top of this section was finalized with about 30 cm of limestone (material from the third region), due to lack of silty material; such layer is not represented on the scheme. Moreover, all transversal schemes of the geophysical test site presented in this Section do not show neither the 10-cm surface layer made in limestone nor the asphalt wearing course.

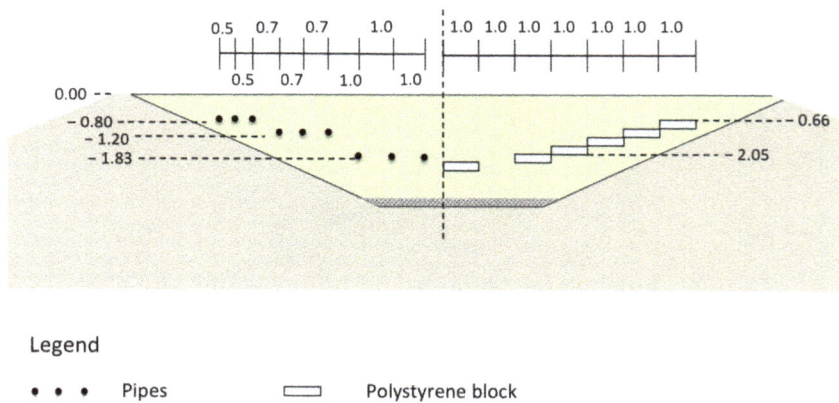

Figure 6. Transversal section of the silt region, showing the embedded targets and their positions. All distances are expressed in meters.

Three pipe layers are present in the silt region, with three pipes per layer: an empty steel pipe, a PVC pipe full of water, and an empty PVC pipe (this is the laying order in all layers, starting from the longitudinal axis of the test site). All pipes are 2.5 m long; the layers are buried at three different depths; the distance between pipes pertaining to the same layer increases with depth. A staircase composed by six expanded polystyrene blocks is also present; the size of the blocks is 1.00 m × 1.00 m × 0.25 m (see the photo in Figure 7).

Figure 7. Placement of one of the polystyrene blocks in the silt section.

3.3.2. Multilayer Region

A scheme of the transversal section of the multilayer region is presented in Figure 8. There are five layers and no targets. Two vertical holes allow carrying out borehole, crosshole and tomography investigations in this region of the test site.

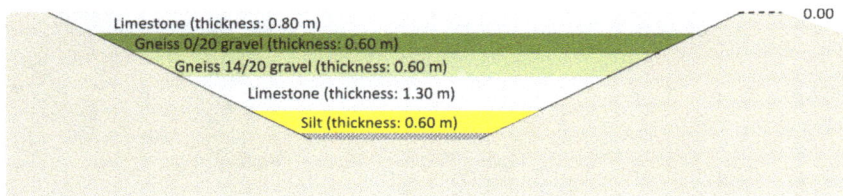

Figure 8. Transversal section of the multilayer region, showing the stratification of materials and the thicknesses of the layers.

3.3.3. Limestone Region

A scheme of the limestone region is presented in Figure 9 (see also Figure 2, for a better understanding of the description).

Here, two series of pipe layers are present, with the same properties as those embedded in the silt region (Figure 10). This permits performing measurements on the same targets, embedded in different host materials. Moreover, in the first series of pipe layers, a large-section electrical cable per pipe layer is present: The cable is orthogonal to the pipe axes, forms a loop, and is taken out at the edge of the site, in order to optionally be able to feed it. The following objects are also present:

- A hemispherical cavity of expanded polystyrene, with a height of 0.50 m and a diameter of 2.50 m;
- Two couples of isolated gneiss blocks, with 300-mm and 500-mm diameters, buried at two different depths; and
- An expanded polystyrene block with size 1.00 m × 1.00 m × 0.25 m.

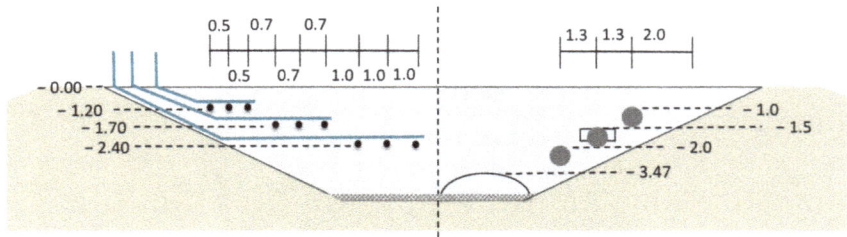

Figure 9. Transversal section of the limestone region, showing the embedded targets and their positions. All distances are expressed in meters.

Figure 10. Placement of one of the pipe series in the limestone section.

3.3.4. Gneiss 14/20 Gravel Region

In this region, the following targets are present (see Figures 11 and 12):

- Three layers of 2.50-m long pipes, same as those described above;
- Two dolmens of about 4 m^3, consisting of 3 or 4 basic blocks and a capstone, located at different depths; and
- A concrete empty pipe, 2-m long, with a diameter of 500 mm (its ends were sealed with polystyrene plates).

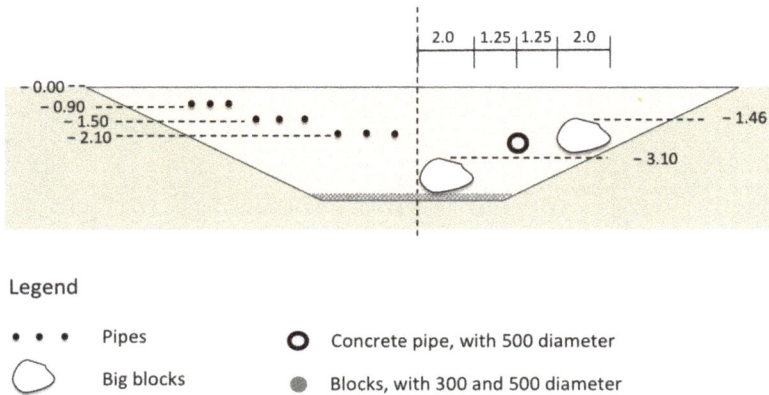

Figure 11. Transversal section of the Gneiss 14/20 gravel region, showing the embedded targets and their positions. All distances are expressed in meters.

Figure 12. Placement of (**a**) one of the pipe series in the gneiss sections, and (**b**) the 500-mm concrete pipe.

3.3.5. Gneiss 0/20 Gravel Region

Schemes of the gneiss region are presented in Figure 13 (see also Figure 2, for a better comprehension, and photos in Figure 14). This region is the widest and occupies four modules of the test site. The following targets are present:

- Two dolmens of about 4 m³, consisting of three or four basic blocks and a capstone, at different depths;
- a concrete empty pipe, 2-m long, with a diameter of 500 mm (its ends were sealed with polystyrene plates)—this pipe is shared with the Gneiss 0/20 gravel region;
- three layers of 2.50-m long pipes, identical to those previously described;
- a masonry wall of parallelepiped hollow blocks;
- another masonry wall of parallelepiped hollow blocks, with three steps and a total height of about 65 cm;
- a steel girder, obliquely buried and crossing the projection of the acquisition line 2 (the girder is not drawn in Figure 13, because it is 'hidden' by the masonry walls);
- a heap of rocky blocks, with diameters from 300 mm to 500 mm, and six isolated blocks, of the same origin as the surrounding Gneiss gravel.

Legend

(a)

(b)

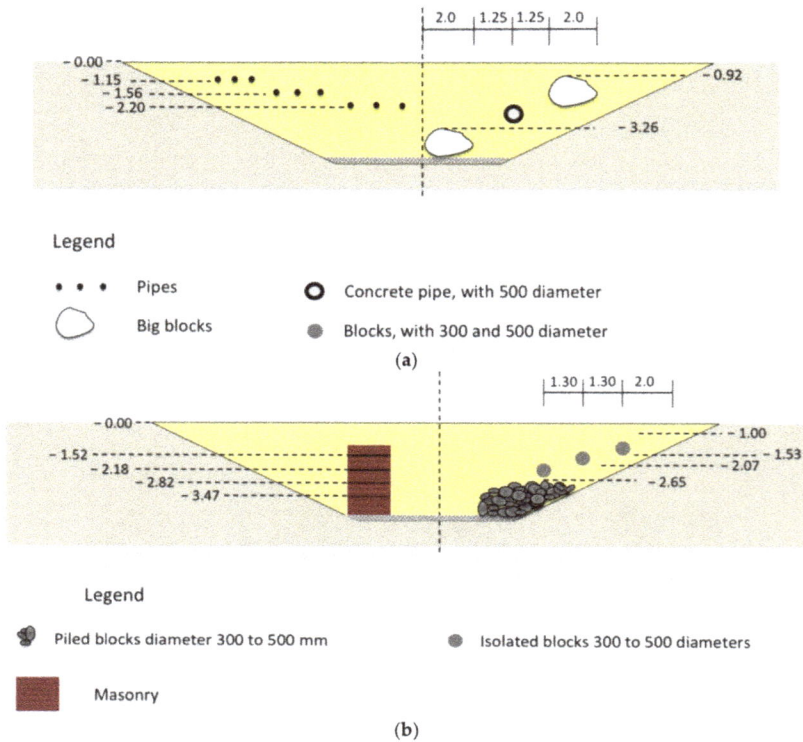

Figure 13. Transversal sections of the Gneiss 0/20 gravel region: (**a**) First part; (**b**) second part. The embedded objects and their positions are shown. All distances are expressed in meters.

Figure 14. Placement of: (**a**) Isolated blocks and a masonry wall; (**b**) a heap of rocky blocks.

3.4. Geolocation of Targets

During the realization of the test site, every embedded object was carefully geolocated, by using a theodolite (visible in Figure 7). Therefore, each target is positioned from several points: 2 points on the upper side of the pipes at their beginnings and ends, 4 points at the upper angles of the polystyrene blocks, and from 3 to 20 points for the rock blocks and dolmens. The surface of the site was georeferenced on the longitudinal axis and on the sides, for every limit of the sections. For each half-section, an interpolation was performed from the 4 surface angles to obtain the height values above the various targets, and then calculate their respective depths (noted in Figures 6, 8, 9, 11 and 13).

4. Equipment and Data Acquisition

Three different GPR systems were used to collect data over the test site, manufactured by GSSI (Geophysical Survey Systems, Inc; Nashua, NH, USA), MALÅ Geoscience (Malå, Sweden), and IDS (Ingegneria dei Sistemi; Pisa, Italy). They were equipped with various antennas, with central frequencies from 200 MHz to 900 MHz. From now on, we will denominate 'GPR 1' the GSSI system, 'GPR 2' the MALÅ system, and 'GPR 3' the IDS system.

GPR 1 was used in bistatic configuration. For measurements performed with 'old' 200-MHz, 400-MHz, 500-MHz, and 900-MHz shielded antennas, the trace spacing was different from one series to another; in some cases, information about the profile length is not available and has to be deducted by analyzing the data. These measurements were performed with antennas manufactured before the Federal Communications Commission (FCC) restrictions on ultra-wideband GPR were emitted. Recently, new measurements were performed by using 200-MHz, 270-MHz, 350 MHz HS (HS stands for hyperstacking), 400-MHz, and 900-MHz last-generation antennas.

For GPR 2, the configuration was bistatic. The distance between the transmitting and receiving antennas was 36 cm with the 250 MHz shielded antenna, 18 cm with the 500 MHz shielded antenna, and 14 cm with the 800 MHz shielded antenna.

With GPR 3, bistatic 200-MHz, 600-MHz, and 900-MHz shielded antennas were tested. Although those antennas were designed before the establishment of the FCC regulation restrictions, they were found suitable for the FCC emission mask. The distance between transmitting and receiving antennas is considered confidential information by the manufacturer, therefore it cannot be provided here.

Data were collected along the eleven red lines shown in Figure 2, as resumed in Tables 3–7; overall, 67 profiles are available. The reason why not all GPR systems and antennas were systematically used over each acquisition line, is that this collection of profiles includes data gathered during different experimental campaigns, carried out by different research teams and for different purposes. Not all the research teams were interested in inspecting all the test site regions.

The file nomenclature is: Freq_Section_ Optional1Position_Optional2.ext, where:

- *Freq*: nominal central frequency;
- *Section*: name identifying the section;
- *Position*: number identifying the acquisition line, which can be 1 (profile recorded 1.25 m from the upstream border of the section—left in Figure 2)/2 (3.75 m)/3 (6.25 m)/4 (8.75 m);
- *Optional1*: "h" letter is added just before the position, when a half-length profile is performed (e.g., h1);
- *Optional2*: "rev" when the profile is done in reverse;
- *ext*: file extension, manufacturer dependent.

For data recorded with GPR 1 and the last-generation 200-MHz, 400-MHz, and 900-MHz antennas, a *b* is added before the extension, to distinguish them from data collected with older antennas working at the same frequencies.

In the silt region (Table 3), 15 profiles were recorded by using all GPR systems, with antennas operating at seven different central frequencies. Note that measurements performed over lines 1 and 2 by using GPR 1 with the 200-MHz antenna are (exceptionally) saved in the same file (200MHz_Silt_h2h1.dzt) instead of being saved in two separate files. In particular, data were collected on the second half of line 2, over the pipes; then, the acquisition continued on the first half of line 1, over the polystyrene blocks; a discontinuity in the data allows understanding where line 2 data end and line 1 data start.

In the multilayer (Table 4), 4 profiles were recorded, by using GPR 1 and GPR 2 and four different central frequencies.

In the limestone region (Table 5), 15 profiles were recorded, by using all GPR systems and nine different central frequencies.

In the Gneiss 14/20 gravel region (Table 6), 15 profiles were recorded, by using all GPR systems and eight different central frequencies.

Finally, in the Gneiss 0/20 gravel region (Table 7), 18 profiles were recorded by using all GPR systems and eight different central frequencies.

Information about the acquisition settings is given in Tables 3–7, for all regions and data; moreover, information on the settings can of course be found in the data file headers. The profiles from 1999–2000 acquired with GPR 1 system were stored after being pre-processed, by using the settings listed in the tables (the used frequency-band filters were Infinite Impulse Response (IIR)); all other data are available in raw format, without application of any filter nor temporal gain. Concerning the 2017 data acquired with GPR 1, two files are available: a .dzt file, which contains raw data; and a .dzx file, which can be opened by using commercial software developed by the radar manufacturer, where pre-processed data are stored, along with pre-filtering information encoded by the operator on the field (in order to see a better image of the radargram on the screen).

Table 3. Information about data collected in the silt region. NA means not available.

Line	GPR System	Year	Antenna Freq (MHz)	File Name	Number of Traces (Profile Length [m])	Scan/m	Sampl./Bits	Range (ns)	Gain (db)	LP (MHz)	HP (MHz)
1	GPR 1	1999	200	200MHz_Silt_h2h1.dzt	1223 (24.44)	50	512/8	110	3/45/70	400	50
		1999	400	400MHz_Silt_1_rev.dzt	1321 (20.02)	66	512/16	70	-1.202×10^{-6}	800	100
		1999	500	500MHz_Silt_1_rev.dzt	1324 (NA)	-	512/16	70	-1.203×10^{-6}	1000	125
		1999	900	900MHz_Silt_1_rev.dzt	1750 (NA)	-	512/16	60	5/30/50/56/56	1800	225
	GPR 2	2002	250	250MHz_Silt_1_rev.rd3	657 (19.90)	33	413/16	80	-	-	-
		2002	500	500MHz_Silt_1_rev.rd3	1313 (25.80)	50	499/16	89	-	-	-
		2002	800	800MHz_Silt_1_rev.rd3	784 (24.20)	32	721/16	89	-	-	-
	GPR 3	2005	200	200MHz_Silt_h1_rev.dt	486 (9.70)	50	1024/16	100	-	-	-
		2005	600	600MHz_Silt_h1_rev.dt	477 (9.52)	50	1024/16	100	-	-	-
		2005	900	900MHz_Silt_h1_rev.dt	475 (9.48)	50	1024/16	100	-	-	-
2	GPR 1	1999	200	200MHz_Silt_h2h1.dzt	1223 (24.44)	50	512/8	110	3/45/70	400	50
		1999	400	400MHz_Silt_2_rev.dzt	1316 (19.94)	66	512/16	70	-1.202×10^{-6}	800	100
		1999	900	900MHz_Silt_2_rev.dzt	1533 (NA)	-	512/16	60	5/30/50/56/56	1800	225
	GPR 3	2005	200	200MHz_Silt_h2_rev.dt	497 (9.92)	50	1024/16	100	-	-	-
		2005	600	600MHz_Silt_h2_rev.dt	474 (9.46)	50	1024/16	100	-	-	-
		2005	900	900MHz_Silt_h2_rev.dt	487 (9.72)	50	1024/16	100	-	-	-

Table 4. Information about data collected in the multilayer. NA means not available.

Line	GPR System	Year	Antenna Freq (MHz)	File Name	Number of Traces (Profile Length [m])	Scan/m	Sampl./Bits	Range (ns)	Gain (db)	LP (MHz)	HP (MHz)
1	GPR 1	1999	400	400MHz_ML.dzt	1198 (NA)	-	512/16	70	-1.202×10^{-6}	800	100
		1999	900	900MHz_ML.dzt	1777 (NA)	-	512/16	60	5/30/50/56/56	1800	225
	GPR.2	2002	250	250MHz_ML.rd3	733 (22.20)	33	415/16	116	-	-	-
		2002	500	500MHz_ML.rd3	1297 (25.50)	50	499/16	89	-	-	-

Table 5. Information about data collected in the limestone region. NA means not available.

Line	GPR System	Year	Antenna Freq (MHz)	File Name	Number of Traces (Profile Length [m])	Scan/m	Sampl./Bits	Range (ns)	Gain (db)	LP (MHz)	HP (MHz)	Stack.	Raw Data (Y/N)
1	GPR 1	1999	400	400MHz_Limestone_1_rev.dzt	1360 (NA)	-	512/16	85	-4.547×10^{-7}	800	100	5	N
	GPR 1	1999	900	900MHz_Limestone_1_rev.dzt	1757 (NA)	-	512/16	60	5/30/50/56/56	1800	225	5	N
		1999	200	200MHz_Limestone_2.dzt	1235 (24.68)	50	512/16	110	3/45/70	400	50	3	N
		2017	270	270MHz_Limestone_h2.dzt	2571 (12.85)	200	1024/32	100	-7.639×10^{-5}	700	75	1	Y
		2017	350	350MHz_Limestone_h2.dzt	2763 (13.81)	200	512/32	100	-9.221×10^{-5}	1095	95	1	Y
	GPR 1	1999	400	400MHz_Limestone_2_rev.dzt	1418 (NA)	-	512/16	85	-4.547×10^{-7}	800	100	5	N
		2017	400	400MHz_Limestone_h2_b.dzt	2581 (12.90)	200	1024/32	100	-8.146×10^{-5}	800	100	1	Y
		1999	900	900MHz_Limestone2_rev.dzt	1861 (NA)	-	512/16	60	5/30/50/56/56	1800	225	5	N
2		2017	900	900MHz_Limestone_h2_b.dzt	2274 (11.36)	200	1024/32	100	10/37/44/51	-	-	-	Y
	GPR 2	2002	250	250MHz_Limestone2_rev.rd3	715 (21.66)	33	415/16	116	-	-	-	1	Y
		2002	500	500MHz_Limestone2_rev.rd3	1172 (23.03)	50	499/16	89	-	-	-	1	Y
		2002	800	800MHz_Limestone2_rev.rd3	825 (25.47)	33	721/16	89	-	-	-	1	Y
	GPR 3	2005	200	200MHz_Limestone_2_rev.dt	1053 (21.04)	50	1024/16	100	-	-	-	1	Y
		2005	600	600MHz_Limestone_2_rev.dt	1038 (20.74)	50	1024/16	100	-	-	-	1	Y
		2005	900	900MHz_Limestone_2_rev.dt	1042 (20.82)	50	1024/16	100	-	-	-	1	Y

Table 6. Information about data collected in the Gneiss 14/20 gravel region. NA means not available.

Line	GPR System	Year	Antenna Freq (MHz)	File Name	Number of Traces (Profile Length [m])	Scan/m	Sampl./Bits	Range (ns)	Gain (db)	LP (MHz)	HP (MHz)	Stack.	Raw Data (Y/N)
1	GPR 1	1999	400	400MHz_Gneiss14-20_1_rev.dzt	1401 (NA)	-	512/16	70	-1.202×10^{-6}	800	100	5	N
	GPR 1	1999	900	900MHz_Gneiss14-20_1_rev.dzt	1473 (NA)	-	512/16	60	5/30/50/56/56	1800	225	5	N
	GPR 2	2002	500	500MHz_Gneiss14-20_1_rev.rd3	1272 (25.00)	50	499/16	89	-	-	-	1	Y
		1999	200	200MHz_Gneiss14-20_2_rev.dzt	1291 (25.80)	50	512/8	110	3/45/70	400	50	3	N
		2017	200	200MHz_Gneiss14-20_2_b.dzt	4922 (24.60)	200	1024/32	100	7/60/68	1000	100	1	Y
		2017	270	270MHz_Gneiss14-20_2.dzt	4775 (23.87)	200	1024/32	100	-0.0001077	540	50	1	Y
		2017	350	350MHz_Gneiss14-20_2.dzt	5168 (25.83)	200	1024/32	100	-5.495×10^{-5}	940	100	1	Y
	GPR 1	2000	400	400MHz_Gneiss14-20_2_rev.dzt	639 (21.27)	30	512/16	90	5/62	665	110	2	N
2		2017	400	400MHz_Gneiss14-20_2.dzt	2462 (24.61)	100	1024/16	100	0/42/61	800	80	1	Y
		2017	500	500MHz_Gneiss14-20_2.dzt	4838 (24.18)	200	1024/16	100	2/40/56/60	1000	100	1	Y
		1999	900	900MHz_Gneiss14-20_2_rev.dzt	1518 (NA)	-	512/16	60	5/30/50/56/56	1800	225	5	N
		2017	900	900MHz_Gneiss14-20_2_b.dzt	4952 (24.75)	200	1024/16	90	4/11/55/60/62	1800	200	1	Y
	GPR 2	2002	250	250MHz_Gneiss14-20_2_rev.rd3	737 (22.32)	33	415/16	116	-	-	-	1	Y
		2002	500	500MHz_Gneiss14-20_2_rev.rd3	1287 (25.29)	50	499/16	89	-	-	-	1	Y
		2002	800	800MHz_Gneiss14-20_2_rev.rd3	820 (25.31)	33	721/16	89	-	-	-	1	Y

Table 7. Information about data collected in the Gneiss 0/20 gravel region. NA means not available.

Line	GPR System	Year	Antenna Freq (MHz)	File Name	Number of Traces (Profile Length [m])	Scan/m	Sampl./Bits	Range (ns)	Gain (db)	LP (MHz)	HP (MHz)	Stack.	Raw Data (Y/N)
1		2017	270	270MHz_Gneiss0-20_h1.dzt	2551 (12.75)	200	1024/32	100	-7.639×10^{-5}	700	75	1	Y
		2017	350	350MHz_Gneiss0-20_h1.dzt	2286 (11.42)	200	512/32	100	-9.221×10^{-5}	1095	95	1	Y
	GPR 1	2017	400	400MHz_Gneiss0-20_h1.dzt	2490 (12.44)	200	1024/32	100	-8.146×10^{-5}	800	100	1	Y
		1999	400	400MHz_Gneiss0-20_1_rev.dzt	1310 (NA)	-	512/16	70	-1.202×10^{-6}	800	100	5	N
		2017	900	900MHz_Gneiss0-20_h1.dzt	2409 (12.04)	200	1024/32	100	4/10/37/44/51	-	-	1	Y
		1999	900	900MHz_Gneiss0-20_1_rev.dzt	1837 (NA)	-	512/16	60	5/30/50/56/56	1800	225	5	N
		2002	250	250MHz_Gneiss0-20_1_rev.rd3	754 (22.84)	33	415/16	116	-	-	-	1	Y
	GPR 2	2002	500	500MHz_Gneiss0-20_1_rev.rd3	1234 (24.25)	50	499/16	89	-	-	-	1	Y
		2002	800	800MHz_Gneiss0-20_1_rev.rd3	801 (24.73)	33	721/16	89	-	-	-	1	Y
		2005	200	200MHz_Gneiss0-20_1_rev.dt	1080 (21.58)	50	1024/16	100	-	-	-	1	Y
	GPR 3	2005	600	600MHz_Gneiss0-20_1_rev.dt	1079 (21.56)	50	1024/16	110	-	-	-	1	Y
		2005	900	900MHz_Gneiss0-20_1_rev.dt	1081 (21.60)	50	1024/16	110	-	-	-	1	Y
2	GPR 1	1999	400	400MHz_Gneiss0-20_2_rev.dzt	1860 (NA)	-	512/16	70	-1.202×10^{-6}	800	100	5	N
		1999	900	900MHz_Gneiss0-20_2_rev.dzt	1458 (NA)	-	512/16	60	5/30/50/56/56	1800	225	5	N
3	GPR 1	1999	400	400MHz_Gneiss0-20_3_rev.dzt	1447 (NA)	-	512/16	70	-1.202×10^{-6}	800	100	5	N
		1999	900	900MHz_Gneiss0-20_3_rev.dzt	1841 (NA)	-	512/16	60	5/30/50/56/56	1800	225	5	N
4	GPR 1	1999	400	400MHz_Gneiss0-20_4_rev.dzt	1672 (NA)	-	512/16	70	-1.202×10^{-6}	800	100	5	N
		1999	900	900MHz_Gneiss0-20_4_rev.dzt	1933 (NA)	-	512/16	60	5/30/50/56/56	1800	225	5	N

5. Results—Maps of All Radargrams Included in the Dataset

In this Section, we present grey-scale maps of the radargrams included in the dataset. All files generated by GPR 1 and GPR 2 were opened and displayed by using the free Matlab-based data processing software MatGPR [70]. However, this software does not open the files generated by GPR 3, which were therefore opened with commercial processing software, exported in ascii format, and then plotted with Matlab. The ascii data files are enclosed to the paper as supplementary material, so that interested researchers can easily open them without needing any commercial tool.

We have chosen to map the radargram data, directly, without applying any editing nor processing procedure. Actually, the purpose of this paper is to present the dataset and provide appropriate information about all profiles included in it, so that the scientific community can use them; our main aim is not to carry out an advanced analysis, comparison and interpretation of the data. Nonetheless, throughout this section we have accompanied the maps with several comments and considerations resulting from direct observation or from a preliminary study of the data.

Depending on the physical characteristics of the host materials, as well as on the nature, distribution and size of the buried objects, the GPR signals recorded along the eleven acquisition lines obviously vary, because they are affected by media lithology and granulometry and by the presence of scatterers. Moreover, for each acquisition line, different radargrams are of course obtained when the spectrum of the signal emitted by the radar changes. In addition, undoubtedly, different radar systems/antennas working at the same central frequency and used over the same acquisition line, provide consistent but not identical results.

Some profiles recorded with GPR 1 were pre-processed by the system before being saved, as pointed out in Section 4; by comparing pre-processed data with raw data acquired at the same frequency and over the same acquisition line, the effects of frequency-band filters application and gain recovery can be clearly appreciated. In particular, due to the geometrical spreading and attenuation of the electromagnetic waves emitted by the radar and transmitted in the subsurface, later trace arrivals in raw data show noticeably lower amplitudes than earlier arrivals; the application of a time-variant gain function allows recovering relative amplitude information (within some inevitable limits of accuracy) and allows seeing on the same grey-scale map the signatures generated by shallower and deeper objects. Moreover, the application of suitable frequency filters improves the signal-to-noise ratio.

5.1. Silt Region

Figures 15–20 show all radargrams gathered over the silt region, as outlined in Table 3 and in the figure captions. The frequency effect of the source is clearly notable in terms of penetration and resolution. As mentioned before, a wearing layer in limestone is present and the relevant interface is clearly visible in the maps, due to the high electromagnetic contrast, for all the central frequencies. Reflections from the lateral boundaries of the test site can be seen, too, and some diffraction signatures from the pipes and polystyrene blocks.

Regarding Figure 15, the second half of the data was collected on half line 1, over the targets, whereas the first half of the data was collected on half line 2, over the targets, too (the data of those two measurements were stored in the same file). By looking at the positions of the target signatures, it can be noticed that the data in Figures 16–20 were collected on a direction opposite to that indicated by the arrows in Figure 2 (as pointed out in Table 3). In Figure 16, the effects of the pipes buried beyond line 2 are visible in the second half of the maps, although the spatial extension of the targets does not reach line 1; such effect is obviously less notable at higher frequencies—a similar phenomenon can be noticed in several figures presented in the following Sections.

In Figure 20, the initial traces of the radargram shown in (a) have to be neglected when analyzing the data. The same is true for various radargrams of the dataset—in general, a proper number of initial traces need (almost) always to be neglected, as they correspond to data recorded above the lateral embankments. Nevertheless, such initial traces are useful for accurately locating the beginning of the test region of interest. Similar comments apply to the final traces of the radargrams, recorded after the end of the test region of interest.

The radargram in Figure 15, recorded with GPR 1, can be compared with the radargrams in Figures 18a and 20a, recorded with GPR 3 at the same central frequency (200 MHz). The data shown in Figure 15 were pre-processed (see Table 3), whereas the data presented in Figures 18 and 20 are raw; for this reason, the signatures generated by targets and interfaces are much more visible in Figure 15. The application of suitable processing steps to the data of Figures 18 and 20 would reveal hidden information and, e.g., enable a comparison between the performances of the two radar systems and antennas. However, this would go beyond our scope and open the perspective of this paper on a number of topics that are not covered here. Analogously, the radargram in Figure 16c can be compared with the radargram in Figure 18c (and similar comments apply): they are recorded over acquisition line 1 by using GPR 1 and GPR 3, respectively, with antennas having a central frequency of 900-MHz. And, the radargram in Figure 19b can be compared with the radargram in Figure 20c: they are recorded over acquisition line 2 by using GPR 1 and GPR 3, respectively, with 900-MHz antennas.

After appropriately processing the available data, it is possible to verify that the silt region appears to be penetrated for no more than a half of its depth; the different responses of the various targets can be examined and geometrical or physical parameters of interest can be extracted (but this is not the objective of the present paper). Moreover, we implemented preliminary FDTD models of this region and by matching synthetic results with the measurements (which allows taking fully into account the wave propagation and scattering phenomena), we found that the relative permittivity of the silt is around $\varepsilon_{r,silt} = 13$, the attenuation of the electromagnetic signal in this region of the test site is between 15 and 45 dB/m, and the penetration is about 1.5 m at the lowest frequency and 1 m at the highest. Such estimations must be taken with care, as the surface limestone layer presents some slight thickness variations, and also because of the compaction performed on the silt above the polystyrene blocks during the test site implementation (which might have compressed the blocks a little). More advanced investigations will be carried out in the future, to obtain more accurate estimations of the various values.

Still concerning the penetration depth, this parameter does not depend only on the material properties and frequency, but also on several other factors (therefore, it is not possible to calculate the conductivity of silt from the attenuation values given above). In particular, the maximum depth that a GPR can reach depends on the radar system dynamics (i.e., the minimum detectable signal-to-noise ratio), the environmental noise, the matching of the antenna to the material (i.e., the ability of the antenna to transmit electromagnetic energy in the material), the shape of the antenna radiation pattern (an antenna may focus more or less than another the electromagnetic energy in the vertical direction), the radar cross section of the targets, their spatial distribution, and the electromagnetic contrast between targets and host material (due to the scattering by the targets, the penetration depth in the test site is smaller than it would be in a target-free scenario).

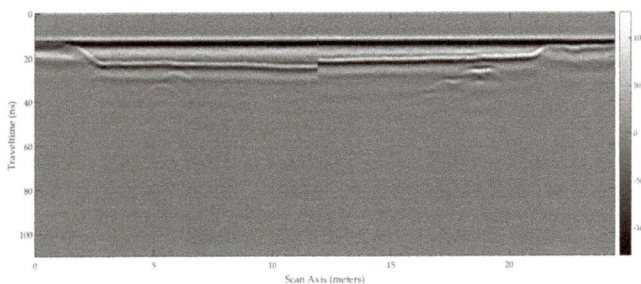

Figure 15. Profile recorded in the silt region by using GPR 1, on acquisition lines 1 and 2, with an antenna operating at the central frequency of 200 MHz.

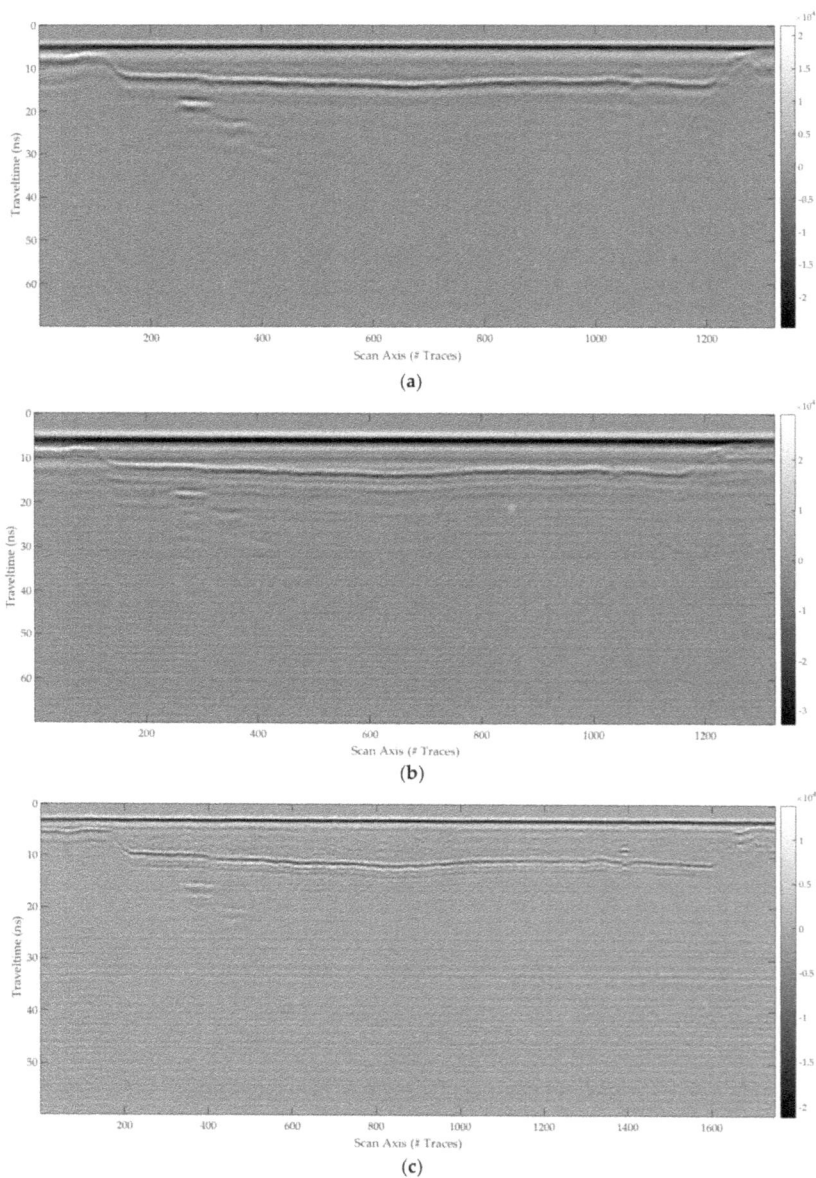

Figure 16. Profiles recorded in the silt region by using GPR 1, on acquisition line 1, with antennas operating at the following central frequencies: (**a**) 400 MHz; (**b**) 500 MHz; (**c**) 900 MHz. Data recorded at 200 MHz are shown in Figure 15.

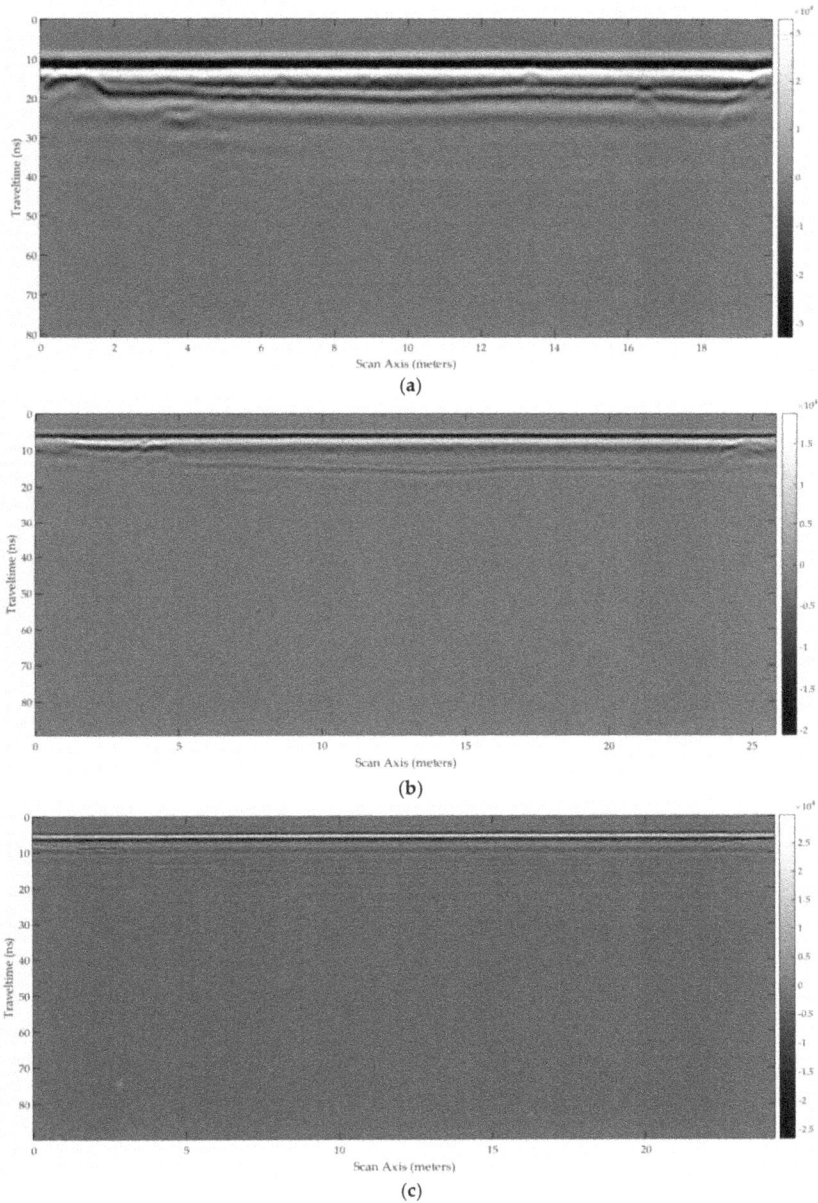

Figure 17. Profiles recorded in the silt region by using GPR 2, on acquisition line 1, with antennas operating at: (**a**) 250 MHz; (**b**) 500 MHz; (**c**) 800 MHz.

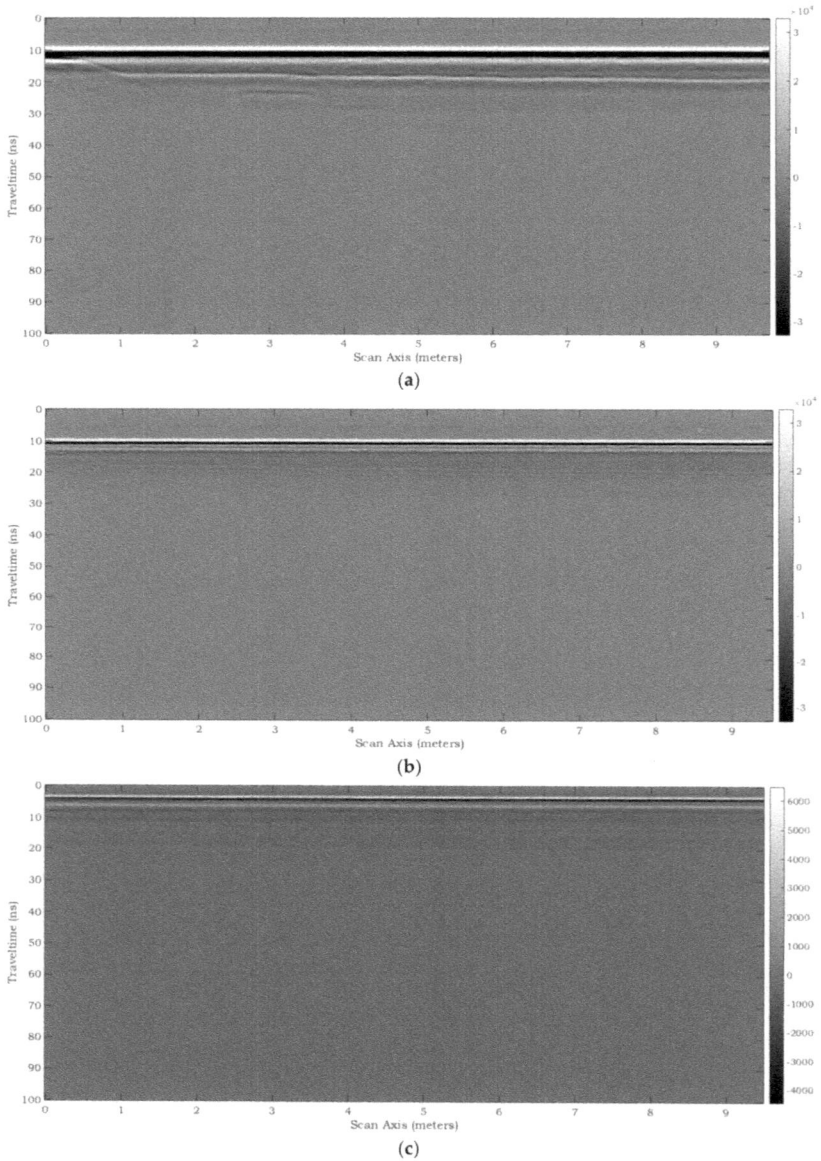

Figure 18. Profiles recorded in the silt region by using GPR 3, on acquisition line 1, with antennas operating at: (**a**) 200 MHz (data collected over a 9.7 m long portion of the acquisition line); (**b**) 600 MHz (9.52 m long portion of the acquisition line); (**c**) 900 MHz (9.48 m long portion of the acquisition line).

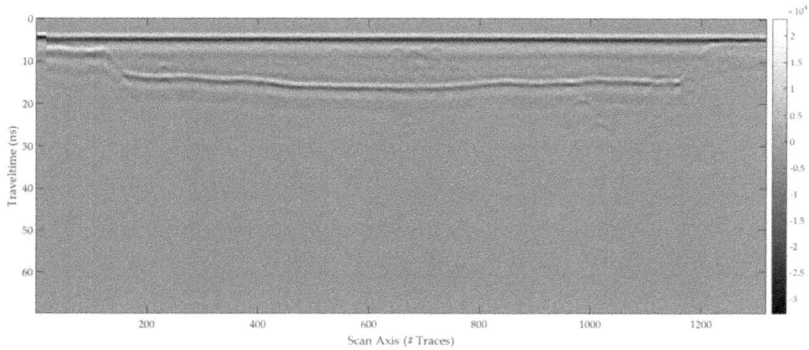

(a)

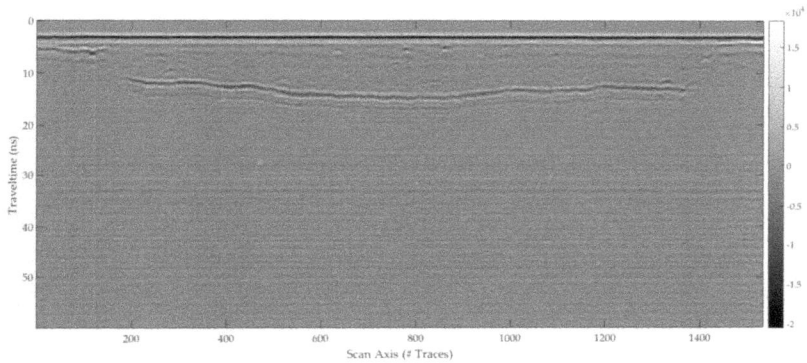

(b)

Figure 19. Profiles recorded in the silt region by using GPR 1, on acquisition line 2, with antennas operating at: (**a**) 400 MHz; (**b**) 900 MHz. Data recorded at 200 MHz are in Figure 15.

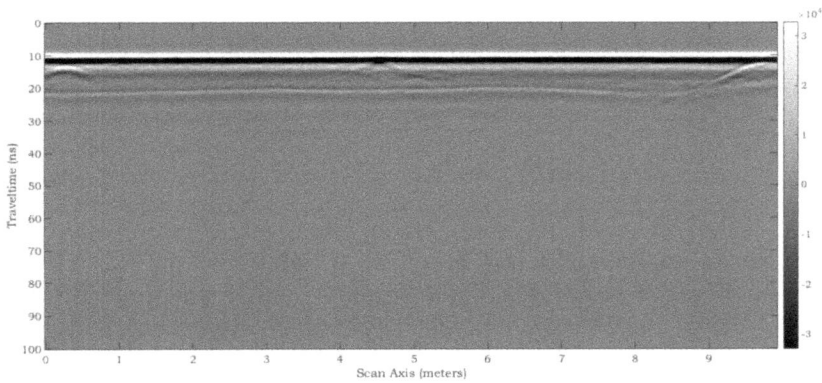

(a)

Figure 20. *Cont.*

(b)

(c)

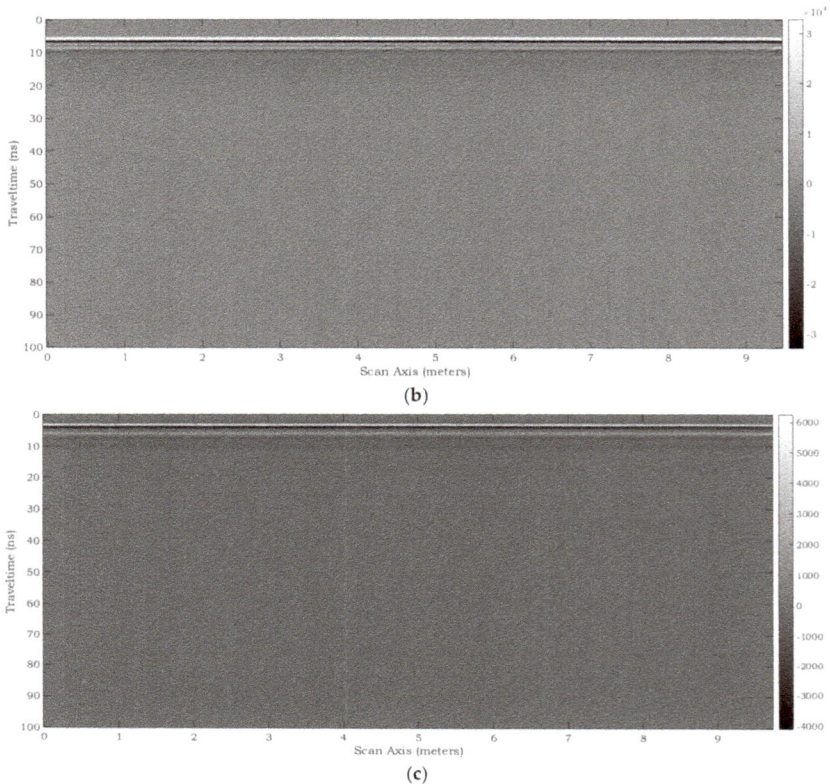

Figure 20. Profiles recorded in the silt region by using GPR 3, on a portion of acquisition line 2, with antennas operating at: (**a**) 200 MHz (9.92 m long); (**b**) 600 MHz (9.48 m long); (**c**) 900 MHz (9.72 m long).

5.2. Multilayer

Figures 21 and 22 show the radargrams collected over the multilayer, as outlined in Table 4 and in the figure captions. The multi-layer section is the only one that was not geo-localized during the implementation of the test site. Then, the thicknesses of every layer remain theoretical.

While studying the GPR profiles (e.g., Figure 21a), several findings can be made. For example, concerning the practical implementation of the layers, despite great precautions to realize homogeneous layers were taken, an elementary ~25-cm implementation of soil sub-layers followed by compaction remains visible due to slight electromagnetic contrasts. A further comment is related to the double travel times in the two layers of gneiss: as the Gneiss 14/20 material presents a high level of porosity (due to the lack of fine and small elements), the corresponding GPR velocity is faster than for Gneiss 0/20 and this should induce different thicknesses for these two Gneiss layers, which are declared similar (60 cm). One possible explanation could be from an unwanted filling of Gneiss 14/20 by fine and small elements of Gneiss 0/20, due to gravity and opened voids in the Gneiss 14/20 skeleton; but this explanation remains only partial. Another observation is associated with GPR scattering phenomena occurring in the upper limestone layer, due to the presence of big coarse aggregates that remained there, although they should have been removed.

The multilayer section remains a challenge for GPR specialists, to estimate the real thicknesses and permittivity values of the various layers.

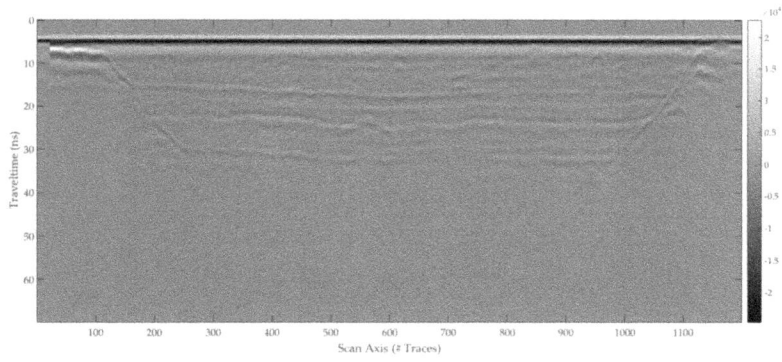

(a)

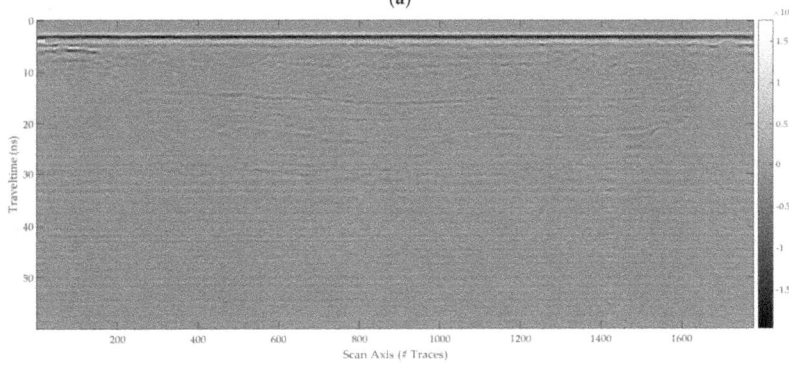

(b)

Figure 21. Profiles recorded in the multilayer by using GPR 1, with antennas operating at: (**a**) 400 MHz; (**b**) 900 MHz.

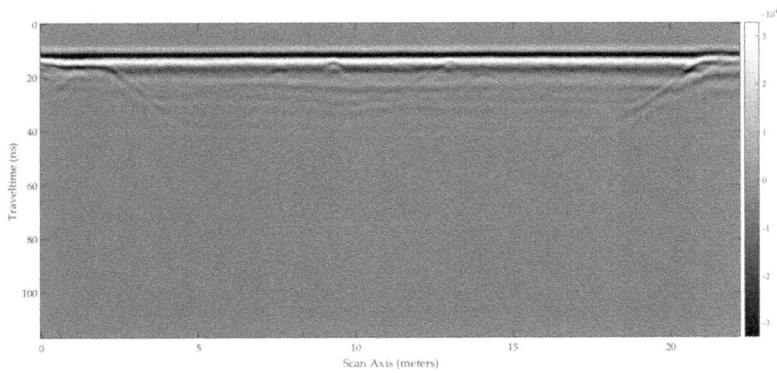

(a)

Figure 22. *Cont.*

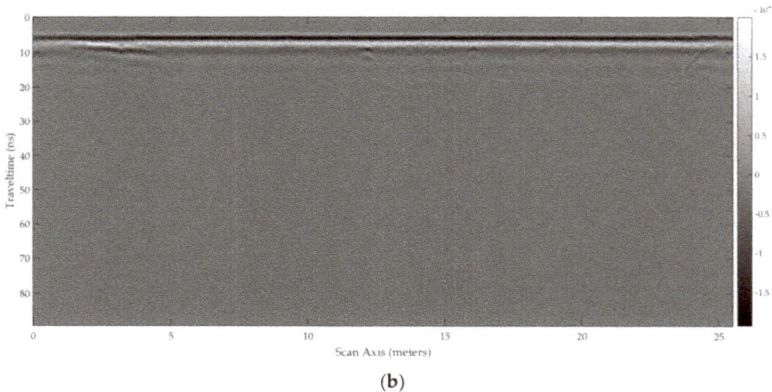

(b)

Figure 22. Profiles recorded in the multilayer by using GPR 2, with antennas operating at: (**a**) 250 MHz; (**b**) 500 MHz.

5.3. Limestone Region

Figures 23–28 show the radargrams gathered over the limestone region, as outlined in Table 5 and in the figure captions. Some radargrams were collected on a direction opposite to that indicated by the arrows in Figure 2, as pointed out in Table 5.

For GPR 1 and acquisition line 2, two different radargrams are available for the central frequency of 400 MHz, collected with an older and a newer antenna (they are shown in Figures 24b and 26a, respectively); the same is true for the central frequency of 900 MHz (Figures 24c and 26b). The radargram in Figure 24a, recorded with GPR 1 over acquisition line 2 at 200 MHz, can be compared with the radargram in Figure 28a, recorded with GPR 3 over the same acquisition line and at the same frequency. Analogously, Figures 24c and 26b can be compared with Figure 28c.

Several observations can be done by studying these maps. This region is particularly interesting for GPR surveys, as the medium is realistic, homogeneous and electromagnetically resistive. As a consequence, one can detect the sides of the test site and its bottom, constituted by a drainage layer and two lateral drainage pipes. The response to the heterogeneities also merits some comments. Every target is clearly detected, such as the polystyrene block (well visible, in Figure 23a–b), which borders generate hyperbolas, or the polystyrene cavity (see Figure 24a–b), and most of the nine pipes. As is well known, the interaction of electromagnetic fields with scatterers strongly depends on their size compared to wavelength. When objects are sufficiently large, such as the polystyrene block and cavity, the interpretation of their signatures in the radargrams is easier and the reflected signal contains complete information about their shape. On the contrary, when objects are small compared to wavelength, they are still detectable but more difficult to characterize, for example it is not trivial to estimate the size of the pipes—especially at lower frequencies. However, it is interesting to observe how pipes of different nature can be distinguished thanks to the differences between their signatures: a steel pipe generates a single hyperbola; a PVC pipe filled with water generates a first hyperbola coming from the upper side of the pipe and successive hyperbolas due to the reflections coming from its bottom (echoes which amplitude is stronger than for the first one); an empty PVC pipe also generates reflections from both its upper side and bottom, but the pattern is different, because the velocity of electromagnetic fields in air is higher than in water, and so the successive echoes are more discernible for the PVC pipe filled with water than for the empty one.

By analyzing the data and implementing preliminary FDTD models of this region, it is reasonable to assume that the relative permittivity of limestone is around $\varepsilon_{r,sand} = 6$, the attenuation of the electromagnetic signal is between 6 and 20 dB/m, the penetration is at least 4.5 m at the lowest frequency (the bottom of the pit is visible) and 2 m at the highest. Further studies are necessary for a more accurate estimation of these values.

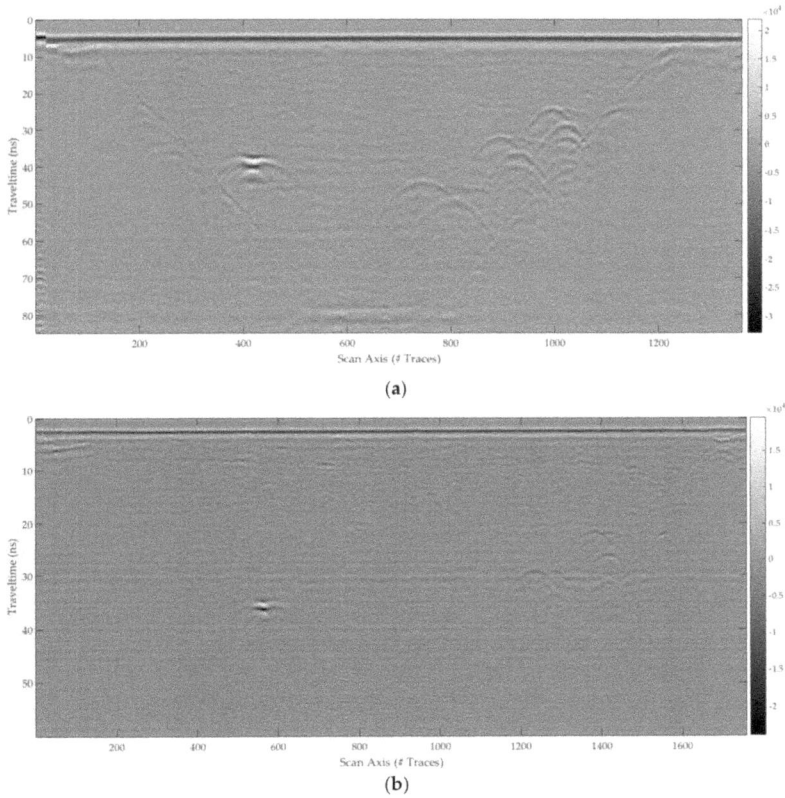

(a)

(b)

Figure 23. Profiles recorded in the limestone region by using GPR 1, on acquisition line 1, with antennas operating at: (**a**) 400 MHz; (**b**) 900 MHz.

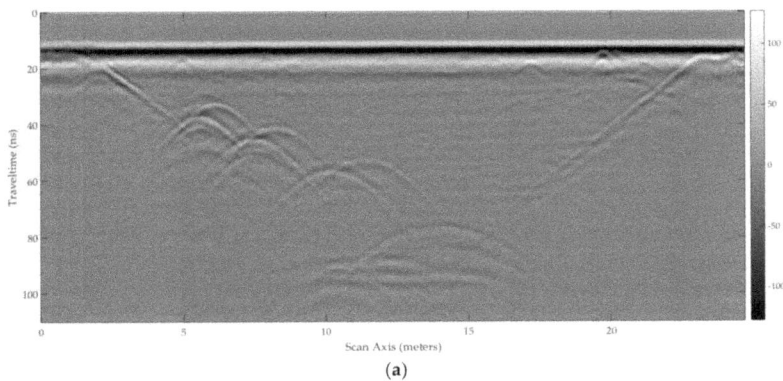

(a)

Figure 24. *Cont.*

(b)

(c)

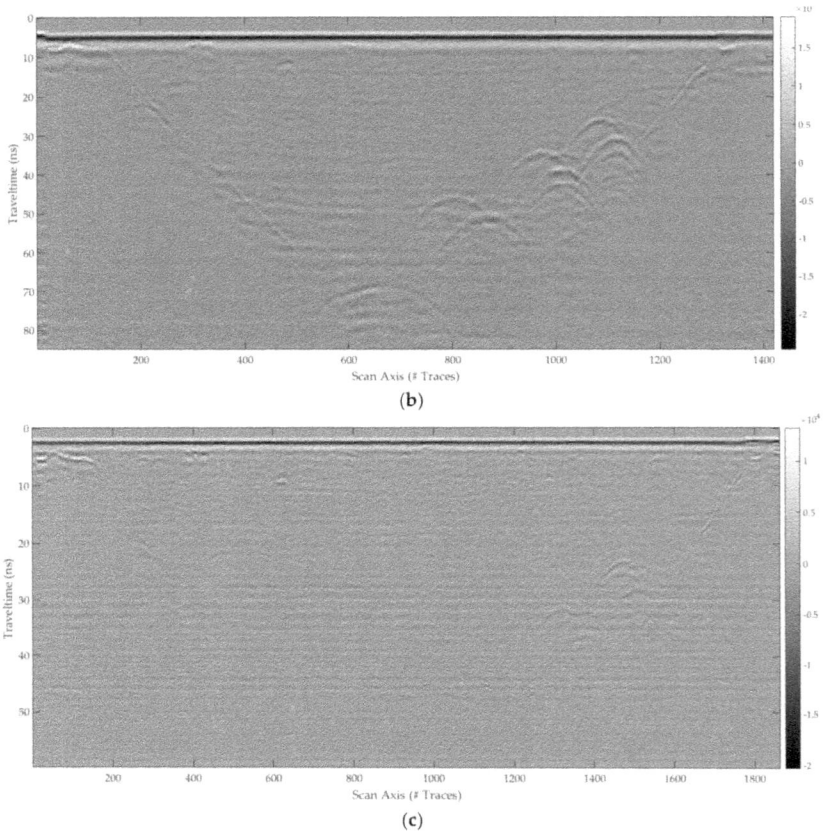

Figure 24. Profiles recorded in the limestone region by using GPR 1, on acquisition line 2, with antennas operating at: (**a**) 200 MHz; (**b**) 400 MHz; (**c**) 900 MHz.

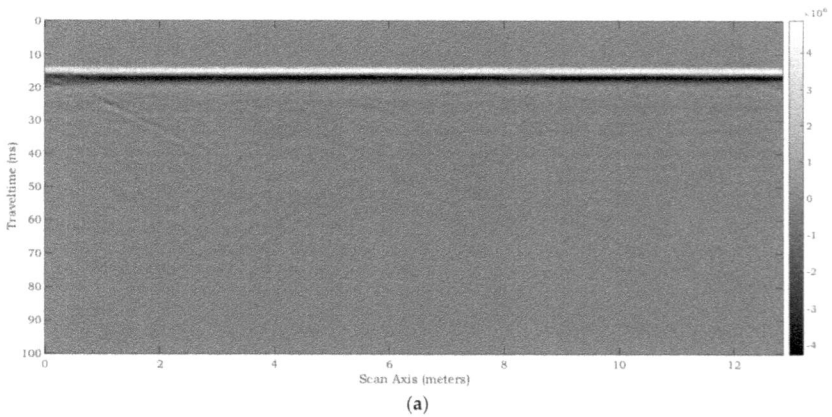

(a)

Figure 25. *Cont.*

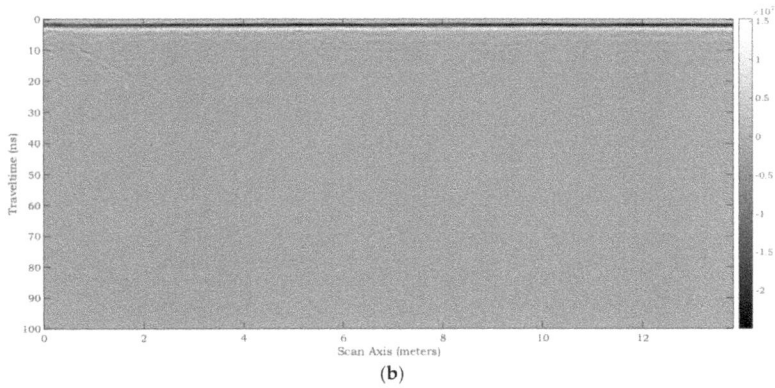

Figure 25. Profiles recorded in the limestone region by using GPR 1, on acquisition line 2, with antennas operating at: (**a**) 270 MHz; (**b**) 350 MHz.

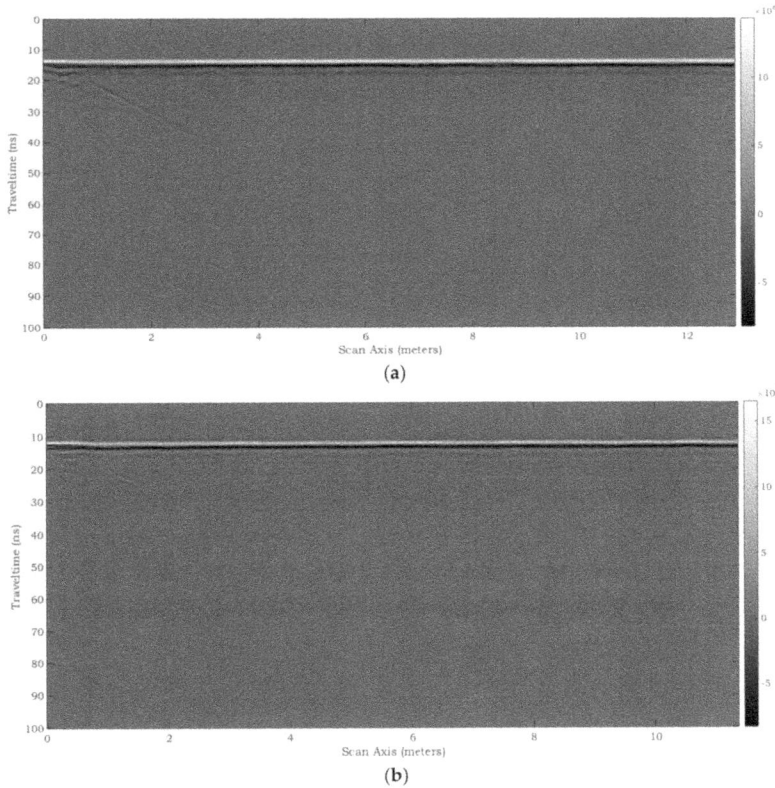

Figure 26. Profiles recorded in the limestone region by using GPR 1, on acquisition line 2, with 'new' antennas operating at: (**a**) 400 MHz; (**b**) 900 MHz.

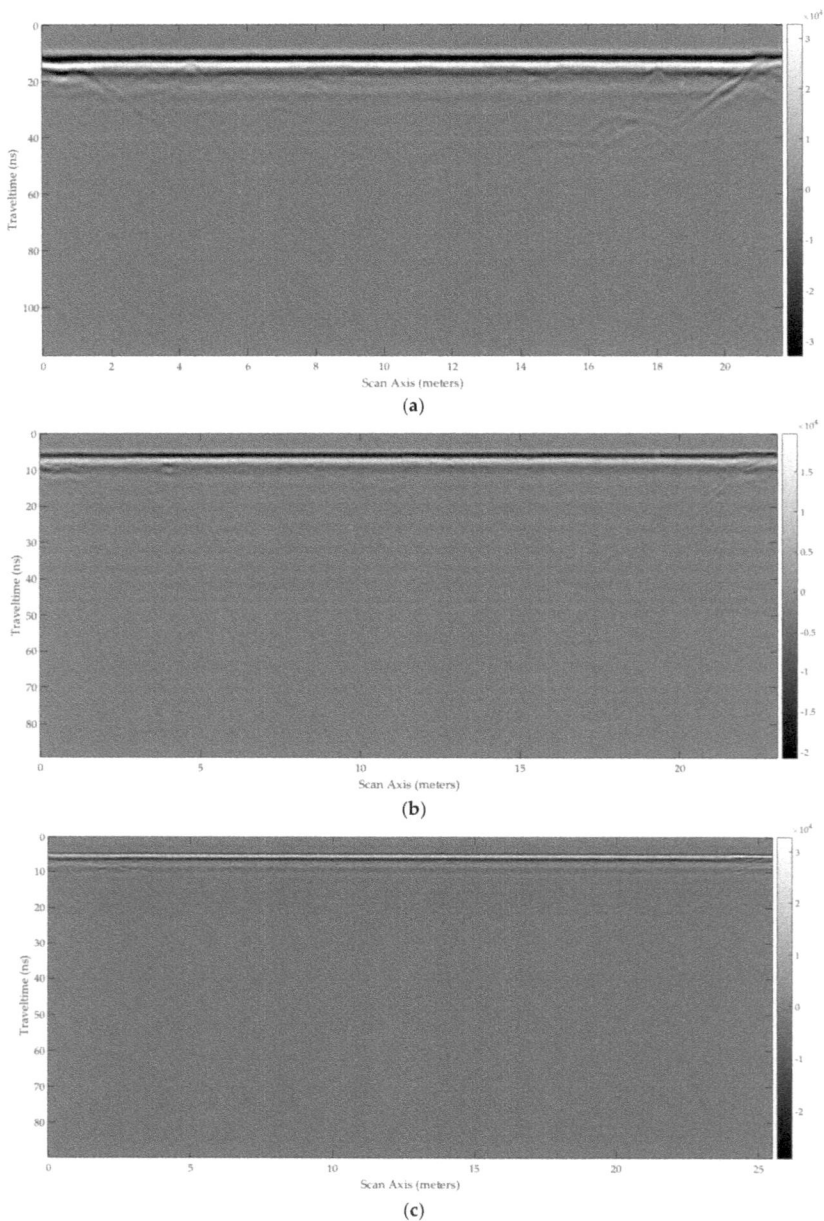

Figure 27. Profiles recorded in the limestone region by using GPR 2, on acquisition line 2, with antennas operating at: (**a**) 250 MHz; (**b**) 500 MHz; (**c**) 800 MHz.

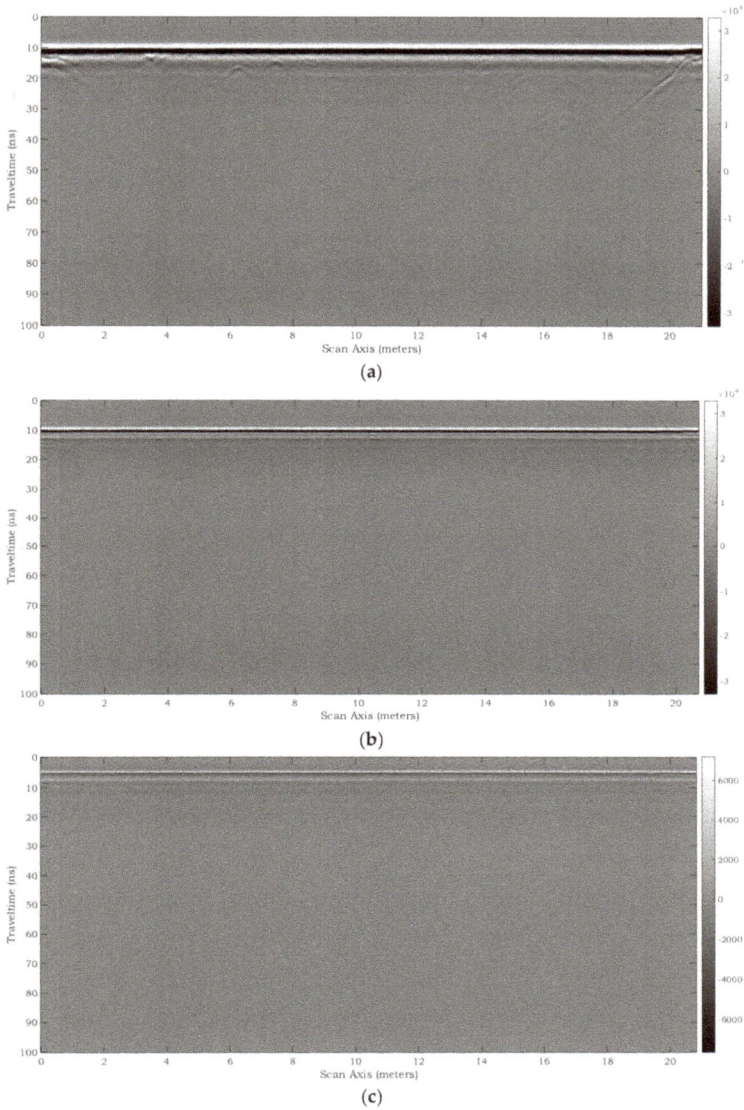

(a)

(b)

(c)

Figure 28. Profiles recorded in the limestone region by using GPR 3, on acquisition line 2, with antennas operating at: (**a**) 200 MHz; (**b**) 600 MHz; (**c**) 900 MHz.

5.4. Gneiss 14/20 Region

Figures 29–34 show the radargrams gathered over the Gneiss 14/20 gravel region, as outlined in Table 6 and in the figure captions. Some radargrams were collected on a direction opposite to that indicated by the arrows in Figure 2, as pointed out in Table 6.

For this section of the test site, the richness of the dataset allows performing various comparisons between radargrams recorded with different equipment working at the same central frequency.

This region was especially designed to propose an artificial lossless medium for GPR-operator beginners in training. Figure 31a is probably the radargram of the dataset that is easiest to read, as it is almost noiseless; this case, along with the results obtained by using the same system at higher frequencies, can be an excellent starting point for testing modeling, imaging, inversion and processing algorithms and it is also particularly good for teaching purposes. The bottom of the pit and all the buried objects are well visible in Figure 31a. The big concrete pipe generates two hyperbolas, which indicate its top and base and allow estimating its size. In the center of the radargram, it is possible to notice several reflections, which are due to the almost-parallel and almost-horizontal interfaces between compaction sub-layers. It is also interesting to observe the two hyperbolas generated by the wedges at the bottom of the site (i.e., where the sloping sides of the pit reach its base and two drains are present). For the signatures generated by the three layers of steel and PVC pipes, comments similar to those written in Section 5.3 apply.

By analysing the data and implementing preliminary FDTD models of this region, it is reasonable to assume that the relative permittivity is around $\varepsilon_{r,g1420} = 3$, the attenuation of the electromagnetic signal is between 1.5 and 4.5 dB/m (this is the region with lowest attenuation), and the penetration is at least 4.5 m at all considered frequencies (meaning that the bottom of the pit can always be seen). Further studies will provide a more accurate estimation of these values.

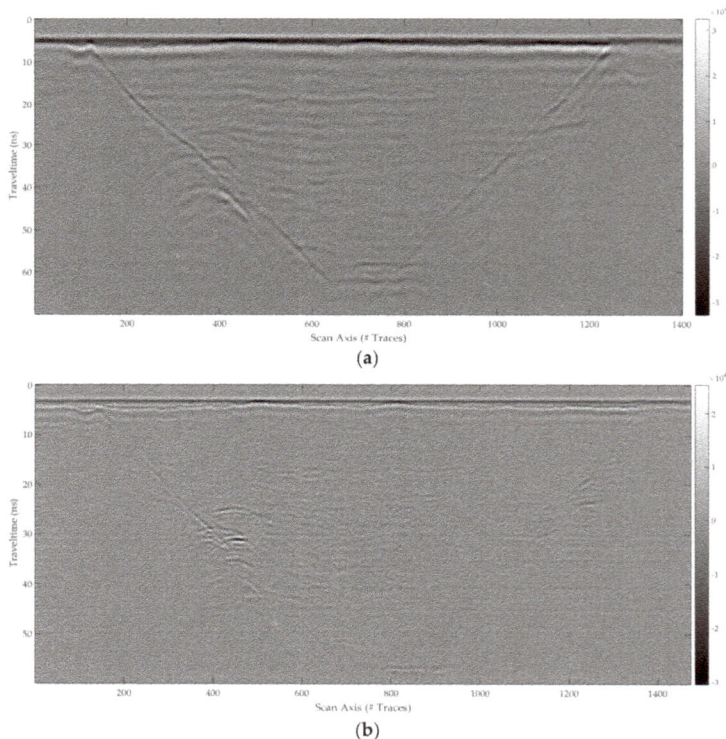

(a)

(b)

Figure 29. Profiles recorded on acquisition line 1 of the Gneiss 14/20 gravel region by using GPR 1 and antennas operating at: (**a**) 400 MHz; (**b**) 900 MHz.

Figure 30. Profile recorded on acquisition line 1 of the Gneiss 14/20 gravel region by using GPR 2 and a 500-MHz antenna.

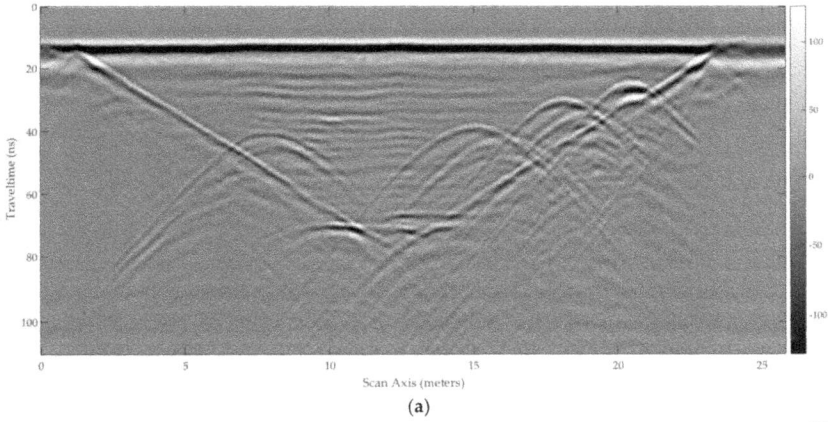

(a)

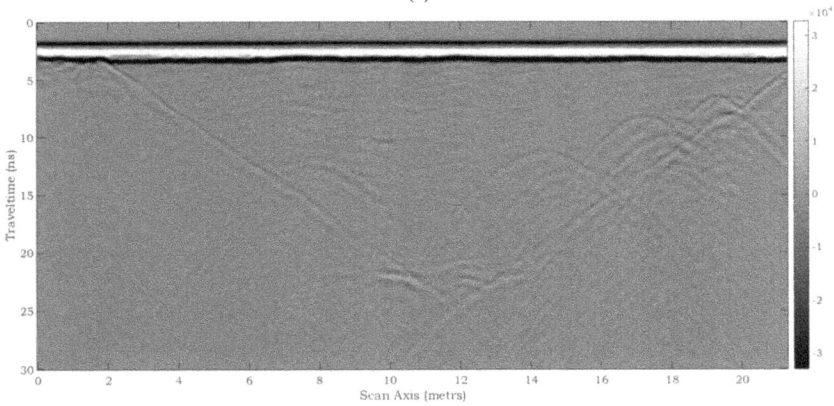

(b)

Figure 31. *Cont.*

(c)

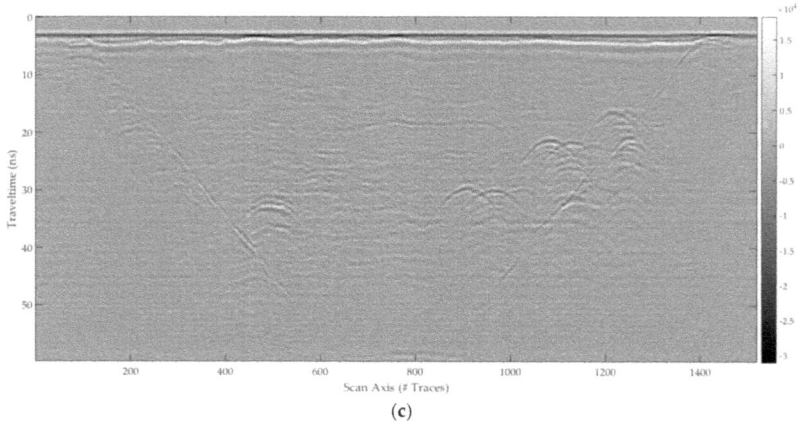

Figure 31. Profiles recorded in the Gneiss 14/20 gravel region by using GPR 1, on acquisition line 2, with antennas operating at: (**a**) 200 MHz; (**b**) 400 MHz; (**c**) 900 MHz.

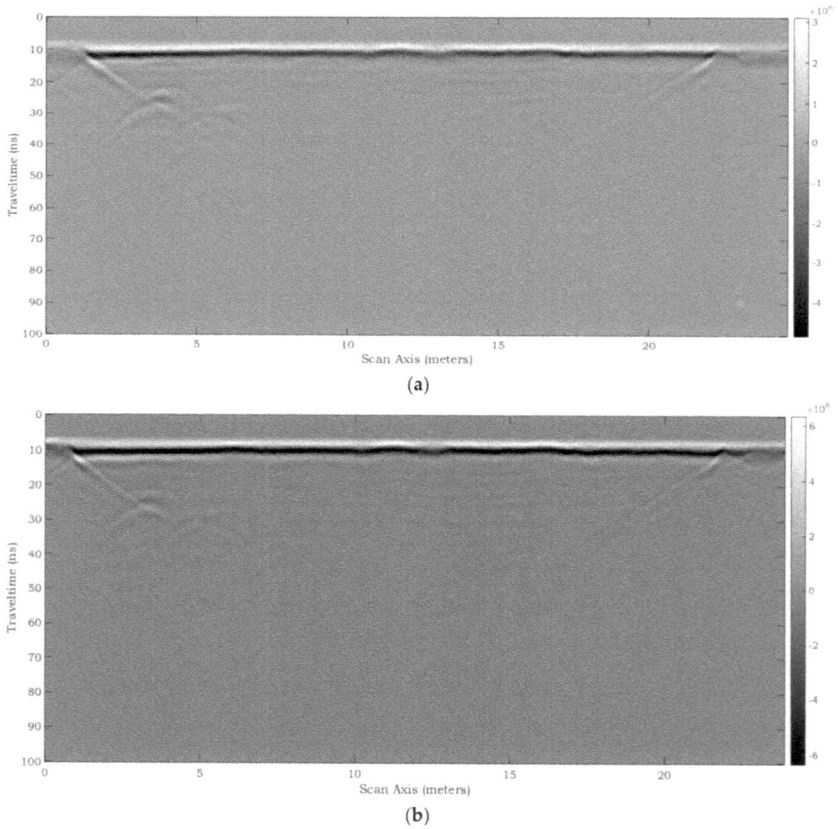

(a)

(b)

Figure 32. *Cont.*

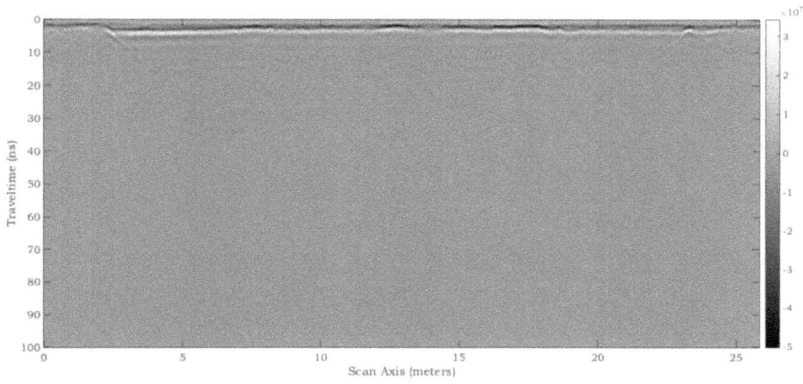

(c)

Figure 32. Profiles recorded in the Gneiss 14/20 gravel region by using GPR 1, on acquisition line 2, with 'new' antennas operating at: (**a**) 200 MHz; (**b**) 270 MHz; (**c**) 350 MHz.

(a)

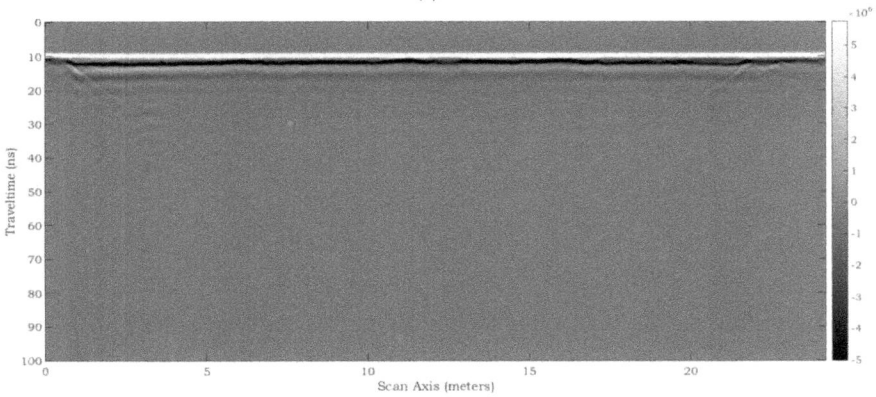

(b)

Figure 33. *Cont.*

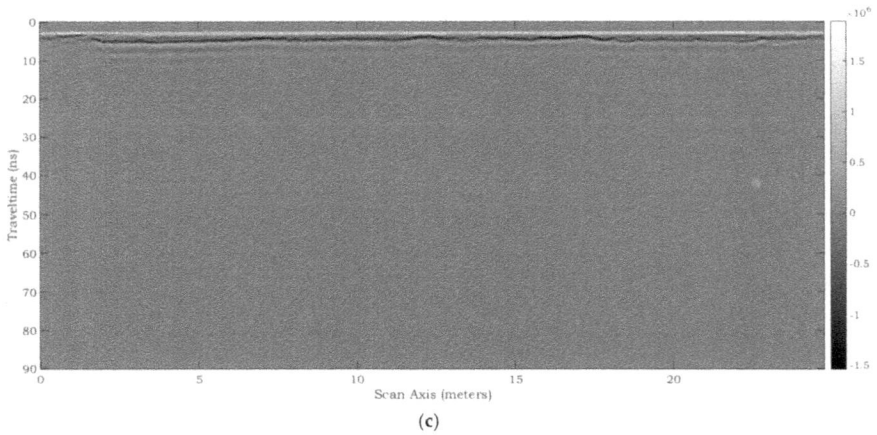

(c)

Figure 33. Profiles recorded in the Gneiss 14/20 gravel region by using GPR 1, on acquisition line 2, with last-generation antennas operating at: (**a**) 400 MHz; (**b**) 500 MHz; (**c**) 900 MHz.

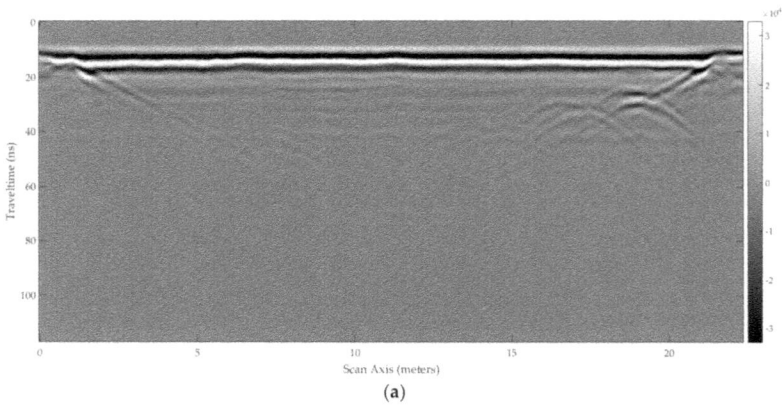

(a)

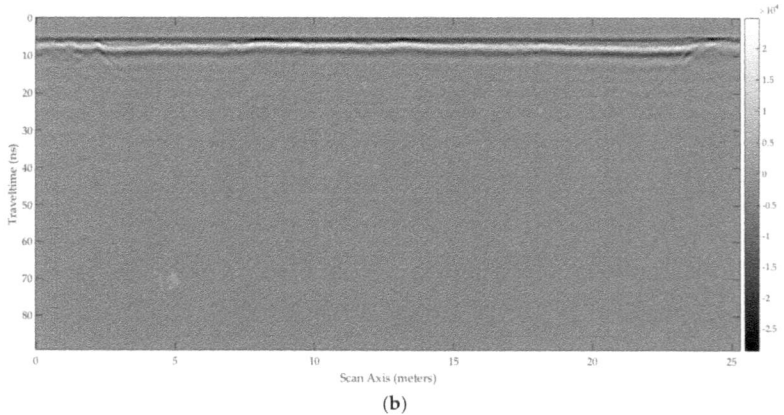

(b)

Figure 34. *Cont.*

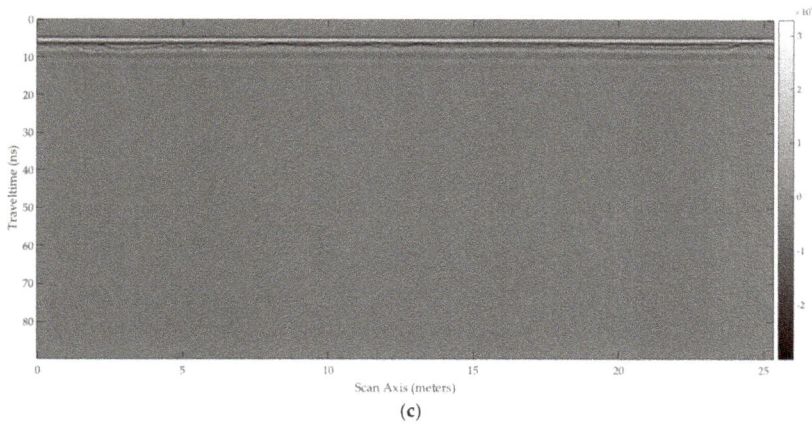

(c)

Figure 34. Profiles recorded in the Gneiss 14/20 gravel region by using GPR 2, on acquisition line 2, with antennas operating at: (**a**) 250 MHz; (**b**) 500 MHz; (**c**) 800 MHz.

5.5. Gneiss 0/20 Region

Finally, Figures 35–42 show the radargrams gathered over the Gneiss 0/20 gravel region, as outlined in Table 7 and in the figure captions.

For acquisition line 1, various comparisons are possible between radargrams collected by using different equipment working at the same central frequency.

The high-density gravel turned out to be so strongly absorbing that the girder is very difficult to detect. A comparison between the radargrams recorded in this region and those recorded in the gneiss 14/20 region, proves how strong is the influence of the fine fractions present on the 0/20 scale: they cause the permittivity of the medium and signal attenuation to be much higher and so prevent any deep penetration of the electromagnetic waves into the ground. Reflections generated by the almost-parallel and almost-horizontal interfaces between compaction sub-layers are visible in the radargrams, same as already commented for the radargrams presented in the previous sub-section: let us now additionally mention that, on one hand, such reflections prove the very high sensitivity of the GPR technique, which is capable to detect interfaces between regions of same material only because they were separately compacted, as well as how large is the amount of information that can be extracted from GPR data; on the other hand, those reflections suggest how difficult the interpretation of GPR data can be, especially in variegated urban scenarios where the measurements can be quite noisy due to the inhomogeneity of the subsurface and to the presence of stones or other scattering elements.

By analyzing the data and implementing preliminary FDTD models of this region, it is reasonable to assume that the relative permittivity is around $\varepsilon_{r,g0020} = 5.5$, the attenuation of the electromagnetic signal is between 9 and 27 dB/m, the penetration is about 2.5 m at the lowest frequency and 1.5 m at the highest. Further studies are necessary for a more accurate estimation of these values.

(a)

(b)

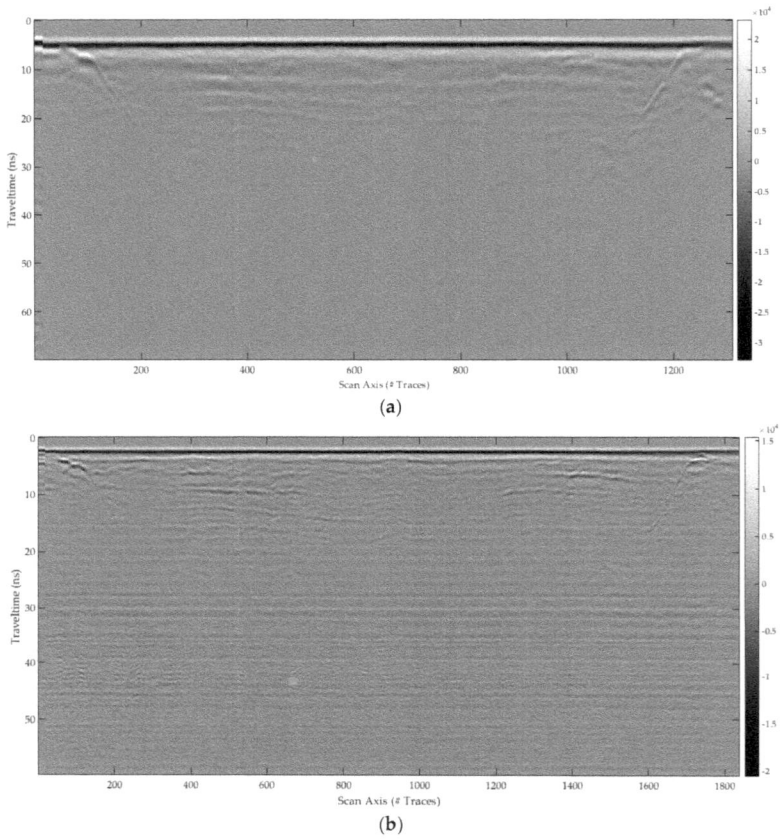

Figure 35. Profiles recorded in the Gneiss 0/20 gravel region by using GPR 1, on acquisition line 1, with antennas operating at: (**a**) 400 MHz; (**b**) 900 MHz.

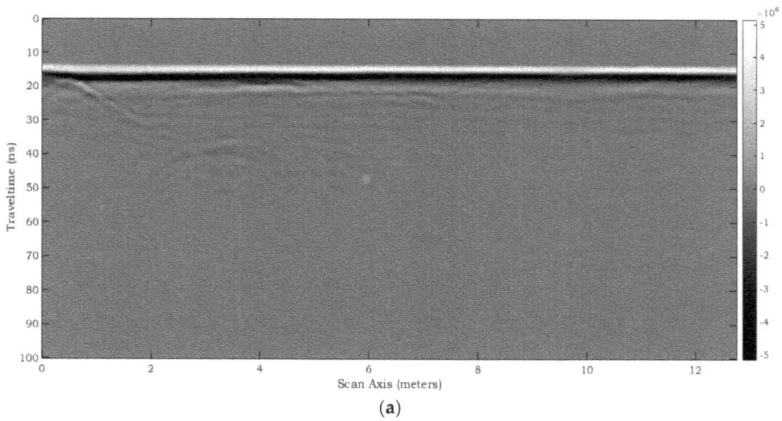

(a)

Figure 36. *Cont.*

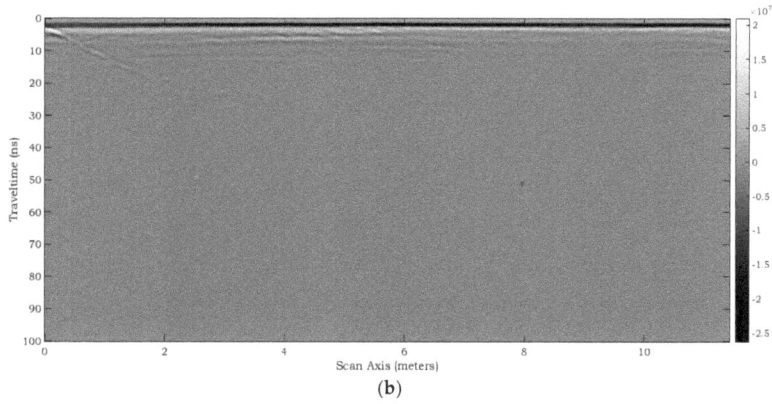

(b)

Figure 36. Profiles recorded in the Gneiss 0/20 gravel region by using GPR 1, on acquisition line 1, with last-generation antennas operating at: (**a**) 270 MHz; (**b**) 350 MHz.

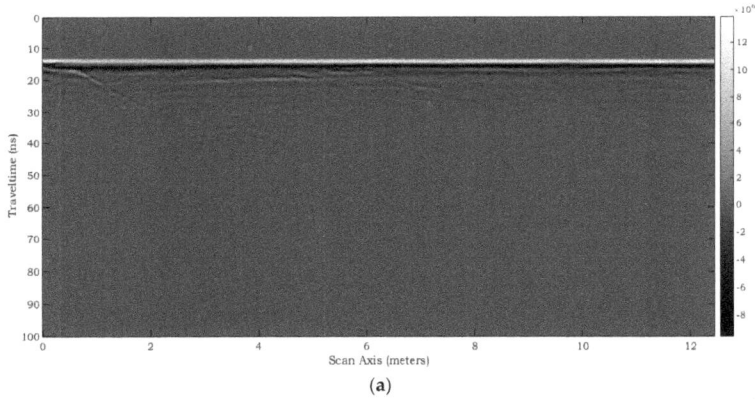

(a)

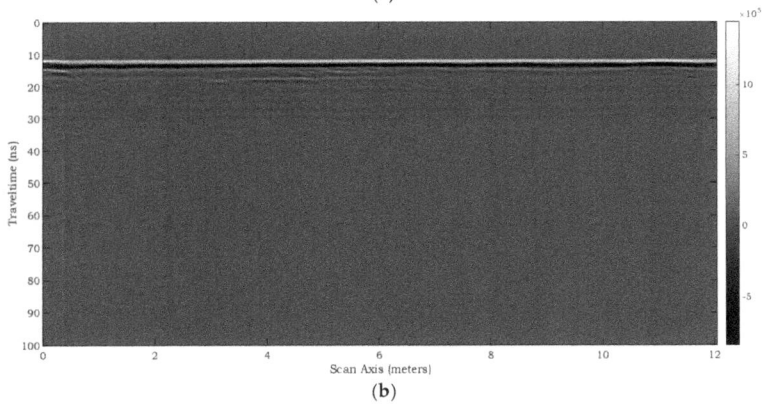

(b)

Figure 37. Profiles recorded in the Gneiss 0/20 gravel region by using GPR 1, on acquisition line 1, with last-generation antennas operating at: (**a**) 400 MHz; (**b**) 900 MHz.

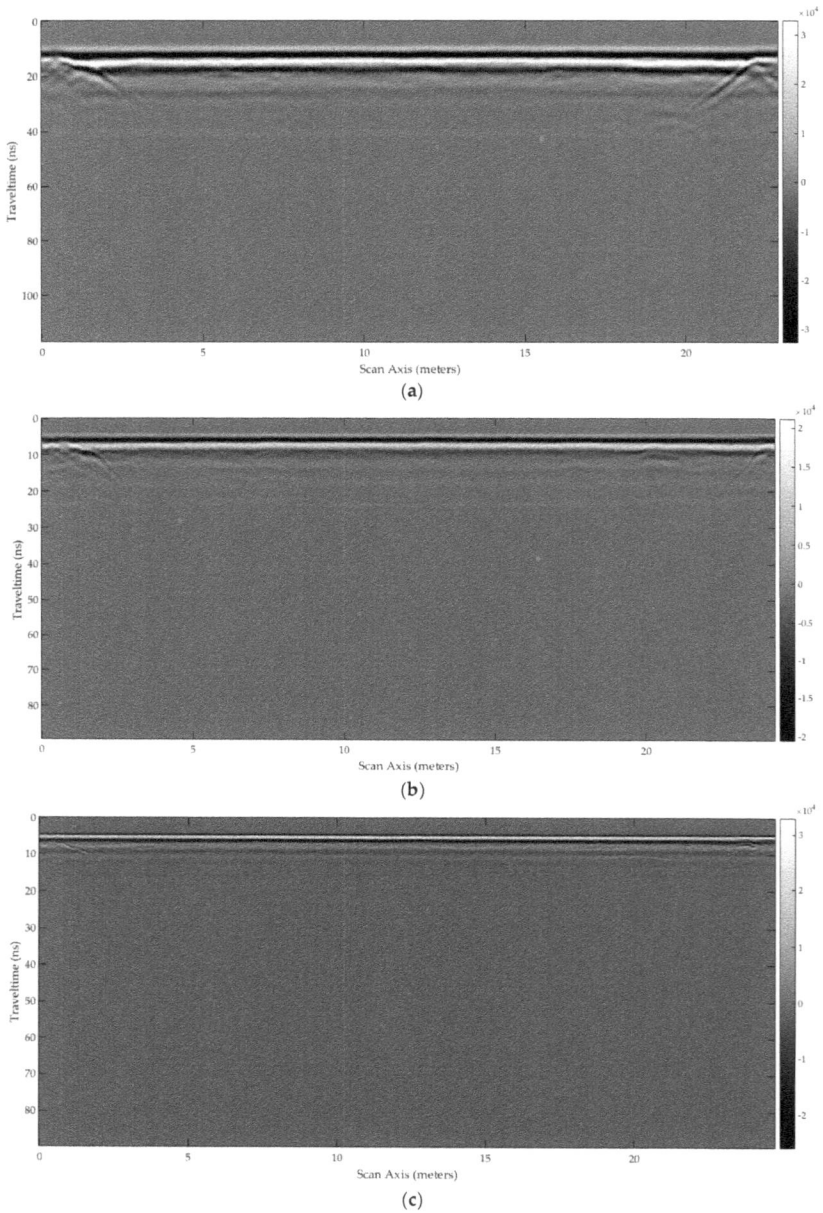

Figure 38. Profiles recorded in the Gneiss 0/20 gravel region by using GPR 2, on acquisition line 1, with antennas operating at: (**a**) 250 MHz; (**b**) 500 MHz; (**c**) 800 MHz.

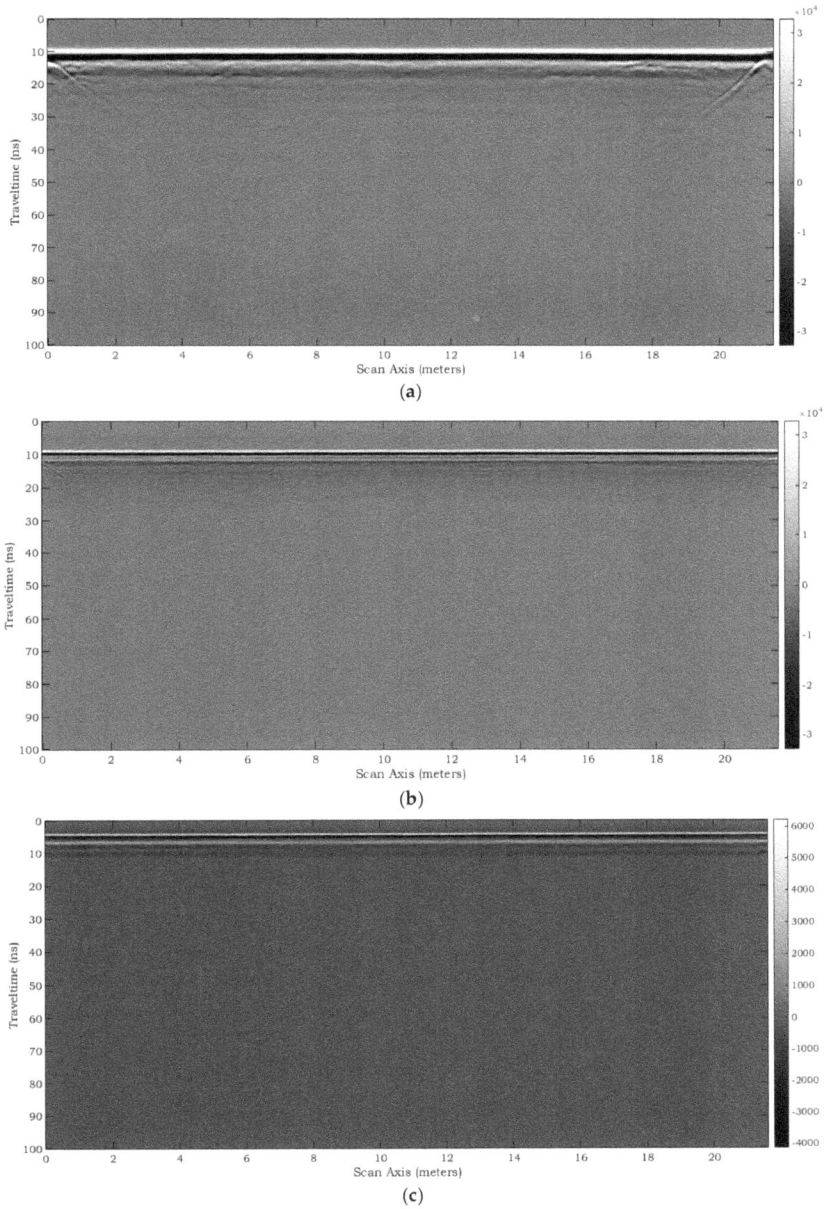

Figure 39. Profiles recorded in the Gneiss 0/20 gravel region by using GPR 3, on acquisition line 1, with antennas operating at: (**a**) 200 MHz; (**b**) 600 MHz; (**c**) 900 MHz.

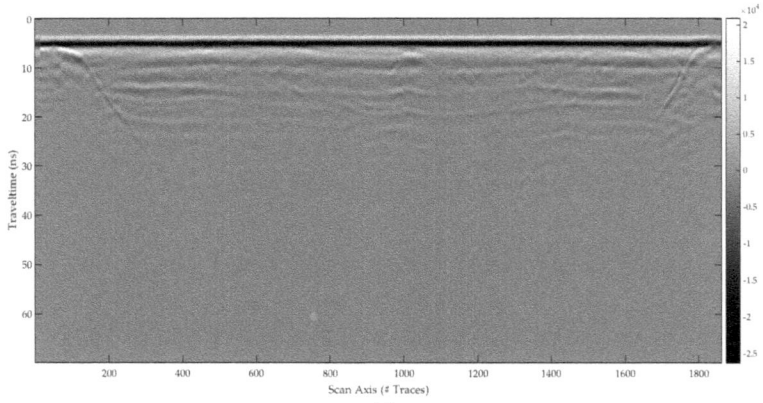

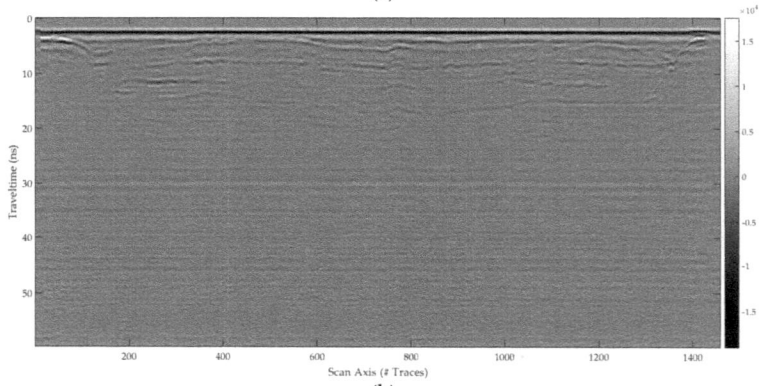

Figure 40. Profiles recorded in the Gneiss 0/20 gravel region by using GPR 1, on acquisition line 2, with antennas operating at: (**a**) 400 MHz; (**b**) 900 MHz.

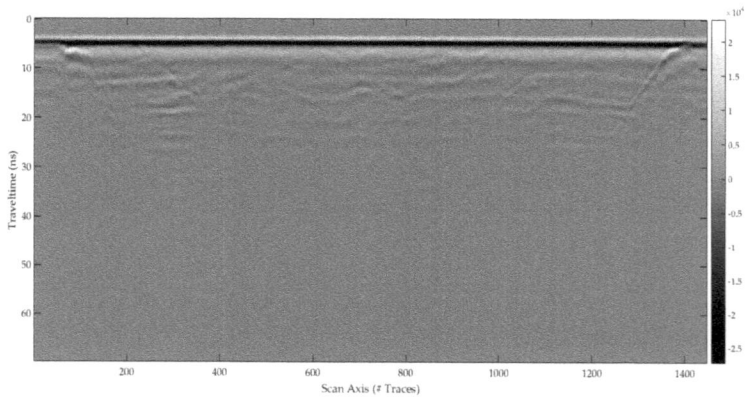

Figure 41. *Cont.*

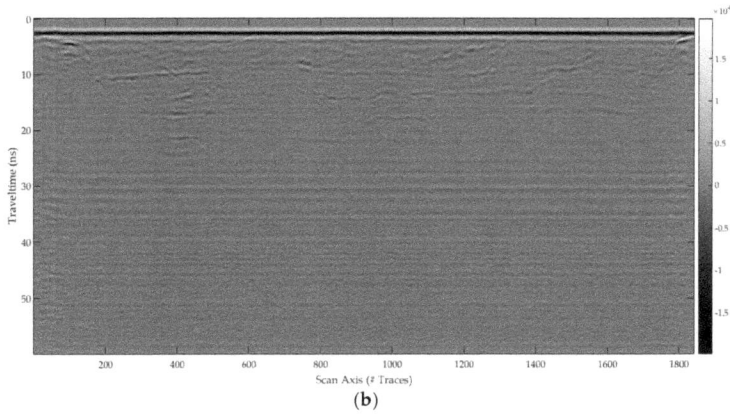

(b)

Figure 41. Profiles recorded in the Gneiss 0/20 gravel region by using GPR 1, on acquisition line 3, with antennas operating at: (**a**) 400 MHz; (**b**) 900 MHz.

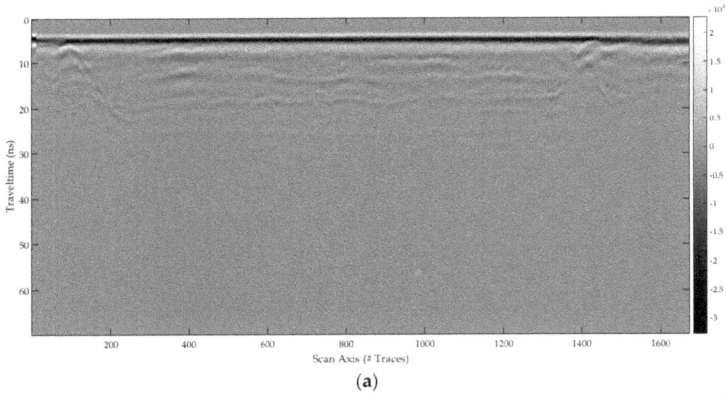

(a)

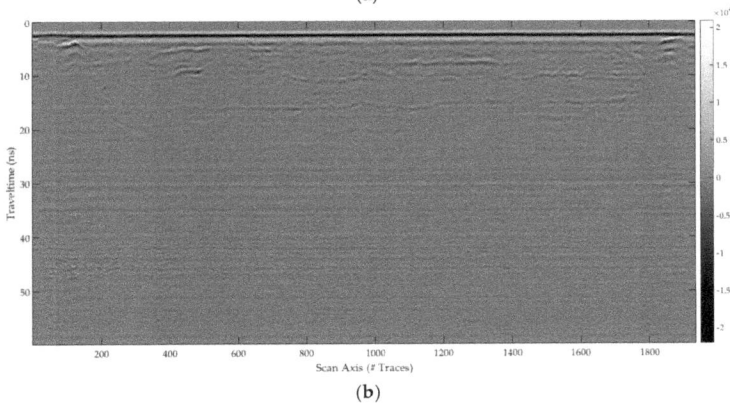

(b)

Figure 42. Profiles recorded in the Gneiss 0/20 gravel region by using GPR 1, on acquisition line 4, with antennas operating at: (**a**) 400 MHz; (**b**) 900 MHz.

6. Conclusions

The geophysical test site, built at the Nantes research center of the French institute of science and technology for transport, spatial planning, development and networks (IFSTTAR), was designed to test new geophysical techniques, innovative methods of measurement, and novel approaches for data analysis. It is proposed on a full-size scale and consists of a pit, 30 m long and 5 m wide at its maximum depth (around 4 m), with sloping sides that extend the width of the site to about 20 m. The pit is divided in five regions filled with different homogeneous soils and hosting georeferenced targets (such as pipes, artificial voids and rocks, masonry walls, etc.). The site was conceived to reserve margins of progress in the detection of heterogeneities by means of ground penetrating radar (GPR) or other geophysical techniques. Actually, some objects buried in the most conductive materials have not yet been detected from the surface.

In this paper, the geophysical test site and its construction process were described for the first time in detail. Then, a large dataset of GPR profiles, recorded by using three pulsed systems equipped with several antennas working in different frequency ranges, was presented. The profiles highlight the performances and limits of the GPR technique in terms of resolution and depth penetration versus soil and/or target. All results were accompanied by comments throughout the paper, and the effects of the application of pre-processing steps before saving the data were pointed out. All data files are found in an archive, enclosed to this paper as 'supplementary material'.

The presented dataset is the core part of the Open Database of Radargrams project of COST (European Cooperation in Science and Technology) Action TU1208 "Civil engineering applications of Ground Penetrating Radar". The idea beyond such initiative is to share with the scientific community a selection of GPR responses to enable an effective benchmark for direct and inverse electromagnetic scattering methods, imaging techniques and signal processing algorithms.

We hope that this dataset will be enriched by the contributions of further users, who are most welcome to visit the geophysical test site and collect new data with their GPR systems. We also hope that the dataset will be made alive by researchers who will process, analyze, invert and interpret the data, or implement electromagnetic models of the test site. It would be very interesting and useful to describe the state of the art of the research in the field by applying different techniques to the radargrams included in this dataset. At last, we hope that the challenge related to the multi-layer section will be achieved.

Supplementary Materials: The supplementary materials are available online at https://zenodo.org/record/1211173#.WsIuC1k0uUk.

Acknowledgments: The authors are grateful to Mercedes Solla for inviting them to submit this paper to the Remote Sensing Special Issue "Recent Advances in GPR Imaging". They thank particularly J. L. Chazelas (Ifsttar) for the design and monitoring of the implementation of the geophysical test site, G. Gugole (Ifsttar) and F. Xie (The Hong Kong Polytechnic University) for performing some GPR measurements, and T. Dubreucq (Ifsttar) for the characterization of the materials. This work is a contribution to COST (European Cooperation in Science and Technology) Action TU1208 "Civil engineering applications of Ground Penetrating Radar" (www.cost.eu, www.GPRadar.eu). The authors thank COST for funding and supporting COST Action TU1208.

Author Contributions: The authors contributed equally to this work.

Conflicts of Interest: The authors declare no conflict of interest.

References

1. Persico, R. *Introduction to Ground Penetrating Radar: Inverse Scattering and Data Processing*; Wiley-IEEE Press: Indianapolis, IN, USA, 2014; p. 392.
2. Benedetto, A.; Pajewski, L. (Eds.) Civil Engineering Applications of Ground Penetrating Radar. In *Springer Transactions in Civil and Environmental Engineering*; Springer International: New Delhi, India, 2015; p. 385. [CrossRef]

3. Chazelas, J.-L. *Création d'un Site-Test Pour les Méthodes Géophysiques Appliquées Aux Travaux Sans Tranchée—Rapport de Présentation Après Réalisation*; Technical Report No. 12; Laboratoire Central des Ponts et Chaussées: Nantes, France, 1998; p. 53. (In French)
4. Official Webpage of COST Action TU1208, Civil Engineering Applications of Ground Penetrating Radar on COST (European Cooperation in Science and Technology) Website. Available online: http://www.cost.eu/COST_Actions/tud/TU1208 (accessed on 28 March 2018).
5. Official Website of COST Action TU1208, Civil Engineering Applications of Ground Penetrating Radar. Available online: http://www.GPRadar.eu (accessed on 28 March 2018).
6. Eide, E.; Våland, P.A.; Sala, J. Ground-Coupled Antenna Array for Step-Frequency GPR. In Proceedings of the 15th International Conference on Ground Penetrating Radar (GPR 2014), Brussels, Belgium, 30 June–4 July 2014; pp. 756–761.
7. Huuskonen-Snicker, E.; Eskelinen, P.; Pellinen, T.; Olkkonen, M.-K. A New Microwave Asphalt Radar Rover for Thin Surface Civil Engineering Applications. *J. RF-Eng. Telecommun.* **2015**, *69*, 377–381. [CrossRef]
8. Ferrara, V.; Troiani, F.; Frezza, F.; Mangini, F.; Pajewski, L.; Simeoni, P.; Tedeschi, N. Design and Realization of a Cheap Ground Penetrating Radar Prototype@ 2.45 GHz. In Proceedings of the 2016 IEEE European Conference on Antennas and Propagation (EuCAP), Davos, Switzerland, 11–15 April 2016; pp. 1–4. [CrossRef]
9. Persico, R.; Leucci, G. Interference mitigation achieved with a reconfigurable stepped-frequency GPR system. *Remote Sens.* **2016**, *8*, 926. [CrossRef]
10. Ferrara, V.; Pietrelli, A.; Chicarella, S.; Pajewski, L. GPR/GPS/IMU system as buried objects locator. *Measurement* **2018**, *114*, 534–541. [CrossRef]
11. Fedeli, A.; Ježová, J.; Lambot, S. Testing of a new lightweight radar system for tomographical reconstruction of circular structures. In *Short-Term Scientific Missions: Years 4 & 5*; Pajewski, L., Rodriguez-Abad, I., Marciniak, M., Eds.; TU1208 GPR Association: Rome, Italy, 2018; pp. 18–44. Available online: http://www.gpradar.eu/resources/books.html (accessed on 28 March 2018).
12. Persico, R.; Provenzano, A.; Trela, C.; Sato, M.; Takahashi, K.; Arcone, S.; Koppenjan, S.; Stolarczyk, L.G.; Utsi, E.C.; Ebihara, S.; et al. *Recommendations for the Safety of People and Instruments in Ground-Penetrating Radar and Near-Surface Geophysical Prospecting*; EAGE: Houten, The Netherland, 2015; p. 66.
13. Pajewski, L.; Persico, R.; Derobert, X.; Balayssac, J.-P.; Ebihara, S.; Grégoire, C.; Ivashchuk, V.; Kind, T.; Krysiński, L.; Wai-Lok Lai, W.; et al. *Catalogue of GPR Test Sites*; COST Action TU1208 Series; TU1208 GPR Association: Rome, Italy, 2017; Available online: http://www.gpradar.eu/resources/books.html (accessed on 28 March 2018). [CrossRef]
14. De Chiara, F.; Fontul, S.; Fortunato, E. GPR Laboratory Tests for Railways Materials Dielectric Properties Assessment. *Remote Sens.* **2014**, *6*, 9712–9728. [CrossRef]
15. Núñez-Nieto, X.; Solla, M.; Novo, A.; Lorenzo, H. Three-dimensional ground-penetrating radar methodologies for the characterization and volumetric reconstruction of underground tunneling. *Constr. Build. Mater.* **2014**, *71*, 551–560. [CrossRef]
16. Sagnard, F.; Norgeot, C.; Dérobert, X.; Baltazart, V.; Merliot, E.; Derkx, F.; Lebental, B. Utility detection and positioning on the urban site Sense-City using Ground-Penetrating Radar systems. *Measurement* **2016**, *88*, 318–330. [CrossRef]
17. Rodríguez-Abad, I.; Klysz, G.; Martínez-Sala, R.; Balayssac, J.-P.; Mené-Aparicio, J. Application of ground-penetrating radar technique to evaluate the waterfront location in hardened concrete. *Geosci. Instr. Methods Data Syst.* **2016**, *5*, 567–574. [CrossRef]
18. Dérobert, X.; Lataste, J.F.; Balayssac, J.-P.; Laurens, S. Evaluation of chloride contamination in concrete using electromagnetic non-destructive testing methods. *NDT E Int.* **2017**, *89*, 19–29. [CrossRef]
19. Reci, H.; Chinh Maï, T.; Mehdi Sbartaï, Z.; Pajewski, L.; Kiri, E. Non-destructive evaluation of moisture content in wood by using Ground Penetrating Radar. *Geosci. Instr. Methods Data Syst.* **2016**, *5*, 575–581. [CrossRef]
20. Pajewski, L.; Fontul, S.; Solla, M. Ground Penetrating Radar for the evaluation and monitoring of transport infrastructures. In *Innovation in Near-Surface Geophysics: Instrumentation, Application and Data Processing Methods*; Persico, R., Piro, S., Linford, N., Eds.; Elsevier: Cambridge, MA, USA, 2018; in press.
21. Frezza, F.; Pajewski, L.; Ponti, C.; Schettini, G.; Tedeschi, N. Through-wall electromagnetic scattering by N conducting cylinders. *J. Opt. Soc. Am. A* **2013**, *30*, 1632–1639. [CrossRef] [PubMed]

22. Poljak, D.; Antonijević, S.; Šesnić, S.; Lalléchère, S.; Drissi, K.E.K. On deterministic-stochastic time domain study of dipole antenna for GPR applications. *Eng. Anal. Bound. Elem.* **2016**, *73*, 14–20. [CrossRef]
23. Warren, C.; Sesnic, S.; Ventura, A.; Pajewski, L.; Poljak, D.; Giannopoulos, A. Comparison of Time-Domain Finite-Difference, Finite-Integration, and Integral-Equation Methods for Dipole Radiation in Half-space Environments. *Prog. Electromagn. Res. M (PIER M)* **2017**, *57*, 175–183. [CrossRef]
24. Prokopovich, I.; Popov, A.; Pajewski, L.; Marciniak, M. Application of coupled-wave Wentzel-Kramers-Brillouin approximation to Ground Penetrating Radar. *Remote Sens.* **2018**, *10*, 22. [CrossRef]
25. André, F.; Lambot, S. Intrinsic Modeling of Near-Field Electromagnetic Induction Antennas for Layered Medium Characterization. *IEEE Trans. Geosci. Remote Sens.* **2014**, *52*, 7457–7469. [CrossRef]
26. De Coster, A.; Tran, A.P.; Lambot, S. Fundamental Analyses on Layered Media Reconstruction Using GPR and Full-Wave Inversion in Near-Field Conditions. *IEEE Trans. Geosci. Remote Sens.* **2016**, *54*, 5143–5158. [CrossRef]
27. Ristic, A.; Bugarinovic, Z.; Vrtunski, M.; Govedarica, M. Point Coordinates Extraction from Localized Hyperbolic Reflections in GPR Data. *J. Appl. Geophys.* **2017**, *144*, 1–17. [CrossRef]
28. Nounouh, S.; Eyraud, C.; Litman, A.; Tortel, H. Near-subsurface imaging in an absorbing embedding medium with a multistatic/single frequency scanner. *Near Surf. Geophys.* **2015**, *13*, 211–218. [CrossRef]
29. Mertens, L.; Persico, R.; Matera, L.; Lambot, S. Automated Detection of Reflection Hyperbolas in Complex GPR Images With No A Priori Knowledge on the Medium. *IEEE Trans. Geosci. Remote Sens.* **2016**, *54*, 580–596. [CrossRef]
30. Salucci, M.; Poli, L.; Anselmi, N.; Massa, A. Multifrequency Particle Swarm Optimization for Enhanced Multiresolution GPR Microwave Imaging. *IEEE Trans. Geosci. Remote Sens.* **2017**, *55*, 1305–1317. [CrossRef]
31. Varela-González, M.; Solla, M.; Martínez-Sánchez, J.; Arias, P. A semi-automatic processing and visualisation tool for ground-penetrating radar pavement thickness data. *Autom. Constr.* **2014**, *45*, 42–49. [CrossRef]
32. Li, J.; Le Bastard, C.; Wang, Y.; Wei, G.; Ma, B.; Sun, M. Enhanced GPR Signal for Layered Media Time-Delay Estimation in Low-SNR Scenario. *IEEE Geosci. Remote Sens. Lett.* **2016**, *13*, 299–303. [CrossRef]
33. Economou, N. Time-varying band-pass filtering GPR data by self-inverse filtering. *Near Surf. Geophys.* **2016**, *14*, 207–217. [CrossRef]
34. Warren, C.; Giannopoulos, A.; Giannakis, I. GPRMax: Open source software to simulate electromagnetic wave propagation for Ground Penetrating Radar. *Comput. Phys. Commun.* **2016**, *209*, 163–170. [CrossRef]
35. Pirrone, D.; Pajewski, L. E²GPR—Edit your geometry, Execute GprMax2D and Plot the Results. In Proceedings of the 2015 IEEE 15th Mediterranean Microwave Symposium, Lecce, Italy, 30 November–2 December 2015; pp. 1–4. [CrossRef]
36. Pajewski, L.; Giannopoulos, A.; Pirrone, D.; Warren, C.; Antonijevic, S.; Doric, V.; Poljak, D. Development of electromagnetic simulators for Ground Penetrating Radar. In Proceedings of the 2017 Conference of the Applied Computational Electromagnetics Society, Florence, Italy, 26–30 March 2017; pp. 1–2. [CrossRef]
37. Meschino, S.; Pajewski, L. SPOT-GPR: A freeware tool for target detection and localization in GPR data developed within the COST Action TU1208. *J. Telecommun. Inf. Technol.* **2017**, *2017*, 43–54. [CrossRef]
38. Meschino, S.; Pajewski, L. A practical guide on using SPOT-GPR, a freeware tool implementing a SAP-DoA technique. *Ground Penetr. Radar* **2018**, *1*, 104–122. [CrossRef]
39. Riveiro, B.; Solla, M. *Non-Destructive Techniques for the Evaluation of Structures and Infrastructure. Structures and Infrastructures*; CRC Press: Boca Raton, FL, USA, 2016; p. 398.
40. Santos-Assuņao, S.; Perez-Gracia, V.; Caselles, O.; Clapes, J.; Salinas, V. Assessment of Complex Masonry Structures with GPR Compared to Other Non-Destructive Testing Studies. *Remote Sens.* **2014**, *6*, 8220–8237. [CrossRef]
41. Persico, R.; D'Amico, S. Use of Ground Penetrating Radar and standard geophysical methods to explore the subsurface. *Ground Penetr. Radar* **2018**, *1*, 1–37. [CrossRef]
42. Solla, M.; Fontul, S. Non-destructive tests for railway evaluation: Detection of fouling and joint interpretation of GPR data and track geometric parameters. *Ground Penetr. Radar* **2018**, *1*, 75–103. [CrossRef]
43. Solla, M.; Lagüela, S. Thermography: Principles and applications. *Ground Penetr. Radar* **2018**, *1*, 123–141. [CrossRef]
44. Núñez-Nieto, X.; Solla, M.; Gómez-Pérez, P.; Lorenzo, H. GPR Signal Characterization for Automated Landmine and UXO Detection Based on Machine Learning Techniques. *Remote Sens.* **2014**, *6*, 9729–9748. [CrossRef]

45. Santos-Assunçao, S.; Perez-Gracia, V.; Salinas, V.; Caselles, O.; Gonzalez-Drigo, R.; Pujades, L.G.; Lantada, N. GPR Backscattering Intensity Analysis Applied to Detect Paleochannels and Infilled Streams for Seismic Nanozonation in Urban Environments. *IEEE J. Sel. Top. Appl. Earth Obs. Remote Sens.* **2016**, *9*, 167–177. [CrossRef]

46. Jezova, J.; Mertens, L.; Lambot, S. Ground-penetrating radar for observing tree trunks and other cylindrical objects. *Constr. Build. Mater.* **2016**, *123*, 214–225. [CrossRef]

47. Ceru, T.; Segina, E.; Gosar, A. Geomorphological Dating of Pleistocene Conglomerates in Central Slovenia Based on Spatial Analyses of Dolines Using LiDAR and Ground Penetrating Radar. *Remote Sens.* **2018**, *9*, 1213. [CrossRef]

48. Pérez-Gracia, V.; Caselles, J.-O.; Clapés, J.; Martinez, G.; Osorio, R. Non-destructive analysis in cultural heritage buildings: Evaluating the Mallorca cathedral supporting structures. *Non Destr. Test. Eval. Int.* **2013**, *59*, 40–47. [CrossRef]

49. Santos-Assunçao, S.; Dimitriadis, K.; Konstantakis, Y.; Perez-Gracia, V.; Anagnostopoulou, E.; Gonzalez-Drigo, R. Ground-penetrating radar evaluation of the ancient Mycenaean monument Tholos Acharnon tomb. *Near Surf. Geophys.* **2016**, *14*, 197–205. [CrossRef]

50. Solla, M.; Asorey-Cacheda, R.; Núñez-Nieto, X.; Conde-Carnero, B. Evaluation of historical bridges through recreation of GPR models with the FDTD algorithm. *Constr. Build. Mater.* **2016**, *77*, 19–27. [CrossRef]

51. Pajewski, L.; Solla, M.; Küçükdemirci, M. Ground-Penetrating Radar for Archaeology and Cultural-Heritage Diagnostics: Activities Carried Out in COST Action TU1208. In *Nondestructive Techniques for the Assessment and Preservation of Historic Structures*; Goncalves, L., Rodrigues, H., Gaspar, F., Eds.; Book Chapter No. 13; CRC Press: Boca Raton, FL, USA, 2017; pp. 215–225.

52. Persico, R.; D'Amico, S.; Rizzo, E.; Capozzoli, L.; Micallef, A. Ground Penetrating Radar investigations in sites of cultural interest in Malta. *Ground Penetr. Radar* **2018**, *1*, 38–61. [CrossRef]

53. Pajewski, L.; Giannopoulos, A. Electromagnetic modelling of Ground Penetrating Radar responses to complex targets. In *Short Term Scientific Missions and Training Schools—Year 1, COST Action TU1208*; Pajewski, L., Marciniak, M., Eds.; Aracne Editrice: Rome, Italy, 2014; pp. 7–45. Available online: http://www.gpradar.eu/resources/books.html (accessed on 28 March 2018).

54. COST Action TU1208. *Proceedings of the 2015 WG Progress Meeting, Edinburgh, UK, April 2015*; Pajewski, L., Ed.; Aracne Editrice: Rome, Italy, 2015.

55. Pajewski, L.; Giannopoulos, A.; van der Kruk, J. Electromagnetic modelling, inversion and data-processing techniques for GPR: Ongoing activities in Working Group 3 of COST Action TU1208. In Proceedings of the 2015 European Geosciences Union (EGU) General Assembly, Vienna, Austria, 12–17 April 2015; abstract ID EGU2015-14988.

56. McGahan, R.V.; Kleinman, R.E. Second annual special session on image reconstruction using real data. *IEEE Antennas Propag. Mag.* **1997**, *39*, 7–32. [CrossRef]

57. McGahan, R.V.; Kleinman, R.E. Third annual special session on image reconstruction using real data: Part 1. *IEEE Antennas Propag. Mag.* **1999**, *41*, 34–51. [CrossRef]

58. McGahan, R.V.; Kleinman, R.E. Third annual special session on image reconstruction using real data: Part 2. *IEEE Antennas Propag. Mag.* **1999**, *41*, 20–40. [CrossRef]

59. Belkebir, K.; Saillard, M. Special section: Testing inversion algorithms against experimental data. *Inverse Probl.* **2001**, *17*, 1565–1571. [CrossRef]

60. Belkebir, K.; Saillard, M. Special section: Testing inversion algorithms against experimental data: Inhomogeneous targets. *Inverse Probl.* **2005**, *21*, S1–S3. [CrossRef]

61. Geffrin, J.-M.; Sabouroux, P.; Eyraud, C. Free space experimental scattering database continuation: Experimental set-up and measurement precision. *Inverse Probl.* **2005**, *21*, S117–S130. [CrossRef]

62. Geffrin, J.-M.; Sabouroux, P. Continuing with the Fresnel database: Experimental setup and improvements in 3D scattering measurements. *Inverse Probl.* **2009**, *25*, 024001. [CrossRef]

63. Versteeg, R. The Marmousi experience: Velocity model determination on a synthetic complex data set. *Lead. Edge* **1994**, *13*, 927–936. [CrossRef]

64. Musumeci, C.; Cocina, O.; De Gori, P.; Patanè, D. Seismological evidence of stress induced by dike injection during the 2001 Mt. *Etna eruption. Geophys. Res. Lett.* **2004**, *31*, L07617. [CrossRef]

65. Solla, M.; Lorenzo, H.; Riveiro, B. Non-destructive methodologies in the assessment of the masonry arch bridge of Traba, Spain. *Eng. Fail. Anal.* **2011**, *18*, 828–835. [CrossRef]

66. Warren, C.; Giannopoulos, A. *gprMax/iwagpr2017-model: v.1.0*; Version v.1.0; Zenodo: Geneva, Switzerland, 2017. [CrossRef]

67. Grandjean, G.; Gourry, J.C.; Bitri, A. Evaluation of GPR techniques for civil-engineering applications: Study on a test site. *J. Appl. Geophys.* **2000**, *45*, 141–156. [CrossRef]

68. Ferber, V. *Déformations Induites Par L'humidification de Sols Fins Compactés, Apport d'un Modèle de Microstructure*; GT80; Laboratoire Central des Ponts et Chausees (LCPC): Bouguenais, France, 2006; 321p, ISBN 2-7208-2464-X.

69. Skempton, A.W. A study of the geotechnical properties of some post-glacial clays. *Géotechnique* **1948**, *1*, 111–124. [CrossRef]

70. Tzanis, A. *MATGPR: A Brief Introduction. In Proceedings of the 2014 Working Group Progress Meeting, Nantes, FR, February 2014*; Pajewski, L., Derobert, X., Eds.; COST Action TU1208 Series; Aracne Editrice: Rome, Italy, 2014; pp. 39–68. Available online: http://www.gpradar.eu/resources/books.html (accessed on 28 March 2018).

remote sensing

MDPI

Article

Reconstructing the Roman Site *"Aquis Querquennis"* (Bande, Spain) from GPR, T-LiDAR and IRT Data Fusion

Iván Puente [1,2,*], Mercedes Solla [1,2], Susana Lagüela [2,3] and Javier Sanjurjo-Pinto [4]

[1]   Defense University Center, Plaza de España s/n, 36920 Marín, Spain; merchisolla@cud.uvigo.es
[2]   Applied Geotechnologies Research Group, University of Vigo, Rúa Maxwell s/n,
      Campus Lagoas-Marcosende, 36310 Vigo, Spain; sulaguela@usal.es or susiminas@uvigo.es
[3]   Department of Cartographic and Terrain Engineering, University of Salamanca, Calle Hornos Caleros 50,
      05003 Ávila, Spain
[4]   PhD Programme in Protection of the Cultural Heritage, University of Vigo, 36310 Vigo, Spain;
      jsanjurjo@jspinto.net
*   Correspondence: ipuente@cud.uvigo.es; Tel.: +34-986-804-848

Received: 7 February 2018; Accepted: 27 February 2018; Published: 1 March 2018

Abstract: This work presents the three-dimensional (3D) reconstruction of one of the most important archaeological sites in Galicia: *"Aquis Querquennis"* (Bande, Spain) using in-situ non-invasive ground-penetrating radar (GPR) and Terrestrial Light Detection and Ranging (T-LiDAR) techniques, complemented with infrared thermography. T-LiDAR is used for the recording of the 3D surface of this particular case and provides high resolution 3D digital models. GPR data processing is performed through the novel software tool "toGPRi", developed by the authors, which allows the creation of a 3D model of the sub-surface and the subsequent XY images or time-slices at different depths. All these products are georeferenced, in such a way that the GPR orthoimages can be combined with the orthoimages from the T-LiDAR for a complete interpretation of the site. In this way, the GPR technique allows for the detection of the structures of the barracks that are buried, and their distribution is completed with the structure measured by the T-LiDAR on the surface. In addition, the detection of buried elements made possible the identification and labelling of the structures of the surface and their uses. These structures are additionally inspected with infrared thermography (IRT) to determine their conservation condition and distinguish between original and subsequent constructions.

Keywords: ground-penetrating radar; terrestrial laser scanning; infrared thermography; archaeology; 3D visualization; toGPRi

1. Introduction

Spain is very rich in cultural heritage, with castles, walls, churches, amphitheaters and monumental buildings of outstanding workmanship spread all over the country. These structures are vulnerable to changing weather patterns and other environmental hazards, thus requiring special protection against these events for the sake of conservation [1]. Their protection and management include material characterization, structural stability analyses and archaeological prospecting. Regarding the latter, the full characterization of archaeological sites using conventional techniques is a long process, while coring and excavations, which are most frequently applied in archaeological assessment, are ground disturbing. The use of non-destructive testing (NDT) is therefore recommended, as these techniques can provide useful information about the conditions of conservation without intrusive intervention [2]. Geophysics, aerial archaeology and other remote sensing techniques can be used to enhance the identification and understanding of the hidden archaeological remains [3].

The application of three-dimensional (3D) technologies to reconstruct archaeological sites has represented a great benefit within the heritage community [4]. In this regard, the use of Terrestrial Light Detection and Ranging (T-LiDAR) for digital recording and documentation of cultural heritage items has increased significantly in recent years [5–7]. This technique generates point clouds with 3D Cartesian coordinates and possibly color information that are useful for the visualization of the as-built realities. Furthermore, due to its capacity to provide fast, dense and accurate measurements, T-LiDAR presents several applications for structural monitoring of cultural heritage structures [8] and civil infrastructure facilities, such as tunnels [9,10], bridges [11], breakwaters [12] and roads [13].

However, buried structures cannot be assessed by visual or optical inspection and consequently, data integration and fusion techniques are needed for the correlation of sub-surface data with its corresponding superficial element. Alternative methods are therefore used for sub-surface inspection, including geophysics, such as magnetic, electrical and electromagnetic methods [14]. Regarding the matter of archaeogeophysics, one of the most noteworthy methods has been ground-penetrating radar (GPR). GPR is a fast data acquisition technique that has been commonly applied for high-resolution imaging in many archaeological applications [15]. However, the understanding of the GPR data has been a long-term challenge among the non-geophysical community. The analysis of the radar signal and the interpretation of the measured data (2D-GPR images) are not straightforward. The use of 3D-GPR data processing and visualization has advanced the acceptance of the use of GPR for archaeological prospecting [16]. The use of 3D imaging techniques and processing software produces more realistic images of buried archaeological remains [17–23], which allows not only the discovery, but also the obtainment of 3D reconstructions of buried structures [24,25]. Furthermore, all the data produced can be combined in a Geographic Information System (GIS) to achieve a more comprehensive archaeological interpretation [26–28].

Infrared thermography (IRT) is a technique for measuring characteristics related to the thermal state of the materials, allowing for the detection of pathologies mainly on their surface [29], but also in the sub-superficial areas if they present a high influence on the thermal state of the object [30,31].

The aim of this paper is to show the results of a morpho-geophysical survey at the archaeological site of "*Aquis Querquennis*" (Bande, Spain) by means of T-LiDAR and GPR, complemented with IRT. The first techniques are powerful tools for the identification and mapping of targeted remains, while IRT is a key technique for the diagnosis and determination of the conservation state of these remains. Thus, the three of them may result in very useful support for archaeologists when dealing with the collection, exploration and preservation of our cultural heritage. GPR data processing is performed using the in-house tool "toGPRi", which has been developed for GPR signal processing that includes the generation of both 2D and 3D imaging. With this objective, the tool implements different filters for signal amplification and noise removal as well as additional topographic correction. In particular, the product generated is a 3D model of the sub-surface and its subsequent XY images or time-slices at different depths, in addition to overlaid images connecting reflections at different levels that provide a better understanding of the occupied underground space. All of these products are georeferenced and can be subsequently imported into a GIS environment and merged with other sources of information for integral data treatment. Thus, the hybrid outcomes derived from the methods of GPR and T-LiDAR, and IRT and T-LiDAR, can be merged into a single image, allowing the reconstruction of the visible reality of the archaeological remains while integrating the unexcavated structures detected in the sub-surface.

2. The Roman Site "Aquis Querquennis"

"*Aquis Querquennis*" was a military camp from Roman times, built in Bande (Galicia, Spain), on the banks of the Limia river. Its occupation dates from the last quarter of the first century to the middle of the second and, according to the findings of the conquest of the site by the legio *septimagemina* detachment, it was a mixed unit of cavalry and infantry. Findings and historical proofs indicate the establishment of the legion at the site during the reign of *Vespasian* (69–79 AD). The encampment was probably built to monitor the construction of the roads that communicated *Bracara Augusta* (Braga, Portugal) and *Asturica Augusta* (Astorga, Spain). The first archaeological excavations were made in the 1920s and new studies were authorized in 1975, centered on the Northwestern area [32].

The settlement presents a classic layout with a rectangular shape and two main orthogonal paths (Figure 1), occupying 3 hectares. The border of the site is surrounded by a wall with softened corners, 3.20 m in height, with semi-cylindrical battlements, which are separated from the interior constructions by 11 m. There is a V-shaped moat outside the wall, 4 m deep and 3 m wide. The encampment had four main doors, corresponding to the ends of the two aforementioned paths, but only two have been excavated. Three barracks (*strigia*) have been identified, each one dedicated to housing two *centuria* with their respective commands. The barracks are composed of two rectangular wings faced around a courtyard (*compluvium*) with a cistern to collect rainwater. Each *centuria* was formed by ten *contubernium*, which is a space for eight soldiers, distributed as six *contubernium* in one wing and the guard quarter and four *contubernium* in the opposite wing, with the suite of rooms for the centurion at the entrance. Each *contubernium* was divided into a small front room (*arma*) for the equipment of the soldiers, and a rear room (*papilio*) where the soldiers slept. The gates and hollows of the barracks are oriented to the south.

Moreover, there were two rectangular granaries raised on stone pillars and delimited by thick walls with buttresses, so the roofs are believed to have been vaulted. A building of almost square shape and rooms arranged around an *impluvium* (a sunken area designed to carry away the rainwater) has been found and is considered to be designated to health care (*valetudinarium*).

The central building of the site was probably used as headquarters, due to its greater dimensions of 34.8 × 32.1 m, the presence of a vestibule flanked by ambulatories, a large central courtyard, a basilica with three doors and a sacral-administrative area or official temple.

Figure 1. Situation of the Roman site *"Aquis Querquennis"* in Galicia (Northwest of Spain) and grids surveyed with the ground-penetrating radar (GPR) technique.

3. Materials and Methods

3.1. Scan Data Acquisition and Processing

A terrestrial laser scanner (T-LiDAR) generates 3D coordinates of an object point by measuring the distance between the scanner's center and the object and both the horizontal and vertical angles. Depending on the number of scans, the distance from the object to the scanner and the minimum angle increment of the system, a very dense point cloud can be achieved. Additionally, the reflectance of the surface can be measured by recording the intensity of the reflected laser beam, and for those scanners with an RGB camera attached, the RGB-value of the reflection is also captured.

A Faro Focus^{3D} X330 scanner (Valladolid, Spain), was used for this work. This is a phase-shift scanner with an effective range of 330 m at 90% reflectivity. Specifications indicate that the ranging noise of a single measurement is 0.3 mm at ranges up to 25 m (90% reflectivity) [33]. The scanner has an integrated 8Mpix HDR camera to provide color to the point clouds and texture maps for the triangulated point cloud data. A topographic tripod was also used for its placement.

The location of the scan positions was planned beforehand in order to minimize the fieldwork and number of scans, while avoiding occlusions, to ensure the full coverage of the site. Actually, a total of 48 scan positions were performed with a base distance between 10 and 15 m, covering the whole monitoring area and obtaining their corresponding point clouds with a spatial resolution of 6–7 mm/10 m distance.

As the locations of the scan positions were not the same in different epochs, the coordinates of identical points sampled in different epochs were not expected to be equal either. Consequently, registration was required to combine multiple data into a single set of range data. The Faro Scene software provides an automatic registration method, using fixed plane targets in common surfaces as control points and is supported by the Iterative Closest Point (ICP) algorithm. Although registration may introduce small misalignment errors, these were in the order of a few millimeters with no significant effect on the final quality of the resulting point cloud [34,35]. Finally, the raw laser point clouds were filtered and segmented in order to remove the points that were not part of the Roman site *"Aquis Querquennis"*.

3.2. GPR Survey

Ground-penetrating radar (GPR) is a geophysical method based on the propagation of very short electromagnetic pulses (1–20 ns) in the frequency band of 10 MHz–2.5 GHz. A transmitting antenna emits the electromagnetic signal into the ground, which is partly reflected when it encounters media with different dielectric properties and is partly transmitted into deeper layers. Then, the reflections produced are recorded upon arrival at the receiving antenna; which is either in a separate antenna box, or in the same antenna box as the transmitter. The strength or intensity of the reflection is provided in terms of amplitude value. The amplitude is higher as the dielectric contrast between two different media is greater. While the antenna is moved along the ground surface, a two-dimensional image (radargram) is obtained, which is an XZ graphic representation of the detected reflections. The *x* axis represents the antenna's displacement along the survey line, and the *z* axis represents the two-way travel time of the pulse emitted (in terms of nanoseconds). If the time required for the electromagnetic pulse to go from the transmitting antenna to the reflector into the ground and return to the receiving antenna is measured, and the velocity of this pulse in the subsurface medium is known, then the position in depth of the reflector can be determined.

3.2.1. GPR Data Acquisition

The GPR survey was conducted using a RAMAC system from MALÅ Geosciences (Malå, Sweden). A 500-MHz antenna was used, because this frequency provides proper vertical resolution as well as sufficient penetration. The data acquisition was carried out using the common-offset mode with the antenna polarization perpendicular to the direction of data collection, and the acquisition parameters were a 2 cm trace-distance interval and a total time window of 75 ns, composed of 512 samples. To complete the trace-distance interval calculation while measuring the profile lengths, the GPR antenna was mounted on a survey cart with an encoder (odometer wheel).

Three-dimensional (3D) GPR methodologies were performed in this work. As illustrated in Figure 1, two grids of approximately 156 m^2 and 787 m^2 were designed to prospect different unexcavated areas of the Roman site. To cover the entire grids, parallel 2D profiles were registered at regular intervals of 20 cm spacing, resulting in a total of 41 and 123 profiles, respectively. All the profile lines were collected in the same direction, starting from the lower left corner along the *x* axis.

3.2.2. GPR Data Processing: "toGPRi" Tool

Data processing was performed using an in-house customized software tool called "toGPRi" [36], which was programmed using GNU Octave programming language [37]. The developed software is compatible with the GNU General Public License v.3 and runs on GNU/Linux. It has been especially created to process data from MALÅ GPR systems in order to generate 3D cubes of data and subsequent georeferenced raster images at different levels of depth representing the underground elements.

As described in Figure 2, the required workflow consists of the following steps: (1) importing the data into the software; (2) checking the data configuration and geometry of the profiles (number of traces, frequency, number of samples, etc.); (3) displaying the 2D images (also called XZ images or radargrams) produced, including pre-processing of the data; (4) applying different filters for data processing, namely to remove noise and to amplify the signal as well as for topographic corrections; (5) exporting the processed 2D data as a series of ".xyz" profile files; (6) creating an Octave 3D matrix for the positioning of the profiles; (7) generating a 3D point cloud as well as XY images (or time-slices) at different levels of depth.

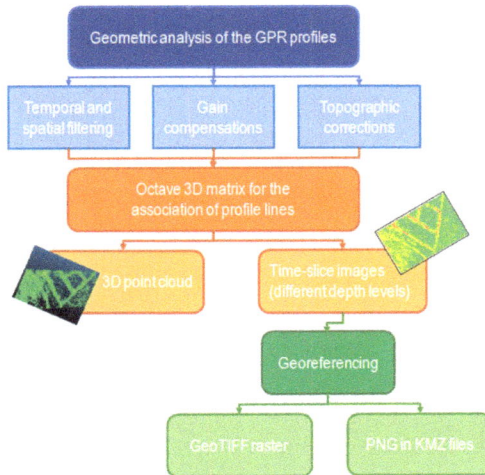

Figure 2. Workflow of the "toGPRi" tool developed for GPR data processing.

The "toGPRi" tool can open and read ".rd3" and ".rad" data files from MALÅ GPR systems, and the radargram is built and displayed using different color maps and contrast levels to optimize its visualization towards interpretation.

Regarding the data processing step, different filters were implemented to reduce clutter or any unwanted noise in the raw-data and to amplify the signal received. The purpose was to enhance the extraction of information from the signal received and to produce better data presentation, to make the data interpretation easier. The filters implemented were time-zero correction, temporal filters (subtract mean trace and vertical smooth), spatial filters (background removal and horizontal smooth), gain application (gain function and temporal constant gain or manual gain), and topographic corrections. Table 1 includes the sequence of filters and parameters considered for the data processing in the particular case study presented in this work.

Table 1. Filters and parameters used for data processing in the case study of the Roman Site *"Aquis Querquennis"*.

FILTER	DESCRIPTION
Time-zero correction	Traces require adjustment to a common time-zero position, since thermal drift, electronic, instability, cable length differences or variations in the antenna air gap can cause "jumps" in the air/ground wavelength's first arrival time. This is usually achieved using some particular criteria such as the air wave first break point or the maximum amplitude peak of the trace.
Subtract mean trace	On some occasions, a continuous or very low frequency component (DC component) appears in the traces recorded by the radar, since the average level of the signal is moved from zero amplitude to a different value. This filter computes the corresponding sample mean value in each pixel of the trace to reduce this component. The filter is applied by selecting a percentage of traces on each side of the target trace to obtain the mean value. The subtract mean trace filter was applied in this work with a percentage of 50%.
Gain function	Signals at greater depths have very low energy due to signal attenuation and geometrical spreading. Gaining consists of amplifying the received signal by multiplying the data using a mathematical function or manually entering gain values. This filter applies a linear and exponential function to each sample. The function is defined as $p * (t/s) * n + eq * (t/s) * n$, where p is the linear factor, q is the exponential factor, t is the time window, s is the number of samples per trace and n is the number of samples filtered. In this work, a gain function was applied from the first sample (once the time-zero was corrected) with a linear factor of 250 and an exponential factor of 10.
Temporal constant gain	This filter allows for the amplification of the signal from a certain depth by multiplying a selected number of samples by a certain factor. It affects all the traces composing the radargram at this temporal distance. This manual gain was applied in this work from sample 1 to sample 158 with a direct factor of 3.
Vertical smooth	Softens data vertically, in a specific horizontal section and time window, to remove high-frequency noise. This filter is applied by selecting a percentage of the mean of the number of samples. Vertical smooth was applied in this work from sample 1 to sample 150 and from trace 100 to trace 380 with a percentage of 35%.
Background removal	The main objective of these filters is to remove potential low-frequency noise which appears in the form of continuous horizontal bands along the recorded traces or only in some parts of them. This noise usually originates by bad coupling between the antenna and the medium and to eliminate ringing. This filter estimates an average of all traces in a window and removes it from every single trace. The main effect in the data is to suppress flat-lying reflectors, emphasizing smaller reflections. For every trace, this filter operates by calculating the mean value of the pixels (positive or negative), inferior to a pre-established percentage. The background removal filter was applied in this work with a percentage of 25%.
Horizontal smooth	Softens data horizontally, in a specific horizontal section and time window, to suppress horizontal reflections coming from the upper and lower interfaces or heterogeneous medium. This filter is applied by selecting a percentage of the mean of the number of traces. Horizontal smooth was applied in this work from sample 75 to sample 130 and from trace 1 to trace 949 with a percentage of 50%.
Topographic corrections	Compensating for topography is often important in improving the accuracy of imaging subsurface features. Features that are not directly underneath the antenna are recorded as if they actually were. The topographic data is loaded through a ".txt" file containing three columns with the XYZ coordinates for several points of the surface line, although only the XZ coordinates are used for correction.

The selection of the filters to be applied and their parameters are configured by the user. The results produced after each filter or after a sequence of different filters can be saved as an Octave binary matrix and the corresponding 8-bit image file (TIFF format). Figure 3 illustrates the results of the data processing described in Table 1.

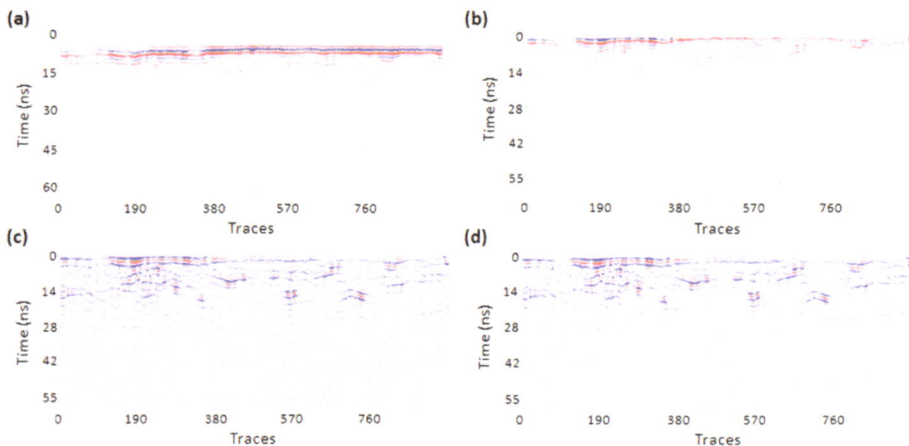

Figure 3. Data processing with the sequence raw-data (**a**), time-zero correction and 50% subtract mean trace (**b**), gain (with p = 250, q = 10, t = 0.075, s = 471) (**c**), smoothing and 25% background removal (**d**).

For the generation of the 3D matrix, a new pattern (.gpr) had to be established regarding the relative position of the profiles. The parameter of distance interval between traces is contained in the ".rad" file for each profile line. The distance between the parallel profile lines needs to be manually introduced. If the distance between profile lines is not regular along the entire grid prospected, the profile lines will be introduced in different phases. Other parameters to be introduced were the total time window, the original maximum number of samples (before the time-zero cut) and the average velocity of propagation of the GPR signals. Amplitude interpolation was used to fill the spaces between consecutive profile lines. The 3D cube obtained was finally saved as a variable named AY3D in a GNU Octave binary file with the extension "3D.b". Once this 3D matrix is built, the next step is to introduce the ".xyz" profile files in order to generate the 3D point cloud in-depth (Figure 4a). In addition, raster images can be generated through horizontal time-slices of the 3D cube and their georeferencing. Regarding the generation of the raster images, these time-slices are directly obtained with the "toGPRi" tool. These images can be obtained for different depth values depending on the length of the time window selected and the slice thickness. Moreover, the user-selectable option "overlaid raster" is also implemented for the computation of the maximum, minimum and mean pixel values of the fusion of layers (Figure 4b). This overlay analysis allows the strongest reflectors at different depths on a single image to be obtained. All the 3D point clouds generated can be exported to be managed in additional software like Cloud Compare (Paris, France) or MeshLab (ISTI-CNR, Pisa, Italia). Furthermore, the raster images produced can be saved in a TIFF image format after the configuration of an alpha channel and a range of transparent pixels (from 0 to 255, according to the 8-bit format) for the establishment of the threshold between black and white. However, if no alpha channel is selected, it will be an indexed color image. The georeferencing of images is also possible given that the "toGPRi" tool includes the geospatial GDAL library tools (libgdal and gdal-bin v2.1.1) [38], from OSGEO (USA), to generate GeoTIFF images and compressed KML files in a WGS84 coordinate system with UTM projection, whose zone can be selected by the user. As an example, Figure 4c shows the superposition of the 3D overlaid image in Figure 4b with the satellite image within Google Earth.

(a)

(b)

Range: 90-130 cm
Velocity: 11.6 cm/ns

(c)

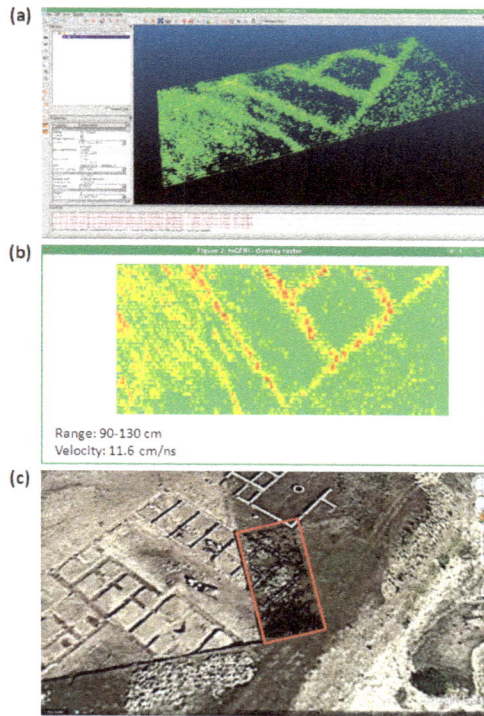

Figure 4. Different GPR results obtained with the "toGPRi" tool: (**a**) visualization of a horizontal slice of the 3D GPR cube in a point cloud viewer; (**b**) raster image generated through the overlay of several time-slices and (**c**) superposition of the 3D overlaid image with the aerial photograph of the site in Google Earth viewer.

3.3. IRT Inspection

Infrared thermography (IRT) is a technique that measures radiation emitted by the bodies in the thermal infrared band of the spectrum. This radiation is a function of the thermal state of the bodies (temperature), and consequently, the surface temperature can be computed from it. Provided that the bodies tend towards thermal equilibrium if no heating is applied, anomalies in the temperature distribution of the bodies are associated with pathologies in their composition or state. Following this fact, a thermographic inspection is performed for the walls of the site, with the aim of providing information on the different stages of construction and determining which part of the remains are original and which belong to posterior reconstructions. This thermographic inspection is performed following the approach of passive thermography; that is, no artificial heating is applied, and the acquisition is performed without direct incidence of the sun on the surfaces. The emissivity of the materials is set to 1, so that the temperatures measured are apparent temperatures, and the differences in temperature in this case are due to differences on the superficial state. This way, temperature differences are associated to (1) different materials and (2) pathologies on the surface, such as cracks, mould or presence of vegetation (Figure 5). In addition, thermographic images are acquired with a 10% overlap between consecutive images and from a position of 10° with perpendicular to the walls. This angle avoids the measurement of radiation reflected from the operator, and the overlap minimizes missing information from any part of the walls.

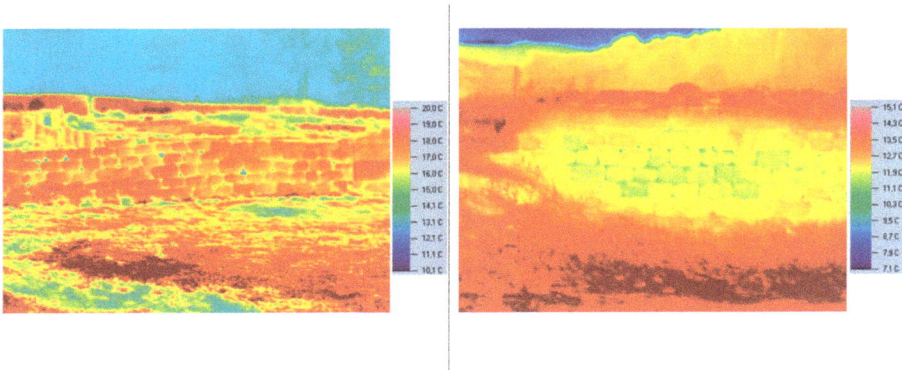

Figure 5. Examples of thermographic images acquired during the inspection of the excavated walls. (**Left**) completely original wall. (**Right**) partially reconstructed wall.

IRT also allows for the detection of sub-superficial anomalies if their effects are high enough to influence the state of the surface. In archaeology, buried elements have been found in [39], where the presence of air chambers under the surface produced temperature differences on the surface from which the presence of buried graves was identified with IRT, and in [40] where the soil coverage was freed from vegetation in order to detect small objects buried next to the surface. In the case of the "*Aquis Querquennis*" site, thermographic tests were performed from a zenithal point of view, in order to cover bigger extensions of the terrain from an angle of approximately 45°. The objective of the tests was to check whether buried walls could be located with IRT.

4. Results and Discussion

4.1. Results of the T-LiDAR Survey

The overall point cloud acquired by the Faro Focus^{3D} X330 system provides a model of the site (Figure 6a) that allows for detailed examination of individual barracks, including the qualitative and quantitative identification of geometric anomalies of structural elements. The walls and openings were all able to be captured, adding to the utility of the model for cultural heritage interpretation. Although there are a few areas inside the three barracks not accessible, this will not affect the overall documentation process.

(a)   (b)

Figure 6. Results of the T-LiDAR survey: (**a**) Detailed view of a colored 3D point cloud of a stone wall and (**b**) orthoimage of the barrack in the Southern area of the Roman site.

Previous surveying of the site involved manual measurements, which required several days to measure and draw in CAD software packages. The resulting model was only accurate at the points where the measurements were taken, therefore leading to an approximate drawing. To overcome this setback, building dimensions were obtained from T-LiDAR orthoimages (Figure 6b). They were generated by applying an affine transformation to each one of the planes of the triangular mesh obtained from the point clouds. The same mathematical model directly relates the pixel coordinates of the registered RGB values from the 8Mpix HDR camera to those of the orthoimages.

The average accuracy of the resulting barrack lengths is typically in the order of one centimeter, which is within the tolerance expected given the minor misalignment errors and the accuracy of single measurements. In addition, thanks to the information about the portion of energy reflected by the surface of the barracks, which depends on its reflectance, it is possible to detect pathologies, measure the size and orientation of rock fissures or compute the damaged area. Lastly, it is noteworthy to mention that T-LiDAR derived measurements can be used as the starting points for advanced studies in the structural analysis of monuments and historical constructions. This fact might help managers, archaeologists, and conservators to design and plan future interventions on this rapidly deteriorating Roman military camp.

4.2. Results of the GPR Prospection

Figure 7a represents the 3D overlaid GPR image obtained in grid 1 (see Figure 1), which corresponds to the unexcavated line of *contubernia* at the right wing of the barrack located in the Southern sector of the Roman camp. This overlaid image was built considering the time-slices (XY images) from 105 cm to 140 cm in depth (assuming a pre-calibrated velocity of propagation of 11.6 cm/ns). Given this image, it was possible to corroborate that, as usual, that sector of the barrack was composed of six *contubernia* of 6 × 3 m, divided into two zones (*papilio* and *arma*) and presenting a guard quarter (8 × 3 m size) at its entrance, its structure being similar to the barracks in the Northwestern sector. Figure 7b allows the validation of the GPR results after a recent excavation conducted in a 2016 campaign.

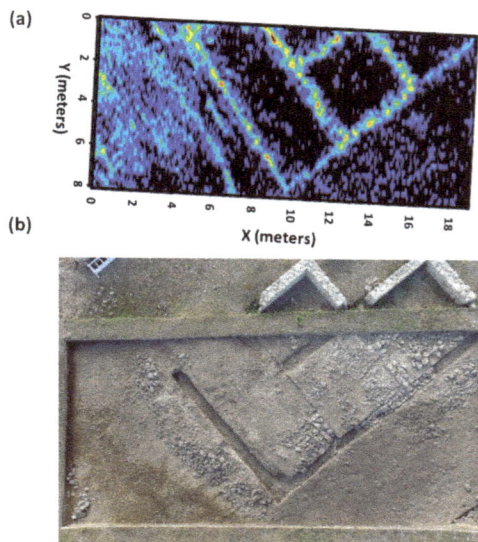

Figure 7. Results of grid 1 in the Southern sector of the camp: (**a**) 3D overlaid image and interpretation, and (**b**) aerial picture of a recent excavation.

Figure 8 presents the results produced for the GPR prospection acquired in grid 2 (see Figure 1). Although the wall reflections begin to appear from 50 cm in depth, the overlaid image in Figure 8 was built considering the time-slices from 70 cm to 150 cm. The purpose was to avoid unwanted reflections from wall falls in order to produce a clear image of the sub-surface structures, which makes the data interpretation easier. To enhance the wall reflections, the image was generated by selecting the maximum amplitude (or intensity) values and these were colored into a black/white palette with an alpha channel. Thus, the white pixels in the imaging represent the maximum amplitude values corresponding to wall reflections—either wanted reflections from sub-surface structures or undesired reflections from fallen walls. Observing the 3D image, it is also important to mention that some of the structure alignments highlighted are interrupted or missed due to the absence of walls by collapse.

The results exhibit the partial structure of a barrack, with two rectangular wings faced around a courtyard. The lower wing presents five *contubernia* (numbers 1–5 in Figure 8) and the guard quarter (number 9) at the entrance of the courtyard, while its counterpart has three *contubernia* (numbers 6–8) and the centurion's suite of rooms (number 10) at the entrance of the courtyard. All *contubernia* have similar dimensions and characteristics, as described for Figure 4. Unfortunately, it was not possible to prospect the complete structure, and the two opposite *contubernia* at the back of the courtyard and central cistern were missed in the GPR imaging. It is important to highlight that although the other excavated barracks present the centurion's suite at the left wing of the entrance (see Figure 1), the barrack under this grid has the centurion's suite at the right wing of the entrance.

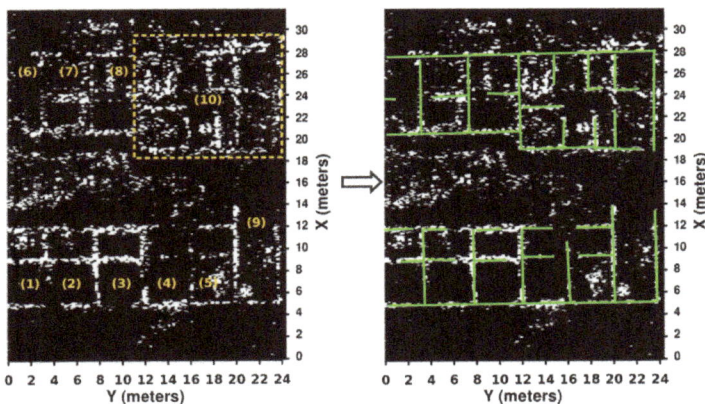

Figure 8. Results of grid 2 in the Northern sector of the camp: 3D overlaid image showing the configuration and interpretation of the *contubernia* identified (1–8), as well as the guard quarter (9) and the centurion's suite (10).

4.3. Results of IRT

The thermographic images acquired from the walls were registered with the T-LiDAR data through six control points in each image and their projective transformation. This way, the thermographic image presents temperature and geometric information, and the visualization of a complete wall in one image is possible in such a way that pathologies appearing in several images can be more clearly identified and interpreted. Figure 9 shows the thermographic orthoimages generated, where the following pathologies are highlighted: stones losing material, lack of union material and presence of vegetation. In addition, the presence of stones from two different times (before and after restoration activities) can be determined from their different temperatures, and because the temperature distribution in the newer stones is homogeneous.

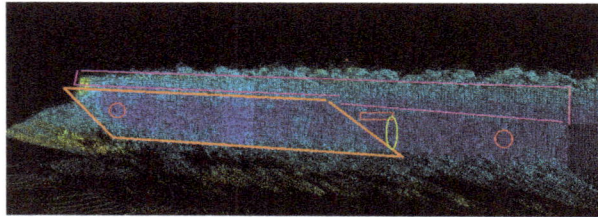

Figure 9. Thermographic image of the walls, registered on the Terrestrial Light Detection and Ranging (T-LiDAR) data for the visualization of complete walls within one product. The following pathologies are highlighted: stones losing material (red circle), lack of union material (red square) and presence of vegetation (green circle). Stones from before and after restoration activities are also marked (orange and purple curves, respectively).

For the test of detection of buried elements, although some images showed a high-temperature grid pattern on the terrain (Figure 10a), closer acquisitions revealed that high temperatures belong to areas with a lack of vegetation cover (Figure 10b). In addition, a partial excavation in the site showed that the current depth of the top of the walls is 21 cm (Figure 10c), while their width is 60 cm. In addition, the walls are made of sedimentary rock (sandstone), with very similar thermal properties (thermal resistance and diffusivity) as the burying sediments. This fact confirms the lack of capability of IRT to detect the buried parts of the wall, since it has been determined that the dimensions of the anomaly should be greater than its depth and that the materials should present different characteristics, in order to be detected with this technique [41]. In the "*Aquis Querquennis*" site, although the width of the walls is greater than its depth, their geology and that of the burying sediments are the same (solid and disaggregated sandstone), not allowing for the existence of different heat fluxes and the consequent appearance of different temperatures. In order to confirm this incapability for detection, the inspection was performed after direct radiation of the sun on the area, trying to maximize the heat excitation of the material and forcing the appearance of a thermal flux in cases where it was possible. The non-appearance of a heat flux as a consequence of the geological equilibrium of the wall and burying materials, acts as a demonstration of the hypothesis.

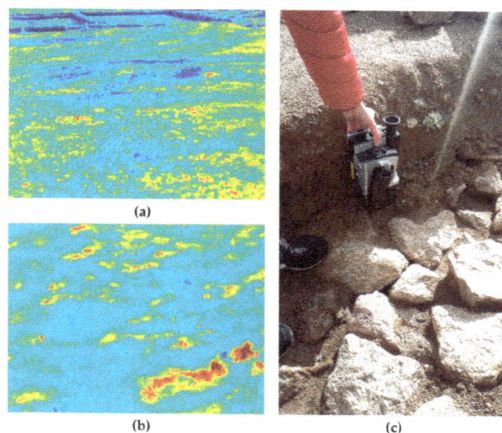

Figure 10. (**a**) Thermographic image of the terrain, showing a possible grid pattern of high temperatures; (**b**) thermographic image of a detail of the terrain, where high temperatures are associated with a lack of vegetation; (**c**) visual demonstration of the depth of the top of the walls, from a wall partially excavated.

4.4. Data Fusion and Visualization of "Aquis Querquennis" T-LiDAR and GPR Results

This final step involves the registration of T-LIDAR data with data from the GPR sensor in order to provide end-users with merged orthoimages of the archaeological site and its surrounding environment. In particular, the data integration process consists of finding corresponding control points in both datasets, comparable to the LiDAR registration process explained in Section 3.1. For this task, eight ground control points were located on the site in each of the four corners of both rectangular grids surveyed with the GPR technique (see Figure 1).

After that, both GPR grids were accurately registered in relation to the 3D geometric model defined by the TLS survey (see Figure 11). This multi-data source approach gives end-users the opportunity to generate orthoimages with different spatial resolutions. In addition, if the point cloud was already georeferenced, the Cartesian representation would enable the user to directly retrieve the coordinates of the points in an absolute coordinate system. In fact, those local coordinates could be transformed into global ones using, for example, a GNSS (Global Navigation Satellite System) base station.

Figure 11. GPR results produced in grids 1 and 2, fused on the T-LiDAR based orthoimage of the Roman site.

5. Conclusions

The application of geomatic techniques at the Roman site *"Aquis Querquennis"*, recognized as a cultural heritage resource, has widely contributed to its documentation and 3D reconstruction. The integration of terrestrial LiDAR and ground-penetrating radar was used for the first time in the encampment in order to support and complement the visual observations and petrophysical characterization carried out by previous restorers. The T-LiDAR method has shown its capabilities for measuring the 3D geometry of the overall settlement, letting users obtain, for example, the lengths, heights and volumes of every building, or the width of a particular stone pillar, without making direct contact with them. In particular, it provides highly accurate and more detailed 3D virtual reconstructions, incomparable with other large-volume measurement systems and their derived products, like the freely-available Google Earth or satellite imagery. Other interesting T-LiDAR derived products, such as 2D plans, sections and orthoimages, can also play important roles for heritage conservation and preservation purposes.

In addition, the use of GPR results revealed hidden constructions or structural details, such as the right wing of one barrack in grid 1 and the partial structure of another barrack in grid 2, which cannot be observed during visual exploration of the surface. With GPR data, the tool "toGPRi" presented in the paper allowed the generation of precise multi-source orthoimages, using the 3D-GPR imaging and the orthoimages from TLS, for a deeper analysis of both visible and underground space. At this point, the tool only supports 3D matrices for the representation of the buried 3D reality and 2D raster layers, but it is expected that for future work, vector layers will be implemented in order to compute parameters such as the volume of buried structures. In the aforementioned case study, the volumetric reconstruction would facilitate the dimensioning of the buildings while providing an intuitive and easily understood layout of the underground space distribution. In particular, the future goal of the "toGPRi" tool is to directly connect the processed results from GPR with LiDAR (both terrestrial and mobile) and DEM (Digital Elevation Model) surface data in geospatial databases with 2.5D and 3D geometries as SQLite/SpatiaLite/RasterLite2 and PostGIS. This way, the tool will be a single application based in a data infrastructure light model.

The inspection was complemented with IRT for the evaluation of the conservation state of the unburied elements. The registration of the thermographic images with T-LiDAR data allows for the direct location of the pathologies on their corresponding positions on the walls, as well as for their geometrical measurement in terms of area, number of stones and depth affected. The fact that buried elements and burying materials are the same (sandstone) has not allowed the detection of the first with IRT under the thermal excitation of the sun, regardless of the depth and dimensions of the buried element to detect.

Regarding novel integrations of techniques, future surveys at the site will deepen the application of infrared thermography with two objectives: first, the thermal characterization of the materials will be performed differentially in order to allow for quantitative diagnosis of the severity of the pathologies; second, information provided by GPR about the elements buried near the surface and their presence and location will be corroborated, through the application of specific thermal excitations and with knowledge of the thermal properties of the buried materials.

Acknowledgments: This study is a contribution to the EU funded COST Action TU-1208 "Civil Engineering Applications of Ground Penetrating Radar". Authors would like to thank the Spanish Ministry of Industry for the support given through human resources grant IJCI-2015-24492. The useful suggestions and support provided by the archaeologist D. Santiago Ferrer Sierra, Director of the Archaeological Center of *"Aquis Querquennis"* in Bande, are also gratefully acknowledged.

Author Contributions: I.P. and M.S. conceived and designed the experiments; I.P., M.S., S.L. and J.S. performed the experiments; I.P., M.S. and S.L. analyzed the data; J.S. contributed to the analysis tools; I.P., M.S. and S.L. wrote the paper.

Conflicts of Interest: The authors declare no conflict of interest. The founding sponsors had no role in the design of the study; in the collection, analyses, or interpretation of data; in the writing of the manuscript, and in the decision to publish the results.

References

1. Vecco, M. A definition of cultural heritage: From the tangible to the intangible. *J. Cult. Herit.* **2010**, *11*, 321–324. [CrossRef]
2. Lin, A.; Novo, A.; Har-Noy, S.; Ricklin, N.; Stamatiou, K. Combining GeoEye-1 satellite remote sensing, UAV aerial Imaging, and geophysical surveys in anomaly detection applied to archaeology. *IEEE J. STARS* **2011**, *4*, 870–876. [CrossRef]

3.   Boschi, F. *Looking to the Future, Caring for the Past. Preventive Archaeology in Theory and Practice*; Bononia University Press: Bologna, Italy, 2016.

4.   Valdelomar, J.T.; Brandtner, J.; Kucera, M.; Wallner, M.; Sandici, V.; Neubauer, W. 4D investigation of Digital Heritage: An interactive application for the auxiliary fortress of Carnuntum. In Proceedings of the Digital Heritage 2015: The 2015 International Congress on Digital Heritage, Granada, Spain, 28 September–2 October 2015; pp. 81–84.

5.   Yastikli, N. Documentation of cultural heritage using digital photogrammetry and laser scanning. *J. Cult. Herit.* **2007**, *8*, 423–427. [CrossRef]

6.   Lerma, J.L.; Navarro, S.; Cabrelles, M.; Villaverde, V. Terrestrial laser scanning and closerange photogrammetry for 3D archaeological documentation: The Upper Palaeolithic Cave of Parpalló as a case study. *J. Archaeol. Sci.* **2010**, *37*, 499–507. [CrossRef]

7.   Puente, I.; Solla, M.; González-Jorge, H.; Arias, P. NDT documentation and evaluation of the Roman Bridge of Lugo using GPR and mobile and static LiDAR. *J. Perform. Constr. Fac.* **2015**, *29*. [CrossRef]

8.   Capra, A.; Bertacchini, E.; Castagnetti, C.; Rivola, R.; Dubbini, M. Recent approaches in geodesy and geomatics for structures monitoring. *Rendiconti Lincei* **2015**, *26*, 53–61. [CrossRef]

9.   Lindenbergh, R.; Uchanski, L.; Bucksch, A.; Van Gosliga, R. Structural monitoring of tunnels using terrestrial laser scanning. *Rep. Geod.* **2009**, *2*, 231–238.

10.   Puente, I.; Akinci, B.; González-Jorge, H.; Díaz-Vilariño, L.; Arias, P. A semi-automated method for extracting vertical clearance and cross sections in tunnels using mobile LiDAR data. *Tunn. Undergr. Space Technol.* **2016**, *59*, 48–54. [CrossRef]

11.   Riveiro, B.; González-Jorge, H.; Varela, M.; Jauregui, D.V. Validation of terrestrial laser scanning and photogrammetry techniques for the measurement of vertical underclearance and beam geometry in structural inspection of bridges. *Measurement* **2013**, *46*, 784–794. [CrossRef]

12.   Puente, I.; Sande, J.; González-Jorge, H.; Pena-González, E.; Maciñeira, E.; Martínez-Sánchez, J.; Arias, P. Novel image analysis approach to the terrestrial LiDAR monitoring of damage in rubble mound breakwaters. *Ocean Eng.* **2014**, *91*, 273–280. [CrossRef]

13.   Cai, H.; Rasdorf, W. Modeling road centerlines and predicting lengths in 3-D using LIDAR point cloud and planimetric road centerline data. *Comput. Aided Civ. Infrastruct. Eng.* **2008**, *23*, 157–173. [CrossRef]

14.   Corsi, C.; Slapšak, B.; Vermeulen, F. *Good Practice in Archaeological Diagnostics, Non-Invasive Survey of Complex Archaeological Sites*; Springer International Publishing: Cham, Switzerland, 2013.

15.   Goodman, D.; Piro, S. *GPR Remote Sensing in Archaeology, Geotechnologies and the Environment*; Springer: Heidelberg, Germany, 2013; Volume 9.

16.   Grasmueck, M.; Weger, R.; Horstmeyer, H. Full-resolution 3D GPR imaging. *Geophysics* **2005**, *70*, K12–K19. [CrossRef]

17.   Leckebusch, J. Ground-penetrating radar: A modern three-dimensional prospection method. *Archaeol. Prospect.* **2003**, *10*, 213–240. [CrossRef]

18.   Novo, A.; Solla, M.; Fenollós, J.L.M.; Lorenzo, H. Searching for the remains of an Early Bronze Age city at Tell Qubr Abu al-'Atiq (Syria) through archaeological investigations and GPR imaging. *J. Cult. Herit.* **2014**, *15*, 575–579. [CrossRef]

19.   Drahor, M.G.; Berge, M.; Öztürk, C. Integrated geophysical surveys for the subsurface mapping of buried structures under and surrounding of the Agios Voukolos Church in Izmir, Turkey. *J. Archaeol. Sci.* **2011**, *38*, 2231–2242. [CrossRef]

20.   Vermeulen, F.; Corsi, C.; De Dapper, M. Surveying the townscape of Roman *Ammaia* in Portugal: An integrated geoarchaeological investigation of the Forum Area. *Geoarchaeology* **2012**, *27*, 123–139. [CrossRef]

21.   Barone, P.M.; Ferrara, C. Geophysics applied to landscape archaeology: Understanding Samnite and Roman relationships in Molise (Italy) using geoarchaeolgical research methods. *Int. J. Archaeol.* **2015**, *3*, 26–36. [CrossRef]

22.   Leucci, G.; De Giorgi, L.; Di Giacomol, G.; Miccoli, I.; Scardozzi, G. 3D GPR survey for the archaeological characterization of the ancient Messapian necropolis in Lecce, South Italy. *J. Archaeol. Sci. Rep.* **2016**, *7*, 290–302. [CrossRef]

23.   Zhao, W.; Forte, E.; Fontana, F.; Pipan, M.; Tian, G. GPR imaging and characterization of ancient Roman ruins in the Aquileia Archaeological Park, NE Italy. *Measurement* **2018**, *113*, 161–171. [CrossRef]

24. Núñez-Nieto, X.; Solla, M.; Novo, A.; Lorenzo, H. Three-dimensional ground-penetrating radar methodologies for the characterization and volumetric reconstruction of underground tunneling. *Constr. Build. Mater.* **2014**, *71*, 551–560. [CrossRef]

25. Malfitana, D.; Leucci, G.; Fragalà, G.; Masini, N.; Scardozzi, G.; Cacciaguerra, G.; Santagati, C.; Shehi, E. The potential of integrated GPR survey and aerial photographic analysis of historic urban areas: A case study and digital reconstruction of a Late Roman *villa* in Durrës (Albania). *J. Archaeol. Sci. Rep.* **2015**, *4*, 276–284. [CrossRef]

26. Neubauer, W.; Seren, S.; Hinterleitner, A.; Löcker, K.; Melichar, P. Archaeological interpretation of combined magnetic and GPR survey of the roman town Flavia Solva, Austria. *ArcheoSciences* **2009**, *33*, 225–228. [CrossRef]

27. Leopold, M.; Plöckl, T.; Forstenaicher, G.; Völkel, J. Integrating pedological and geophysical methods to enhance the informative value of an archaeological prospection—The example of a Roman villa rustica near Regensburg, Germany. *J. Archaeol. Sci.* **2010**, *37*, 1731–1741. [CrossRef]

28. Papadopoulos, N.G.; Sarris, A.; Salvi, M.C.; Dederix, S.; Soupios, P.; Dikmen, U. Rediscovering the small theatre and amphitheatre of ancient Ierapytna (SE Crete) by integrated geophysical methods. *J. Archaeol. Sci.* **2012**, *39*, 1960–1973. [CrossRef]

29. Garrido, I.; Lagüela, S.; Arias, P.; Balado, J. Thermal-based analysis for the automatic detection and characterization of thermal bridges in buildings. *Energy Build.* **2018**, *158*, 1358–1367. [CrossRef]

30. Sfarra, S.; Marcucci, E.; Ambrosini, D.; Paoletti, D. Infrared exploitation of the architectural heritage: From passive infrared thermography to hybrid infrared thermography approach. *Mater. Constr.* **2016**, *66*, 1–16. [CrossRef]

31. Sfarra, S.; Regi, M. Wavelet analysis applied to thermographic data for the detection of sub-superficial flaws in mosaics. *Eur. Phys. J. Appl. Phys.* **2016**, *74*, 1–11. [CrossRef]

32. Rodríguez Colmenero, A. *Aquae Querquennae, Roman Headquarters and Inn. A Guide to the Archaeological Site*; Dirección Xeral de Turismo: Xunta de Galicia, Spain, 2016.

33. FARO Technologies. *FARO Focus 3D X330*; Product specifications; FARO Technologies, Inc.: Stuttgart, Germany, 2014.

34. Men, H.; Gebre, B.; Pochiraju, K. Color point cloud registration with 4D ICP algorithm. In Proceedings of the 2011 IEEE International Conference on Robotics and Automation (ICRA), Shanghai, China, 9–13 May 2011; pp. 1511–1516.

35. Jost, T.; Hügli, H. Fast ICP Algorithms for Shape Registration. In *Joint Pattern Recognition Symposium*; Springer: Heidelberg/Berlin, Germany, 2002; pp. 91–99.

36. ToGPRi Tools. Available online: https://savannah.nongnu.org/projects/togpri (accessed on 25 May 2017).

37. GNU Octave. Available online: https://www.gnu.org/software/octave/ (accessed on 28 September 2017).

38. Open Source Geospatial Consortium. Available online: http://www.gdal.org/ (accessed on 15 June 2017).

39. Santulli, C. IR thermography for the detection of buried objects: A short review. *Non-Destruct. Test. Online* **2007**, *12*, 12.

40. Lunden, B. Aerial thermography: A remote sensing technique applied to detection of buried archaeological remains at a site in Dalecarlia, Sweden. *Geogr. Annal.* **1985**, *67*, 161–166.

41. Carlomagno, G.; Di Maio, R.; Fedi, M.; Meola, C. Integration of infrared thermography and high-frequency electromagnetic methods in archaeological surveys. *J. Geophys. Eng.* **2011**, *8*, 93–105. [CrossRef]

remote sensing

MDPI

Article

Utilization of Integrated Geophysical Techniques to Delineate the Extraction of Mining Bench of Ornamental Rocks (Marble)

Julián Martínez [1], Violeta Montiel [2], Javier Rey [3,*], Francisco Cañadas [2] and Pedro Vera [2]

[1] Department of Mechanical and Mining Engineering, Higher Polytechnic School of Linares, University of Jaén, 23700 Linares, Spain; jmartine@ujaen.es

[2] Department of Telecommunication Engineering, Higher Polytechnic School of Linares, University of Jaén, 23700 Linares, Spain; vmontiel@ujaen.es (V.M.); fcanadas@ujaen.es (F.C.); pvera@ujaen.es (P.V.)

[3] Department of Geology, Higher Polytechnic School of Linares, University of Jaén, 23700 Linares, Spain

* Correspondence: jrey@ujaen.es

Received: 24 October 2017; Accepted: 8 December 2017; Published: 15 December 2017

Abstract: Low yields in ornamental rock mining remain one of the most important problems in this industry. This fact is usually associated with the presence of anisotropies in the rock, which makes it difficult to extract the blocks. An optimised planning of the exploitation, together with an improved geological understanding of the deposit, could increase these yields. In this work, marble mining in Macael (Spain) was studied to test the capacity of non-destructive geophysical prospecting methods (GPR and ERI) as tools to characterize the geology of the deposit. It is well-known that the ERI method provides a greater penetration depth. By using this technique, it is possible to distinguish the boundaries between the marble and the underlying micaschists, the morphology of the unit to be exploited, and even fracture zones to be identified. Therefore, this technique could be used in the early stages of research, to estimate the reserves of the deposit. The GPR methodology, with a lower penetration depth, is able to offer more detailed information. Specifically, it detects lateral and vertical changes of the facies inside the marble unit, as well as the anisotropies of the rock (fractures or holes). This technique would be suitable for use in a second stage of research. On the one hand, it is very useful for characterization of the texture and fabric of the rock, which allows us to determine in advance its properties, and therefore, the quality for ornamental use. On the other hand, the localization of anisotropy using the GPR technique will make it possible to improve the planning of the rock exploitation in order to increase yields. Both integrated geophysical techniques are effective for assessing the quality of ornamental rock and thus can serve as useful tools in mine planning to improve yields and costs.

Keywords: ground penetrating radar; electrical resistivity imaging; quarry; marble

1. Introduction

Mine planning increasingly requires prior knowledge of the characteristics and structures of the rocks to be mined. The competitive ornamental rock industry, which has an increasingly internationalized market, needs to optimize the costs of extracting commercial-size blocks for subsequent sale and manufacturing in the cutting workshop.

Reducing operating costs in the ornamental rock sector is closely linked to the extraction techniques and to the yield of the extracted block (percentage of useful stone relative to the total extracted volume). The yield from ornamental rock quarries is usually very low (approximately 20%), which is mainly due to the presence of fractures, joints, and other anisotropies that control the extraction of commercial size blocks. The low yields are also affected by the dimensions of the mining benches, which are usually defined in a standardized way without determining the mechanical characteristics of the rock massif.

Our goal was to apply non-destructive research techniques to analyse the geometry of the body to be mined, the presence of fractures, and the rock mass quality, which should be known to develop the optimal extraction plan for maximum profitability. To do this, we selected two geophysical methods: ground penetrating radar (GPR) and electrical resistivity imaging (ERI).

GPR is a geophysical prospecting technique that has been applied in geological studies [1–5]. However, few studies have used GPR to characterize ornamental rock. Sigurdsson [6] performed a GPR study in limestone quarries and was able to differentiate zones with textural variations. Porsani [7] used these techniques in granite quarries in Brazil. Several other studies have used this technique in carbonate quarries to detect fractures [8–10]. Other studies show that GPR techniques provide complementary and high-resolution information for determining the textures and fabrics of sedimentary rocks [11], which offers great potential for delimiting the geological structure to be exploited in a quarry. The critical fractures and quality of an ornamental stone deposit have been examined [12]. In this research, fracture status was evaluated by an in-situ GPR test. The resulting 3D GPR model allowed an exploration of the extension, shape, and orientation of the detected fractures surfaces.

ERI techniques have also been used to support geological studies [13–15]. However, few studies have been performed on lithological characterization and fracturing in ornamental rocks.

There are very few publications in which GPR and ERI are used for lithological characterization, fracture and cavity identification, and rock quality determination of ornamental rock. Recently, the results of using GPR and ERI in three quarries were presented at SUSTAMINING [16] in Verona (Italy). ERI was used in two marble quarries in Turkey and Italy and a granite quarry in Spain.

Because of the few previous investigations, this study analyses the possibility of using GPR and ERI as non-destructive analysis techniques in ornamental rocks (marbles) to identify the presence of fractures and karstification, lithologic contacts, and variations in the massif that will limit the final size of the blocks to be extracted and thus the design of the exploitation bench and extraction systems.

This study was conducted in the region of Macael (Figure 1), which is currently the most important centre of marble production in Spain and the second most important in Europe after Italy. With marble reserves of more than 70 Mm^3, peak production was reached in 2002 with the extraction of 2×10^6 Tm. Fieldwork was conducted in the most important quarry mining area of Macael (the Macael Sur concession) on a marble mining bench with adequate dimensions and geometry for the study.

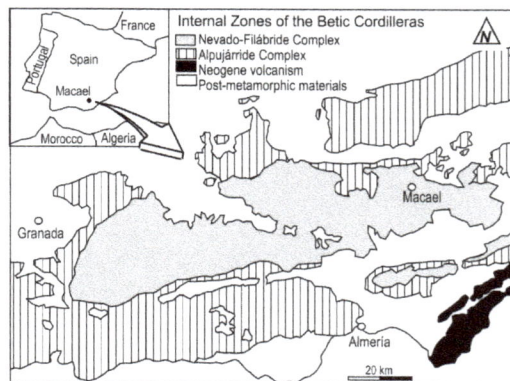

Figure 1. Simplified geological map of the inner zones of the Baetic System showing the location of the study region ([17]).

2. Background to the SITE

2.1. Geological Setting

The rocks that outcrop in the quarry belong to the internal areas of the Baetic Range and specifically to the Nevado-Filabride complex [18] (Figure 1), which consists of three tectonic units: the Nevado-Lubrin, Bedar-Macael, and Almocaizar formations.

The characteristic materials of the Nevado-Filabride Formation are Triassic carbonate rocks, which sometimes alternate with micaschists, calcareous micaschists, quartzite micaschists with garnet, and quartzite micaschists with amphibole [17,19]. The marbles are located in the Bedar-Macael Formation [19] and are composed of layered white, grey, and yellow marbles with thicknesses ranging between 10 and 30 m.

The white marble, which is commercially known as "Macael White", is granoblastic and ranges from equigranular with large grains to heterogranular with medium-fine grains. It also contains small amounts of quartz and isolated crystals of muscovite and feldspar. The marbles generally contain a foliation that is associated with grain size variations, certain orientations of calcite crystals, and the orientations of phyllosilicates.

2.2. Extraction System

Open-pit mining is commonly used in the Macael mining district. Hillside quarries are mined by means of descending benches (Figure 2a). Clearing is performed by drilling and blasting and sometimes creates up to 100 m of rubble, which is removed to a dump by mine trucks. The rubble/ore ratio is 6–7 m^3/Tn.

Once the primary mining bench is excavated and prepared (Figure 2b,c), it is subjected to successive stages of subdivision perpendicular to the front. Two main cutting methods are used. The first is drilling and blasting (using mining powder or detonating cord) combined with mechanical methods, although the use of explosives is increasingly falling out of favour because it damages the rock. The second and most common method is the use of mechanical techniques to cut the marble, such as a diamond wire cutter and rarely a ranging arm shearer. To do this, three perpendicular holes that intersect each other at a point are drilled on the primary bench (Figure 2b), through which a diamond wire cutter is passed to obtain a "secondary" bench. These benches, which have standard dimensions of $5 \times 15 \times 15$ m in Macael, are in turn subdivided into five "tertiary" benches of $3 \times 15 \times 5$ m by cuts with a diamond wire (Figure 2c). In these tertiary benches, the commercial blocks are extracted using loaders or excavators with hydraulic drilling hammers or gunpowder. Overturning the marble bench in these quarries is unusual, although the quarries usually contain a sand bed, so segregated blocks are not damaged by falling (Figure 2c). These cutting and extraction operations are important for the optimization of the quarry because the blocks can be broken due to cracks and fractures. The blocks and rock fragments that are extracted are squared on the quarry floor using a cutting wire for the blocks (Figure 2d) and drills and wedges for the fragments (Figure 2e), before being sent to the cutting factories. Finally, the rejects from the block extraction and squaring are used in the micronized stone industry (Figure 2f).

Figure 2. (**a**) Overview of the quarry; (**b**) Details of the three perpendicular holes that delineate the secondary bank; (**c**) Cutting of tertiary blocks with a diamond wire (detail of the block is outlined). Excavator extracting blocks over the sand bed. Note the breakage along a fracture plane; (**d**) Squaring a block with a diamond wire; (**e**) Squaring a block with a drill and wedges along a crack; (**f**) Excavator with a hydraulic hammer reducing the size of the stone intended for micronized carbonate.

The presence of lithological alternations, fractures, joints, and occasional karstification in the marble level poses a major problem in mining (Figure 3a,b). It is important to identify these planes of weakness to design and plan the appropriate size and cutting orientations of the secondary and tertiary benches. Hence, using the most appropriate and economical techniques for exploitation will reduce losses and considerably lower costs.

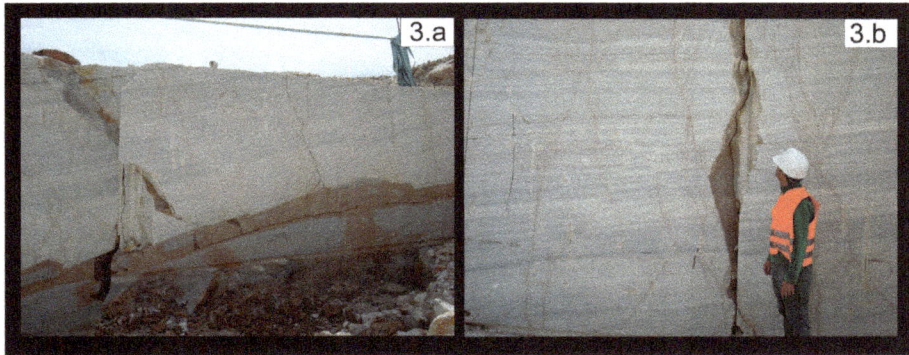

Figure 3. (**a**) Contact of the marble with the micaschist and fracturing; (**b**) Fracture, karstification, and banded marble.

3. Methods

For this GPR and ERI study, we selected a quarry bench that is 48 m long and 5 m high. The bench is composed of marble (Macael white), which is characterized by a granoblastique texture. There are some pelitic intercalations (more or less abundant), with a lepidoblastic texture, which allow the original stratification (f_0) to be identified (Figure 3a,b). On the wall of the carbonates and at both ends of the bench, the underlying micaschists were identified by a dipping brecciated contact that forms a gentle syncline (Figure 3a). Vertical closed fractures dip to the northeast in both the marble and the micaschist. Several undeveloped karstic cavities (Figure 3b) are also located on the face of the bench.

3.1. GPR Data Acquisition and Processing

This study used a Pro-Ex model RAMAC/GPR system manufactured by MALÅ GeoScience (Malå, Sweden). The depth/resolution requirements and the soil conditions determine the choice of antenna frequency. On one side, the depth penetration is controlled by the centre frequency and the conductivity of the material. In low-loss materials, the approximate maximum penetration depth for frequencies of 200–250 MHz is approximately 8–20 m, whereas, when using 100 MHz antenna, it is about 25 m [20,21]. On the other side, resolution depends on the frequency of the antenna. Taking a velocity of 100 mm/ns, by using antennas of 250 and 100 MHz, the wavelength will be 0.4 m and 1 m, Δv will be 0.1 m and 0.25 m, and $\sqrt{r \cdot 0.2}$ m and $\sqrt{r \cdot 0.5}$ m, respectively, where r is the distance between the antenna and the target [22]. In this work, a 250 MHz antenna is selected because it provides the best resolution with a penetration depth of 20 m that satisfies our survey requirements. The test was performed using the following operating parameters: nominal frequency of 250 MHz, time window between 189.8 ns (approximately 10 m) and 359.8 ns (approximately 18 m), distance interval of 0.05 m, sampling frequency of 2.54 GHz, number of stacks of two, and the velocity of 100 mm/ns was estimated during the data processing. Specifically, the velocity was estimated using the popular method of hyperbola fitting [22], which is included in ReflexW software [23], when the data obtained were analysed.

Several profiles were measured in the quarry, which are parallel and perpendicular to the main bench of stone exploited (Figures 4 and 5). The first profiles were obtained at the top of the quarry (Figure 4a,b and Figure 5) in the longitudinal direction (profile P01; 48 m long) and transversal directions (profiles P02–P05; 20 m long). Three profiles (P06–P08) were obtained in the lower part of the quarry in two directions (40 m long for P07 and P08 and 48 m long for P06). The last profile (P09) was obtained directly at the face of the quarry (36 m long due to difficulties in measuring the entire face).

Figure 4. (**a**) GPR profile performed on the marble bench; (**b**) Detail from the 250 MHz antenna on the bank; (**c**) Position of the ERI profile on the primary bank; (**d**) Detail of a copper electrode and clay.

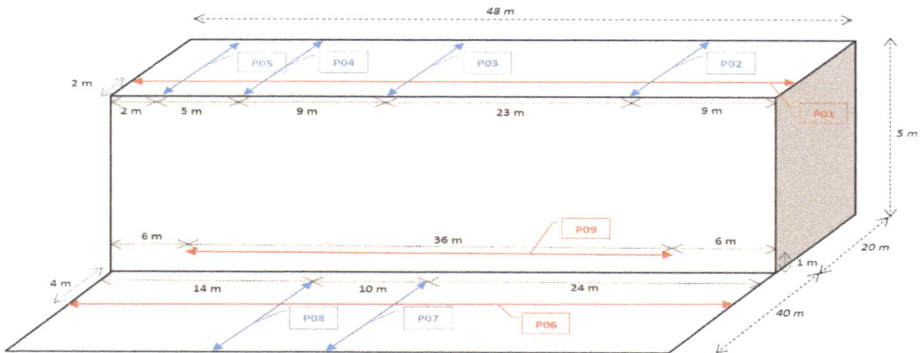

Figure 5. Schematic showing the locations of the GPR profiles.

Figure 6a shows details of the structure of the quarry. The marble can be differentiated from the micaschist in both the first face of the quarry (approximately 5 m high) and in the second face (3 m deep). The measurements from profile P09 were collected at the first face of the quarry (P09), where two people are shown in the figure.

A set of standard pre-processing techniques [24], using software developed with Matlab, was applied to the radargrams. First of all, DC removal subtracts the mean from the centre value of a window of 40 ns and 75 ns (smallest and largest profile, respectively). Secondly, amplitude correction is applied using trace equalization, where every trace is normalized according to the maximum absolute value of the amplitudes in a window of 160 ns and 300 ns (smallest and largest profile, respectively). Then, band-pass filtering, with 125 MHz and 325 MHz as the lower and upper cut-off frequency points, removes the unwanted frequencies. Finally, automatic trace subtraction, as the noise background removal method [25], removes the mean of several traces within a window of 22 ns. As a result, the effects of clutter (energy that is unrelated to the real target) are reduced [26] and the signal-to-noise (SNR) ratio is enhanced.

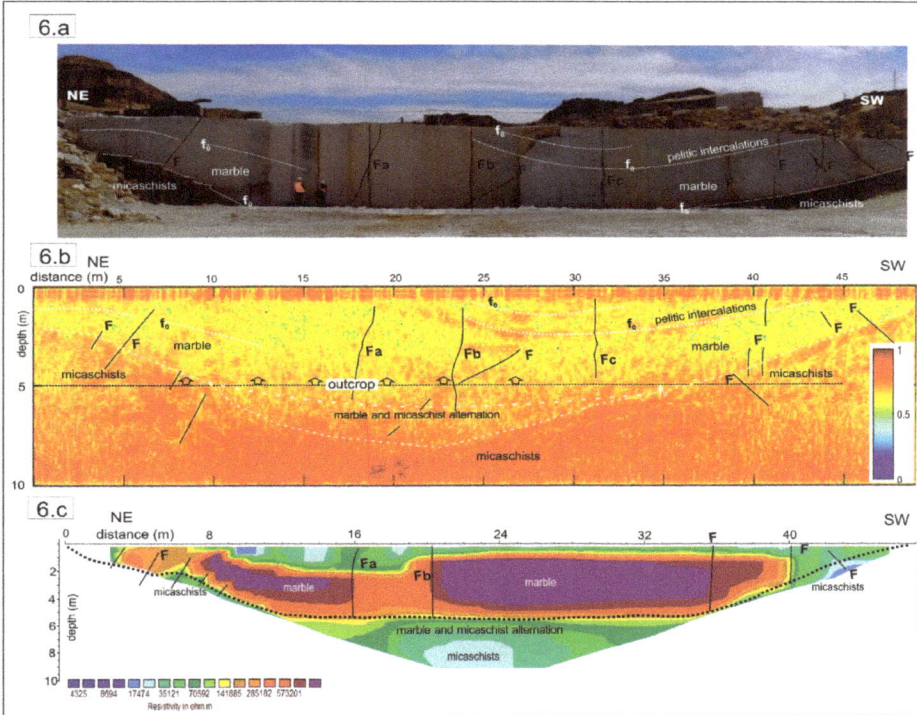

Figure 6. Overview of the front of the marble mining bank. Stratification(f_0) (defined by the presence of pelitic intercalations) and observed fractures (F) are outlined on the outcrop. Some of these fractures are correlated with Figure 6a,b and Figure 7a–c. (**a**) Radargram of the bank corresponding to profile P01. The amplitude scale is plotted in a logarithmic scale (**b**) Interpretation of the ERI profile of real resistivity sections.

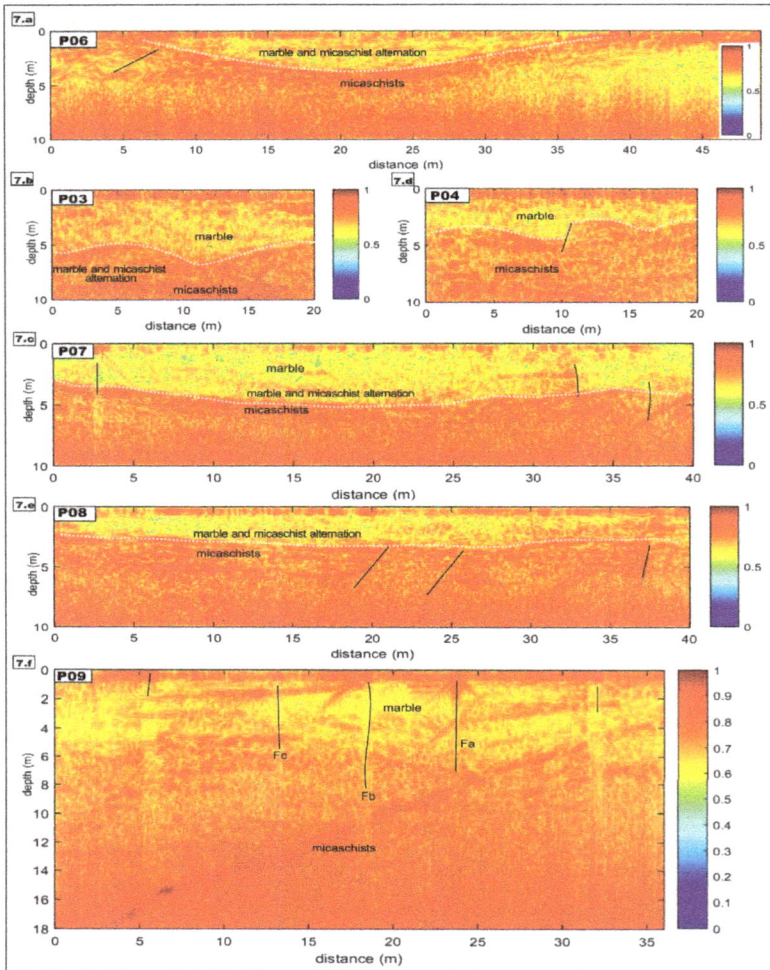

Figure 7. Pre-processed radargrams collected along different faces of the quarry (see Figure 4a). The intersection between the different profiles and the fractures is outlined (Figure 4). P06 (**a**); P03 (**b**); P07 (**c**); P04 (**d**); P08 (**e**); P09 (**f**). The amplitude scale is plotted in a logarithmic scale.

3.2. ERI Data Acquisition and Processing

ERI can be used to determine the distribution of a characteristic physical parameter in the subsurface based on a very large number of measurements collected at the ground surface. We measured the electrical resistivity of certain materials, which is the resistance that is encountered by an electric current when passing through a material [27,28]. The variations in the geoelectric behaviour allow 2D profiles to be constructed, which makes ERI one of the most effective non-destructive tools for studying and characterizing discontinuities in the subsurface [28,29].

The ERI method involves implanting numerous electrodes along profiles (Figure 4c). The electrode spacing is dependent on the required resolution, depth, and objectives; the shorter the distance between the electrodes, the greater the resolution, whereas the greater the distance between the electrodes, the greater the depth at which readings can be taken [29]. In this study, we performed one electrical resistivity imaging profile using a Wenner-Schlumberger array. This array remains stable and behaves well in response to variations in both the vertical and horizontal resistivities, which makes it useful when investigating horizontal layers that may contain lateral facies changes and/or vertical structures [30]. We used the RESECS electrical tomography equipment made by Deutsche Montan Technologie (DMT). The electrodes were connected to measurement equipment, and we used a specific sequential programme for each objective, which defines the electrodes that should be used at any given moment and their layout. We measured the apparent resistivity at each electrode position, which was then attributed to a particular geometric point in the subsurface. We avoided drilling holes in the marble bench to insert the electrodes. Copper electrodes were composed of a 5 × 5 cm plate and a 2 cm rod where the electrical connection was made and were used for the measurements (Figure 4d). After cleaning the surface of the bench, the electrodes were placed on a base of wet clay with brine to obtain better current transmission to the ground. Because the marble behaves like a large resistor, it was necessary to use the maximum current that the instrument could deliver (up to 800 V peak-to-peak).

The profile on the bench parallel to the front was 48 m long and used an electrode spacing of 1 m (Figure 4a–c, Figures 5 and 6a). This electrode spacing provides sufficient penetration without losing information about the structure of the rock. A total of 463 measurements were taken, which reached a depth of 9.5 m.

The electrical tomography profiles were interpreted based on the apparent resistivity values that were obtained during the fieldwork, which were processed using the RES2DINV software for the resistivity and induced polarization [31]. This programme is based on the least squares method with smoothing constraints modified by the Quasi-Newton optimization method. The inversion method constructs a model of the subsurface using rectangular prisms and calculates the resistivity values for each while minimizing the differences between the observed and calculated apparent resistivities [32,33]. The apparent resistivity data were refined before the modelling to remove the extrema (bad data points); 61 points were removed. Because of the wide variation in the measured resistivity values and the abrupt variations between neighbouring points, robust inversion and the combined Marquardt and Occam inversion method were used, which appear to provide better results in resolving compact structures.

4. Results and Discussion

Figure 6b shows the radargram for profile P01. The effectiveness of the GPR technique is demonstrated by the correlation of the geological data (observed in the field measurements in Figure 6a) with the GPR results. Two units are identified: an upper unit that corresponds to the marble (beginning at a depth of 2 m, deepening to approximately 5 m in the middle of the profile, and rising again at the end of the quarry face) and a lower unit that is related to the micaschists. The lower unit is composed of a marble and micaschist alternation (upper part) and micaschist. This smooth variation can be observed in Figure 6b,c.

The presence of mica-rich layers (lepidoblastic texture) in marbles is shown in the radargrams. Detection of these layers is relevant from two points of view: firstly, they can identify anisotropy layers and therefore can offer information related to their weakness and breaking of the rock. On the other hand, the presence of this banding in ornamental rock can lower its market value.

The radargram demonstrates the suitability of GPR for detecting discontinuities, including stratification (f_0 in Figure 6a,b), fractures, and cavities, such as at horizontal distances from 10 to 25 m and a depth of 3 m. This information is relevant for cutting blocks by adjusting their size based on the encountered fractures.

Figure 7 shows other radargrams. Figure 7a shows profile P06, which contains a noisy area from 5 m to the end of the profile. Figure 7b,c show profiles P03 and P07, respectively, and Figure 7d,e show profiles P04 and P08, respectively. The upper profiles show that the micaschist is at depths of 6–8 m, whereas the lower profiles show depths of 2–4 m. These results demonstrate that the resulting information is consistent. Furthermore, the radargrams from the lower part of the quarry (Figure 7a,c,e) show areas with higher energy than the other profiles, so the upper part of the quarry can be expected to have better quality rock. Several fractures can also be identified in these radargrams. Finally, Figure 7f shows profile P09. The first part of the radargram (0–6 m) is interpreted to have higher quality rock, and several fractures are located at distances of 18 m, 25 m, and 32 m along the profile. We cannot identify the micaschists because the profile was measured in the middle part of the face (from 6 m to 42 m; Figures 5 and 6a).

Besides, GPR can be considered as a promising tool to distinguish the type of fracture because, in addition to detecting fractures, it can provide information on their spatial orientation. As previously mentioned, Figure 7f shows this utility in which three hyperbolas correspond to Fa, Fb, and Fc vertical fractures. However, pelitic intercalations identified, which can be seen as quasi-horizontal fractures, provide different GPR responses (horizontal pattern in the radargram). Then, GPR could be used as a fracture classification technique.

The ERI profile reaches a depth of 9.5 m (Figure 6c), as expected with the array configuration used: the Wenner-Schlumberger array in which the penetration depth usually results in a $1/5$ of the total length [34]. Therefore, longer profiles could provide a greater penetration depth in these materials, being crucial in the preliminary stage of the research in order to identify the presence of layers to be exploited.

Figure 6c shows a profile where three zones can be distinguished from the electrical responses. An upper zone with low resistivities runs along the bench between 9 and 15 m from the origin of the profile and reaches up to 2 m deep, which we interpret as a marble decompression zone. An intermediate zone with very high resistivities that correspond to the marbles that are the target for extraction is identified below the upper zone; it reaches a depth of 5 m in the central part of the profile. Finally, a lower zone with lower resistivities is well defined due to the contacts at the bottom and laterally with the underlying micaschists (Figure 6a,c).

Although the marble in the intermediate zone is fairly homogeneous along the bench, several discontinuities can be identified. Low resistivity zones are observed within the marble from the origin to 7 m, between 16 and 21 m, and from 40 to 47 m, which are associated with the most highly fractured zones. We consider these areas to be zones of weakness with a high risk of breakage during cutting and removal of the blocks. Accordingly, the areas on the bench between 7 and 16 m and between 21 and 40 m appear to be intact and would allow for further optimization of cutting and removing the blocks from the extraction bench.

By means of ERI, the boundaries between the two units of the rock (marble and micaschists) are clearly identified. In addition, the increase of the profile length could allow a greater penetration depth. However, textural changes could not be detected in the marble, only more or less altered areas. Therefore, ERI is a suitable tool that can be used as a preliminary stage of the research, which requires an overall view of the deposit. In a second stage of the research, GPR could offer more detailed information on the anisotropies of the rock to be exploited.

5. Conclusions

This study showed that GPR and ERI provide additional information for determining the quality of ornamental rocks "in situ" in quarries in terms of their lithology and the presence of fractures and contacts. The techniques provide consistent results.

ERI provides more general and basic information about the lithological unit, so it should be applied in the first stages of the quarry research. In this example, the electrical responses allowed us to identify three levels in the extraction bench and to identify the areas within the marble with the most fractures. Thus, this technique can be used at a preliminary stage for the estimation of the reserves of the deposit. However, it does not provide information on the texture or fabric of the rock, so it does not give us much information on the quality of the rock to be extracted (commercial value or market price). A large number of small fractures, but capable of breaking a block at the time of extraction,

Remote Sens. **2017**, *9*, 1322

can be completely undetected using this method. Therefore, the methodology does not offer much potential for planning the exploitation in order to increase yields.

GPR was used to obtain images of the quarry, in which the marble units can be clearly identified. Unlike ERI, the results demonstrated the usefulness of this technique for evaluating the quality of the marble (it allows one to define the stratification and even the fabric and the texture of the rock) and the structural anomalies (fractures, holes, . . .). Thus, GPR can be used as a tool to characterize the unit to mine in much more detail, as well its commercial value.

In ornamental rock mining, where operations are systematic and repetitive and the mechanical work is optimized using a standardized cutting routine, it is necessary to have a good knowledge of the extraction bench to schedule the cutting sequences. Thus, zones could be delineated that allow commercial-size blocks to be cut with techniques such as diamond wire cutting or the use of ranging arm shearers. In addition, zones with fractures that will generate small blocks and/or rejects can be identified so they can be removed by light blasting with mining powder, which is much more economical. The greater complexity of the different extraction plans for each area of the quarry would be offset by the improved yields.

Therefore, the combined use of tools such as GPR and ERI can help in the a priori identification of fractures and planes of weakness in ornamental rock mines, which will lead to efficient mine planning.

Acknowledgments: This study was performed using funds provided by the Government of Andalusia (Project TIC-7278).

Author Contributions: The fieldwork of ERI and the processing of the information was carried out by Julián Martínez. The fieldwork of GPR and signal processing was carried out by Violeta Montiel and Pedro Vera. Francisco Cañadas and Javier Rey participated in the interpretation of the results and writing of the manuscript.

Conflicts of Interest: The authors declare no conflict of interest.

References

1. Gómez-Ortiz, D.; Martín-Crespo, T.; Martín-Velázquez, S.; Martínez-Pagán, P.; Higueras, H.; Manzano, M. Application of ground penetrating radar (GPR) to delineate clay layers in wetlands. A case study in the Soto Grande and Soto Chico watercourses, Doñana (SW Spain). *J. Appl. Geophys.* **2010**, *72*, 107–113. [CrossRef]
2. Rey, J.; Martínez, J.; Hidalgo, C. Investigating fluvial features with electrical resistivity imaging and ground-penetrating radar: The Guadalquivir River terrace (Jaén, Southern Spain). *Sediment. Geol.* **2013**, *295*, 27–37. [CrossRef]
3. Pueyo-Anchuela, O.; Luzón, A.; Gil Garbi, H.; Pérez, A.; Pocoví, A.; Soriano, M.A. Combination of electromagnetic, geophysical methods and sedimentological studies for the development of 3D models in alluvial sediments affected by karst (Ebro Basin, NE Spain). *J. Appl. Geophys.* **2014**, *102*, 85–91. [CrossRef]
4. Comas, X.; Terry, N.; Slater, L.; Warren, M.; Kolka, R.; Kristiyono, A.; Sudiana, N.; Nurjaman, D.; Darusman, T. Imaging tropical peatlands in Indonesia using ground-penetrating radar (GPR) and electrical resistivity imaging (ERI): Implications for carbon stock estimates and peat soil characterization. *Biogeosciences* **2015**, *12*, 2995–3007. [CrossRef]
5. Steelman, C.M.; Kennedy, C.S.; Parker, B.L. Geophysical conceptualization of a fractured sedimentary bedrock riverbed using ground-penetrating radar and induced electrical conductivity. *J. Hydrol.* **2015**, *521*, 433–446. [CrossRef]
6. Sigurdsson, T.; Overgaard, T. Application of GPR for 3-D visualization of geological and structural variation in a limestone formation. *J. Appl. Geophys.* **1998**, *40*, 29–36. [CrossRef]
7. Porsani, J.L.; Sauck, W.A.; Junior, A.O.S. GPR for mapping fractures and as a guide for the extraction of ornamental granite from a quarry: A case study from southern Brazil. *J. Appl. Geophys.* **2006**, *58*, 177–187. [CrossRef]
8. Grandjean, G.; Gourry, J.C. GPR data processing for 3D fracture mapping in a marble quarry (Thassos, Greece). *J. Appl. Geophys.* **1996**, *36*, 19–30. [CrossRef]
9. Kadioglu, S. Photographing layer thicknesses and discontinuities in a marble quarry with 3D GPR visualization. *J. Appl. Geophys.* **2008**, *64*, 109–114. [CrossRef]
10. Rey, J.; Martínez, J.; Vera, P.; Ruiz, N.; Cañadas, F.; Montiel, V. Ground-penetrating radar method used for the characterisation of ornamental stone quarries. *Constr. Build. Mater.* **2015**, *77*, 439–447. [CrossRef]

11. Rey, J.; Martínez, J.; Montiel-Zafra, V.; Cañadas, F.; Ruiz, N. Characterization of the sedimentary fabrics in ornamental rocks by using GPR. *Near Surf. Geophys.* **2017**, *15*, 1–9.

12. Elkarmoty, M.; Colla, C.; Gabrielli, E.; Bondu, S.; Bruno, R. A Combination of GPR Survey and Laboratory Rock Tests for Evaluating an Ornamental Stone Deposit in a Quarry Bench. *Procedia Eng.* **2017**, *191*, 999–2017. [CrossRef]

13. Maillet, G.M.; Rizzo, E.; Revil, A.; Vella, C. High resolution electrical resistivity tomography (ERT) in a transition zone environment: Application for detailed internal architecture and infilling processes study of a Rhône River paleo-channel. *Mar. Geophys. Res.* **2005**, *26*, 317–328. [CrossRef]

14. Martínez, J.; Benavente, J.; García-Aróstegui, J.L.; Hidalgo, M.C.; Rey, J. Contribution of electrical resistivity tomography to the study of detrital aquifers affected by seawater intrusión-extrusion effects: The river Vélez delta (Vélez- Málaga, southern Spain). *Eng. Geol.* **2009**, *108*, 161–168. [CrossRef]

15. Rey, J.; Martínez, J.; Hidalgo, C.; Rojas, D. Heavy metal pollution in the Quaternary Garza basin: A multidisciplinary study of the environmental risks posed by mining (Linares, southern Spain). *Catena* **2013**, *110*, 234–242. [CrossRef]

16. SUSTAMINING. *Selective and Sustainable Exploitation of Ornamental Stones Based on Demand*; SME-2012-2-314926; European Commission: Brussels, Belgium, 2015.

17. Torres-Ruiz, J.; Pesquera, A.; Sánchez-Vizcaíno, V. Chromian tourmaline and associated Cr-bearing minerals from the Nevado-Filábride Complex (Betic Cordilleras, SE Spain). *Miner. Mag.* **2003**, *67*, 517–533. [CrossRef]

18. IGME. *Mapa Geológico de España. Escala 1/50.000. Hoja nº 1013 (Macael)*; Servicio de Publicaciones Instituto Geológico y Minero de España: Madrid, Spain, 1975.

19. López Sánchez-Vizcaino, V. Evolución Petrológica y Geoquímica de las Rocas Carbonáticas en el Área de Macael-Cóbdar, Complejo Nevado-Filábride, SE España. Ph.D. Thesis, University of Granada, Granada, Spain, 1994.

20. Pro Ex Manual, Mala. 2017. Available online: http://www.guidelinegeo.com/ (accessed on 12 April 2017).

21. Bristow, C.S.; Jol, H.M. *Ground Penetrating Radar in Sediments*; Geological Society of London: London, UK, 2003; Volume 211.

22. Jol, H.M. *Ground Penetrating Radar Theory and Applications*, 1st ed.; Elsevier Science: Atlanta, GA, USA, 2008.

23. ReflexW Software, Sandmeier Geophysical Research. 2017. Available online: http://www.sandmeier-geo.de/reflexw.html (accessed on 30 November 2017).

24. Szymczyk, M.; Szymczyk, P. Preprocessing of GPR data. *Image Process. Commun.* **2014**, *18*, 83–90. [CrossRef]

25. Solimene, R.; Cuccaro, A.; DellAversano, A.; Catapano, I.; Soldovieri, F. Ground Clutter Removal in GPR Surveys. *IEEE J. Sel. Top. Appl. Earth Obs. Remote Sens.* **2014**, *7*, 792–798. [CrossRef]

26. Daniels, D.J. *Ground Penetrating Radar*, 2nd ed.; The Institution of Electrical Engineers: London, UK, 2004.

27. Telford, W.M.; Geldart, L.P.; Sheriff, R.E. *Applied Geophysics*; Cambridge University Press: Cambridge, UK, 1990.

28. Store, H.; Storz, W.; Jacobs, F. Electrical resistivity tomography to investigate geological structures of earth's upper crust. *Geophys. Prospect.* **2000**, *48*, 455–471. [CrossRef]

29. Sasaki, Y. Resolution of resistivity tomography inferred from numerical simulation. *Geophys. Prospect.* **1992**, *40*, 453–464. [CrossRef]

30. Dahlin, T.; Zhou, B. A numerical comparison of 2D resistivity imaging with 10 electrode arrays. *Geophys. Prospect.* **2004**, *52*, 379–398. [CrossRef]

31. Griffiths, D.H.; Barker, R.D. Two-dimensional resistivity imaging and modelling in areas of complex geology. *J. Appl. Geophys.* **1993**, *29*, 211–226. [CrossRef]

32. Loke, M.H.; Barker, R.D. Rapid least-squares inversion of apparent resistivity pseudosections by a quasi-Newton method. *Geophys. Prospect.* **1996**, *44*, 131–152. [CrossRef]

33. Loke, M.H.; Dahlin, T. A comparison of the Gauss-Newton and quasi-Newton methods in resistivity imaging inversion. *J. Appl. Geophys.* **2002**, *49*, 149–162. [CrossRef]

34. Pazdirek, O.; Blaha, V. Examples of resistivity imaging using ME-I00 resistivity field acquisition system. In Proceedings of the EAGE 58th Conference and Technical Exhibition Extended Abstracts, Amsterdam, The Netherlands, 3–7 June 1996.

remote sensing

MDPI

Article

Geomorphological Dating of Pleistocene Conglomerates in Central Slovenia Based on Spatial Analyses of Dolines Using LiDAR and Ground Penetrating Radar

Teja Čeru [1,*], Ela Šegina [2] and Andrej Gosar [1,3]

[1] Faculty of Natural Sciences and Engineering, University of Ljubljana, Aškerčeva 12, 1000 Ljubljana, Slovenia; andrej.gosar@gov.si
[2] Independent Researcher, Miklošičeva 4a, 1230 Domžale, Slovenia; ela.segina@gmail.com
[3] Slovenian Environment Agency, Seismology and Geology Office, Vojkova 1b, 1000 Ljubljana, Slovenia
* Correspondence: teja.ceru@ntf.uni-lj.si; Tel.: +386-40-752-084

Received: 23 October 2017; Accepted: 17 November 2017; Published: 24 November 2017

Abstract: On Kranjsko polje in central Slovenia, carbonate conglomerates have been dated to several Pleistocene glacial phases by relative dating based on the morphostratigrafic mapping and borehole data, and by paleomagnetic and [10]Be analyses. To define how the age of conglomerates determines the geomorphological characteristics of karst surface features, morphometrical and distributive spatial analyses of dolines were performed on three test sites including old, middle, and young Pleistocene conglomerates. As dolines on conglomerates are covered by a thick soil cover and show a strong human influence, the ground penetrating radar (GPR) method was first applied to select dolines appropriate for further morphometrical and distributive analyses. A considerable modification of natural morphology was revealed for cultivated dolines, excluding this type of depression from spatial analyses. Input parameters for spatial analyses (doline rim and deepest point) were manually extracted from the 1 × 1 m grid digital elevation model (DEM) originating from the high-resolution LiDAR (Light Detection and Ranging) data. Basic geomorphological characteristics, namely circularity index, planar size, depth, and density index of dolines were calculated for each relative age of conglomerates, and common characteristics were determined from these data to establish a general surface typology for a particular conglomerate. The obtained surface typologies were spatially extrapolated to the wider conglomerate area in central Slovenia to test the existent geological dating. Spatial analyses generally confirmed previous dating, while in four areas the geomorphological characteristics of dolines did not correspond to the existing dating and require further revision and modification. Doline populations exhibit specific and common morphometrical and distributive characteristics on conglomerates of a particular age and can be a reliable and fast indicator for their dating.

Keywords: doline; karst; land cultivation; morphometrical analysis; distributive analysis; conglomerate; LiDAR; digital elevation model (DEM); ground penetrating radar (GPR); Kranjsko polje

1. Introduction

Consolidated and unconsolidated clastic deposits in the Ljubljana basin (central Slovenia) have been related to Quaternary alternation of glacial and interglacial periods by several relative and absolute dating methods. The specific morphological characteristics of a surface linked to these different ages of deposits have already been noticed and briefly described by several authors [1–3]. Šifrer [1] had noticed that conglomerates are karstified to different degrees depending on their age.

He observed that dolines lost their typical forms and are elongated and shallower on the old Günz conglomerate compared to Mindel conglomerate, where dolines are circular and deep [1]. He also noticed that the youngest Riss conglomerate is almost unaffected by karstification and that the surface is only slightly undulating.

The linkage between the morphometrical properties of doline populations and a particular age of carbonate conglomerate terraces was noticed for late-Miocene conglomerate in Italy [4]. However, the morphometrical and distributive characteristics of a particular relative age of a conglomerate have not been quantitatively observed and typified yet. Karst surface morphology, in particular doline population properties, also has not been applied as a quantified tool to date carbonate conglomerate terraces.

New methodologies for capturing relief data are more available nowadays and offer new extensions to modern spatial analyses. The increasingly available high-resolution LiDAR (Light Detection and Ranging) method for obtaining elevation data and creating digital elevation models (DEMs) is still in early stages of ability and reliability testing compared to traditional topographic and aerial surveying methods in modern karst geomorphology [5]. Researches applying high-resolution DEMs are mainly focused on automation of dolines detection and delineation [5–8], rather than on interpretations of the acquired geomorphological data [4,9].

In order to extract relevant spatial information from high resolution input data, a careful selection of reliable test samples is required. Dolines have traditionally been subject to intensive anthropogenic reshaping, thus the human impact should be taken into consideration when selecting dolines for further spatial analyses. People have used dolines for agricultural land purposes and have transformed them in order to gain the surface suitable for cultivation [10]. Nowadays, many dolines are no longer used for traditional activities, such as farming, gardening, pasturing, and water supply. They are abandoned or completely filled with various kinds of unknown material [11]. The morphological difference between uncultivated and cultivated dolines can be indicated already from a visual examination of high-resolution digital elevation models (DEMs). Cultivated dolines in the study area seem to be shallower and leveled compared to those that were uncultivated. As the quality of morphometrical and distributive analyses essentially depends on the relevance of the input data, ground penetrating radar (GPR) was employed to define the representative doline sample.

Dolines and their morphometrical and distributive properties have been mostly studied on compact carbonate rocks (limestones, dolomites). Ferrarese and Sauro [4] studied the conglomerate karst of Montello in Italy, which they denote as "the classical karst of the conglomerate rock" because of the similarities to the classical karst on limestones in Slovenia (Kras). Due to the particularity of karstification in carbonate conglomerates [4,12] the results of this study could not be directly transmitted to karst surfaces on carbonate rocks such as limestones and dolomites.

Spatial relief characteristics were quantified using high-resolution LiDAR data, and the typification of karst surfaces on conglomerates of different relative ages was generated. Types of karst surfaces were specified by morphometrical and distributive characteristics of dolines. They served as a tool for dating wider conglomerate terraces in the area.

2. Geological Settings of Study Area and the Age of Quaternary Deposits

The Ljubljana basin was formed along the Sava fault and filled with glacial, glaciofluvial, and fluvial sediments deposited in glacial periods in Quaternary [1,2]. Quaternary deposits overlie impermeable, poorly lithified Oligocene mudstone ("sivica"), which covers underlying Tertiary limestones (Figure 1). Pebbles that constitute the conglomerates are mostly carbonate, and the cement mostly consists of calcite [2,13,14]. The non-carbonate parts (around 10%) are quartz pebbles [13] and pebbles of mainly sedimentary rocks [3]. Uneven surfaces of the cave walls, where the pebbles protrude out, show that the carbonate cement dissolves faster than pebbles in conglomerate [12].

Karst and contact-karst surface features such as dolines, shafts, blind and pocket valleys have developed in conglomerates of the wider area. Several small caves have been found in conglomerates of all ages. Four general types of eogenetic caves were recognized on the older conglomerate terrace Udin Boršt: linear stream caves, shelter caves, breakdown caves, and vadose shafts [15,16]. Most of them are narrow horizontal passages that developed at the contact between the permeable carbonate conglomerates and the impermeable Oligocene mudstone "sivica" [12].

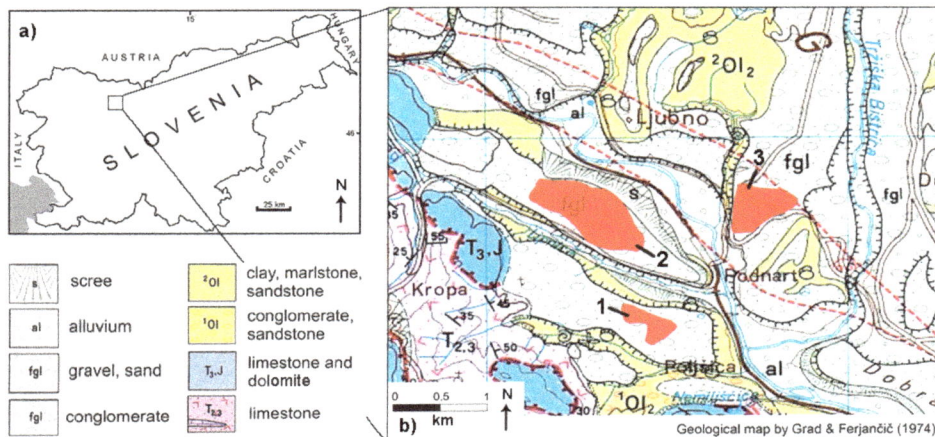

Figure 1. (**a**) The location of the study area; (**b**) A geological map [17] of the study area with three test sites (1: Poljšica; 2: Dobrava; 3: Podbrezje).

The first relative dating techniques were based on the comparison of the relative elevation of the surfaces, the induration of conglomerates (the degree of the hardening and cementation), as well as the degree of surface degradation in terms of erosion, karstification, and soil thickness [1,2,18,19]. Four major morphostratigraphic units were distinguished in the Ljubljana Basin (Table 1): the Older conglomerate fill of Günz age, the Middle conglomerate fill of Mindel age, the Younger conglomerate fill of Riss age, and Gravel fill of Würm age [1,2,19,20]. Würm gravel fill is related to the latest maximum cold period, based on the palynological analysis of a lacustrine sediments core [21].

Absolute age estimations of conglomerates were done by cosmogenic-nuclide burial dating using ^{10}Be (cosmogenic radionuclides dating) and paleomagnetic methods [22–24]. Absolute dating (cosmogenic-nuclide burial dating using ^{26}Al and ^{10}Be) gave a burial age of 1.86 ± 0.19 Ma for the oldest conglomerate terrace Udin Boršt [25]. The chronological and relative ages of gravel and conglomerate fills are quite well accepted, starting from Penck and Brückner [26], while absolute dating is still ambiguous.

All three conglomerates are covered with a thick soil layer. The thickness of soil varies from 1 m to more than 8 m, depending on the age of the conglomerate. However, soil thickness differs between areas of each individual conglomerate fill due to multiple periods of gravel deposition, or due to the initial differences of the deposits [27]. Soil classification range from Mollisols on the youngest Würm conglomerate, to Alfisols on the Riss, and Ultisols on the Mindel and Günz [22,27]. Soil thickness, the thickness of Bt horizons, the amount and continuity of clay coatings, and the amount of Fe and Mn concretions increase with soil age [27]. The variability of soil properties is generally higher within subareas than between areas of the individual conglomerate fill, except for soil thickness.

Study sites of three different conglomerates, named young, middle, and old in this study, correspond to the Younger, Middle, and Older conglomerate infills named by Žlebnik. Visual analyses of the Ljubljana basin conglomerate areas indicated a high uniformity of doline properties on particular ages of conglomerates. Locations of study sites (Podbrezje, Dobrava, Poljšica) were selected on the basis of geological maps done by Žlebnik [2] and Pavich and Vidic [22] on locations with accordant dating of both studies that are located close together for easier visual representation on the map. The borders of study sites were defined so as to include only forested dolines.

Table 1. The relative and absolute dating methods and inferred ages for the three study sites. The names of glaciations are used as morphostratigraphic and not chronostratigraphic terms.

Locations of Test Sites		Relative Dating [2]	Absolute Dating [22,24]	
		Related Glaciations	Estimated Age (ka)	Uncertainty Intervals (ka)
Gravel	/	Würm	Würm I, II, III (62, 44, 32)	(50–70, 40–50, 20–35)
Young conglomerate	Podbrezje	Riss	Riss, 450	435–515
Middle conglomerate	Dobrava	Mindel	Mindel I, II (960, 980)	(780–1000, >780)
Old conglomerate	Poljšica	Günz	Günz 1800	>1000

3. Methods

3.1. LiDAR and Morphometrical Analyses

New methods for automated doline detection and delineation as well as numerous tests of their reliability are lately subject to considerable expansion [5,7,8,28,29]. Additionally, more and more precise input data (LiDAR) for such researches has recently been made available, but issues such as understanding the concept of surface karstification and the pitfalls of new methods should be considered when trying to obtain results of the same quality as those expected from high-resolution data.

Šušteršič [30] developed The Pure Karst Model with which he recognized the fundamental processes of karstification having a vertical direction of outflow and resulting in basic elements of karst surface, which are centrically organized depressions and intermediate elevations. The configuration of such a surface does not follow the principles of fluvial morphology that are based on surface runoff and result in a connected drainage system consisting of valleys and intermediate ridges. This should be considered at data interpolation (some methods favor the hydrologically correct DEMs creating a connected drainage structure—for example the interpolation tool "Topo to raster" by ArcGIS) and when detecting and delineating karst depressions (some methods detect hydrologically closed depressions that are based on surface runoff). Due to high spatial heterogeneity of rock properties and vertical drainage, the karst surface is vertically irregular and therefore depressions often do not have a hydrologically closed upper rim. Such characteristics of the karst surface prevent researchers from employing tools for automated doline detection and delineation that are based on fluvial laws of surface runoff. Obu and Podobnikar [29] presented the low precision of the automated recognition of karst depressions obtained from DEMs (Digital Elevation Models). They employed a 12.5 m resolution DEM, but did not even detect depressions with a 40 m diameter. Following some recommendations, the point density of LiDAR data required for the detection and delineation of dolines is established to be at least 5–12 pt/m^2 [31].

The importance and the problem of the doline upper rim definition have already been discussed by several authors [32–34]. Despite this, in most studies the doline upper rim is simply treated as the uppermost closed contour [5,35]. Even though it is the most reasonable definition of the doline perimeter, the principle of "an abrupt change in the surface slope" [33] has seldom been used in practice due to its fieldwork requirements. Newly available high-resolution LiDAR data allows manual delineation of dolines using the same method with a less time-consuming procedure.

In this research, the criteria for detecting a doline was "a depression of any size and depth recognizable on the 1 × 1 m grid DEM". Doline rims were, with some prior field examination, determined visually and digitized manually from the shaded relief with the principle of "an abrupt change in the surface slope" [32]. Such a highly subjective method of doline delineation was applied as it was established to be more reliable than automated methods [36]. In the case of very shallow depressions, detection of dolines on DEM proved more reliable than field detection. Due to the inexpressive rims of those dolines, field determination of the perimeter was more problematic than remote determination. Field and remote data were more accordant in the case of deep depressions. On the basis of these findings, all the input data (presence, location, and shape of dolines) and the attributes of dolines (circularity, planar size, depth) that were employed in the following spatial analyses, were acquired only from the 1 × 1 m grid DEM generated from the high-resolution LiDAR data, by the tool "Point to raster" by ArcGIS, which directly assigns the point value to the raster pixel. The highly circular nature of the depressions comprised within the given dataset prompted the use of a circularity index that described those features most accurately [36]. The density index was calculated to include the size of dolines, since dolines treated as points can give misleading results. When big dolines are close to each other, the points representing their centers are as far apart as their radiuses (Figure 2d). In this case, the distance between the centers of depressions 1 and 2 is the same as the distance between the centers of depressions 2 and 3, while the distance between the depressions as entities is different. Thus the density of a depression as the ratio between the area of the depression and the area of the belonging Voronoi polygon was calculated (i.e., density index). Morphometrical and distributive analyses (Table 2) were carried out on a total of 279 depressions.

Spatial analyses were performed using a 1 × 1 m grid DEM generated from LiDAR data [37]. The point density of LiDAR data for the study area was 5 pt/m^2 with up to 30 cm horizontal and up to 15 cm vertical precision [38]. LiDAR data was filtered with a gLiDAR tool [39] by extracting the most-contrasted connected components in order to remove the non-ground objects [40,41].

Table 2. The investigated characteristics of depressions on conglomerates and methods employed in morphometrical and distributive analyses.

	Doline Characteristic	Parameter	Method
Morphometrical analyses	The circularity of the planar shape	Circularity index (Ic)	$\dfrac{Pcc}{Pd}$ where Pcc = The circumference of the circumscribed circle Pd = The perimeter of the doline (Figure 2a)
	The size of the planar shape	A	The area of the doline planar shape (m^2) (Figure 2b)
	Depth	h	The vertical distance (m) between the highest elevation of the doline rim and the lowest elevation of the doline bottom (Figure 2c)
Distributive analyses	Density including the size of dolines	Density index (Id)	$\dfrac{Ad}{Av}$ where Ad = The area (m^2) of the doline planar shape Av = The area (m^2) of the zone where any location is closer to its associated doline than to any other doline (Voronoi polygon)

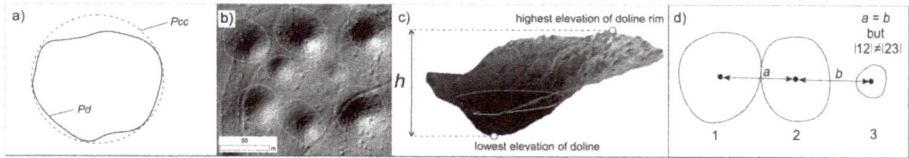

Figure 2. (**a**) The parameters of a doline circularity index; (**b**) The area of a doline planar shape, visually determined from LiDAR data; (**c**) The acquisition of the depth parameter; (**d**) The influence of doline size on density calculation.

3.2. Ground Penetrating Radar

Ground penetrating radar is a non-invasive technique that has engineering, environmental, glaciological, and archaeological applications. Numerous studies describe the successful application of different geophysical techniques to characterize the internal geometry of dolines, providing better understanding of their formation and evolution. Among them, the GPR method provides the highest resolution images. It has been widely used to image the structure of dolines and to locate the potential collapse dolines [42–48].

GPR can be an appropriate tool in many new sub-disciplines that have emerged over the last 15 years, such as forensic geophysics, bio-geophysics, and agro-geophysics [49]. Recently, many studies describe the application of GPR to investigate soil moisture/water content and to determine the extent and lateral variations of soil horizons and their properties [50–55]. Since the thickness of soil cover at the study area exceeds 5 m, knowledge of soils and soil properties is essential to understand the effectiveness of GPR. The chemical, physical, mineralogical, and electromagnetic properties influence the propagation velocity, attenuation, and penetration depth of electromagnetic energy [49].

The electrical conductivity of soils increases with increasing water, soluble salt, and/or clay contents and is governed not only by the amount of clay particles, but also by the types of clay minerals [56]. In the granitic terrain, it was established that the mineralogical composition and the abundance of the minerals strongly influence the depth of penetration where the increased biotite content in regolith restricts GPR performance [54]. According to all mentioned factors, depth penetration in different soils varies from 30 m in clay-free sands to less than 0.5 m in wet clayey soils [57,58].

The boundaries which separate soil horizons are usually associated with differences in moisture contents, physical (texture and bulk density) and/or chemical properties. GPR can therefore be used to detect the boundaries between subsurface horizons [49]. However, if electrical properties are similar, different horizons and sometimes even boundaries between soil materials and bedrock are indistinguishable on radar profiles [58]. Some diagnostic surface pedological horizons, such as the argillic horizon (Bt), have a distinctly higher clay content than the overlying horizon and radar signals are rapidly attenuated at the boundary. Therefore, GPR can be a successful tool for estimating the depth to Bt horizons, which have well-defined upper boundaries that display abrupt increases in bulk density and illuviated silicate clays ([49] and references therein). Due to the increase of clay content and bulk density, a Bt horizon generally provides smooth, continuous reflectors on the GPR data that occur at uniform depths [49].

The GPR Survey and Data Processing

The ProEx (MALA Geoscience, Sweden) ground penetrating radar with an RTA (Rough Terrain Antennas) unshielded 50 MHz bistatic antenna was used with a common offset technique. The length of the GPR system with a flexible snake-like design is 9.25 m, and the spacing between the receiver and the transmitter is 4 m [59]. In addition, a 250 MHz bistatic shielded antenna was used.

GPR profiles were recorded at all three test sites (Figure 1) over cultivated and uncultivated dolines to quantify the anthropogenic shape modifications. Profiles were acquired at cultivated areas Poljšica (old conglomerate) and Dobrava (middle conglomerate) to reveal the anthropogenic impact on the depth of the dolines (Profiles 1 and 2). Profile 3 was recorded on the young conglomerate above a roadcut to obtain the contact between the soil and conglomerate bedrock where the soil cover is significantly thinner than the soil cover on the old and middle conglomerate, and mainly does not exceed 2 m. This profile was recorded as a testing profile where the contact is visible on the terrain to correlate the results obtained from Profile 4. On the young conglomerate at Podbrezje, dolines are not clearly expressed on cultivated areas; only the profile over the uncultivated doline in the forest was measured to obtain the primary shape of the doline (Profile 4).

All profiles were measured in two directions perpendicular to each other with a 50 MHz and 250 MHz antenna. It turned out that the unshielded 50 MHz antenna gave more useful results due to better depth of penetration, especially on the old and middle conglomerate, where soil cover is thicker than 5 m. The profiles measured in the forest contain too much noise from the trees due to the unshielded antenna. Here, the shielded 250 MHz antenna provided more useful information, mostly on the young conglomerate, where soil cover is relatively thin and the depth of penetration is around 6–7 m. Selected profiles and their basic data are summarized in Table 3.

Table 3. Basic data of the presented GPR profiles acquired with the 50 MHz unshielded and 250 MHz shielded antennas.

Profile	Location	Type	Length (m)	Antenna Frequency
Profile 1	Poljšica (old conglomerate)	cultivated doline	46.4	50 MHz
Profile 2	Dobrava (middle conglomerate)	cultivated doline	57.4	50 MHz
Profile 3	Podbrezje (young conglomerate)	soil/conglomerate	79.5	250 MHz
Profile 4	Podbrezje (young conglomerate)	uncultivated doline	64.4	250 MHz

The profiles were processed with the ReflexW program. The following processing steps were applied: subtract-mean (dewow), time zero correction, background removal, manual gain, bandpass filtering, and topographic correction. The velocity used to convert the two-way travel time into depth was different depending on the expected prevailing material and the depth of penetration for each frequency. In Profiles 1 and 2, the signal did not reach the underlying conglomerate due to several meters of soil, thus the velocity for average soil was used according to the values in the literature ($\varepsilon_r = 16$; $\upsilon = 0.075$ m/ns; [49,60]). In Profile 4, several hyperbolic diffractions occur below the soil in the conglomerate basement. By hyperbola fitting, velocities between 0.10–0.08 m/ns were observed. An average velocity $\upsilon = 0.09$ m/ns was used to convert two-way travel time in Profiles 3 and 4. In the literature, there is no standard value for velocity in such material, but the expected velocity for a carbonate conglomerate should be lower than in carbonates due to the heterogeneous composition causing diffraction and signal scattering. The velocity obtained in Profile 4 could represent the relevant velocity for conglomerate.

4. Study Area and Test Sites

4.1. GPR Results and Defining the Appropriate Test Dolines for Morphometrical Analyses

The DEM derived from the LiDAR data already indicates the morphological difference between cultivated and uncultivated dolines (Figure 3). Cultivated dolines are shallower and their primary form was reshaped by anthropogenic tilling. On old conglomerate, the depth of uncultivated dolines is 6–8 m, while the depth of cultivated dolines is 4–5 m. The slopes of the cultivated dolines are less steep and smoother due to anthropogenic tilling.

Laterally continuous strong reflectors are visible in Profiles 1 and 2 (Figures 4 and 5). They most likely belong to the argillic horizon (Bt) related to clay accumulation, because its depth corresponds to the data observed by a previous study [22]. Following the inventory of the complete soil profiles obtained on conglomerates of different ages in the area, the Bt horizon occurs at an approximate depth of 0.3–0.6 m below the surface and underneath the pedological A or B horizon [22]. The thickness of this horizon amounts to 2.5 m on young conglomerate and more than 3 m on middle and old conglomerates [22].

The difference in the dielectric constant between the upper horizon (generally A or AB horizon according to Pavich and Vidic [22]) and the Bt horizon due to the clay and moisture content in the Bt horizon leads to the strong reflectivity at the boundary between horizons. The Bt horizon causes strong signal attenuation and presents the limit of depth penetration.

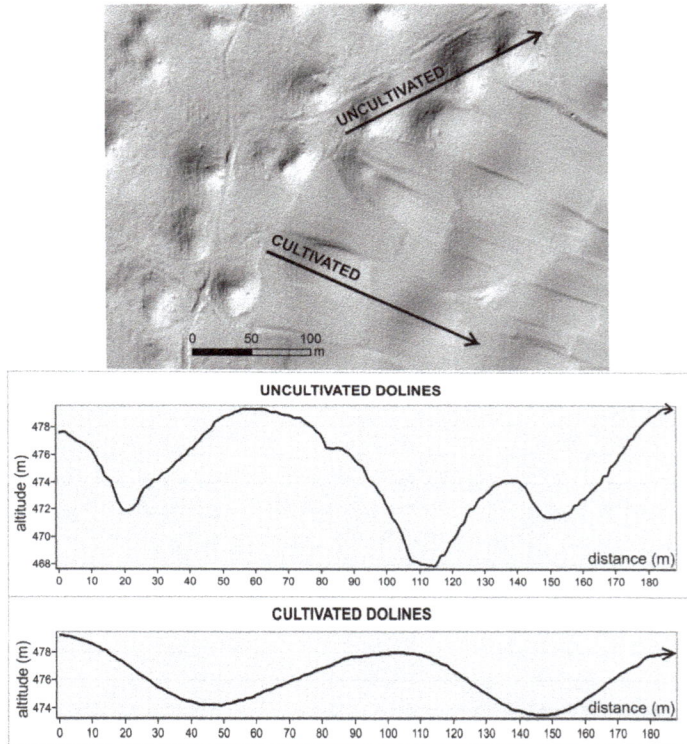

Figure 3. Topographic profiles across cultivated and uncultivated dolines at Poljšica field site [37].

Subsurface horizons are normally parallel and uniform so their shape can be an indirect indicator of agricultural changes in dolines. If the horizons within a doline appear at the same depth as in the surrounding area, the shape of the Bt horizon can provide information about the thickness of the redeposited material, as well as the shape and depth of the doline before cultivation. A slightly undulating reflector within a doline indicates that the primary doline was deeper (Figures 4 and 5). On Profiles 1 and 2, the depth of the redeposited material above the Bt horizon can be roughly evaluated. The maximum thickness of the redeposited material is 2.5 m, considering the dielectric constant for the average soil. This corresponds to the difference between the depths of cultivated and uncultivated dolines derived from DEM. Soil from the slopes was partially terraced and partially moved to the bottom of the dolines to acquire a larger area and to lower the inclination of the slopes. Based on the

GPR data, the redeposited soil was mixed with more sandy material because the signal penetrates well through it.

Another explanation for the irregular upper boundary of the Bt horizon can be attributed to the underlying dissolution features that are associated with karst processes in conglomerate basement [53]. However, this interpretation is unlikely because the reflector representing the Bt horizon is too smooth and continuous for its shape to be the consequence of the karstification processes under dolines. Furthermore, the estimated thickness of the redeposited material corresponds well with the depths derived from DEM.

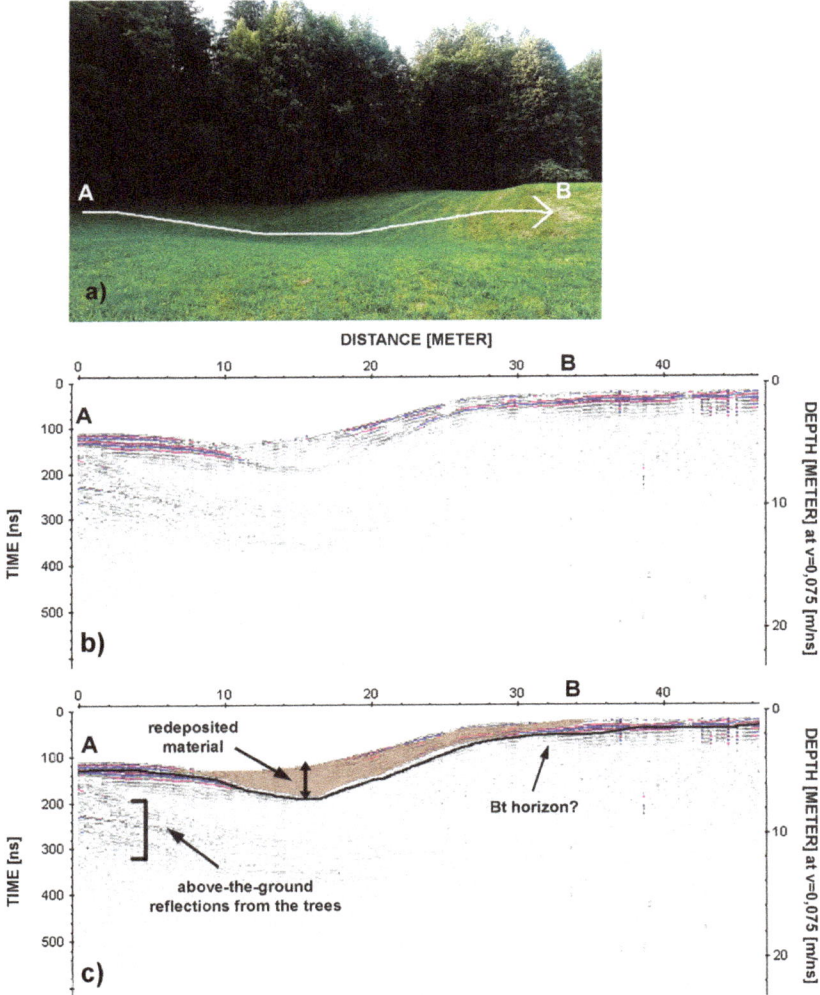

Figure 4. (**a**) The direction of Profile 1 acquired over a cultivated doline on old conglomerate (Poljšica). The right side of the doline is slightly terraced; (**b**) A processed and topographically corrected radargram; (**c**) An interpreted radargram.

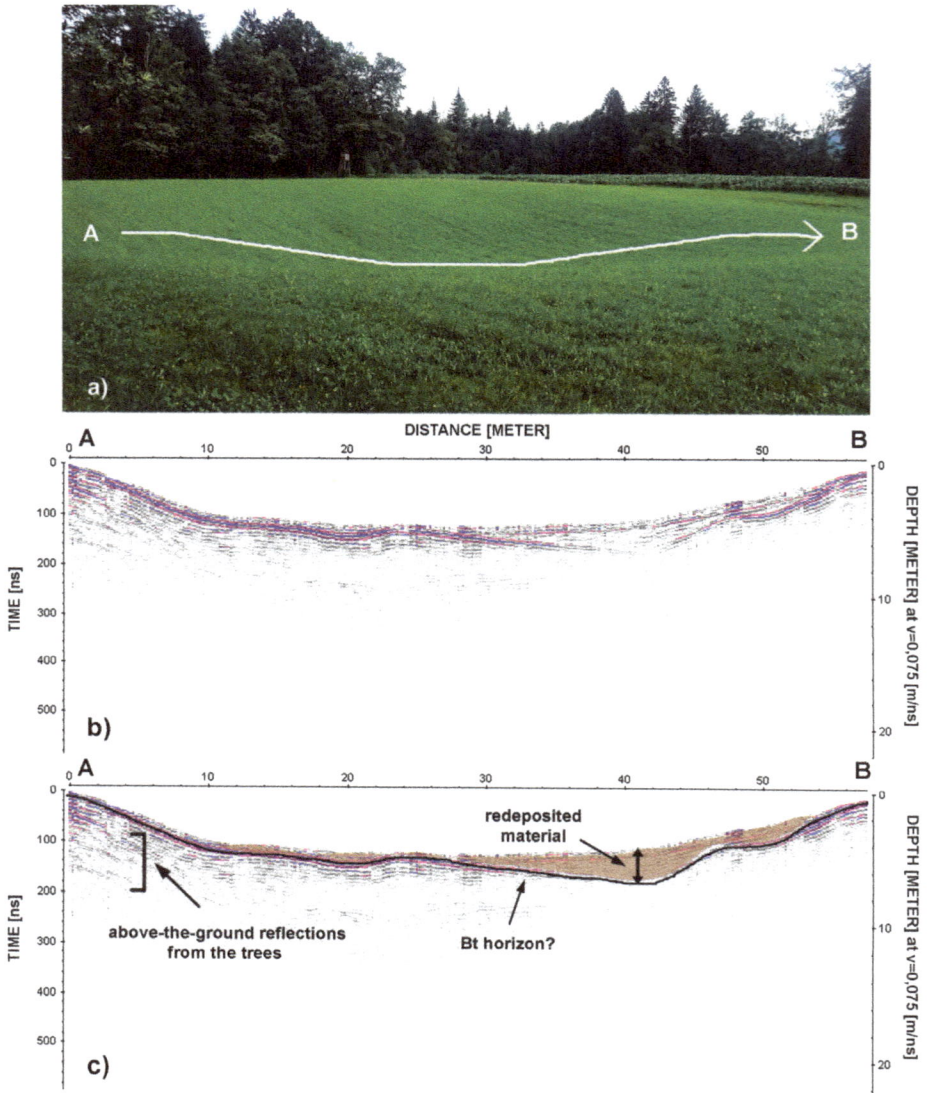

Figure 5. (a) The direction of Profile 2 acquired over a cultivated doline on middle conglomerate (Dobrava); (b) A processed and topographically corrected radargram; (c) An interpreted radargram.

The contact between the soil cover and conglomerate basements is mostly very irregular, forming pockets and bulges (Figure 6A). Profile 3 was measured to observe the contact between the soil cover and the conglomerate bedrock on young conglomerate over a roadcut, where the soil cover is relatively thin (up to 2 m). The boundary between the soil and conglomerate is unclear on the GPR data due to the uneven contact and insufficient difference in dielectric properties related to the gradual transition of soil into the conglomerate basement (Figure 6). If the contrast in dielectric properties is too low or the difference is not sudden enough, no distinct reflections are found on the GPR data.

In order to confirm the primary shape of cultivated dolines acquired with GPR, an uncultivated doline on young conglomerate was also measured (Profile 4, Figure 7). The soil cover on young conglomerate is significantly thinner, so the contact between the soil and conglomerate can provide some information about the primary shape of the doline.

The radargram of Profile 4 shows that no distinctive reflector is visible that could represent the Bt horizon as in Profiles 1 and 2. Similarly, the soil/bedrock interface over the uncultivated doline is not clearly expressed as high-amplitude reflections, but is distinguishable in the form of occurrences of hyperbolic diffractions (Figure 7). They are the consequence of cavities and fractures within a karstified carbonate conglomerate. The soil/bedrock interface is unclear due to the presence of coarse fragments or blocks in the overlying soil, irregular bedrock surfaces, and fracturing that are characteristic of karst on carbonates and on conglomerates.

The boundary soil/bedrock (highlighted by the dashed line in Figure 7) is defined by hyperbolic anomalies. The thickness of the soil cover varies between 1 and 3 m, which corresponds with the known soil thickness on young conglomerate [22].

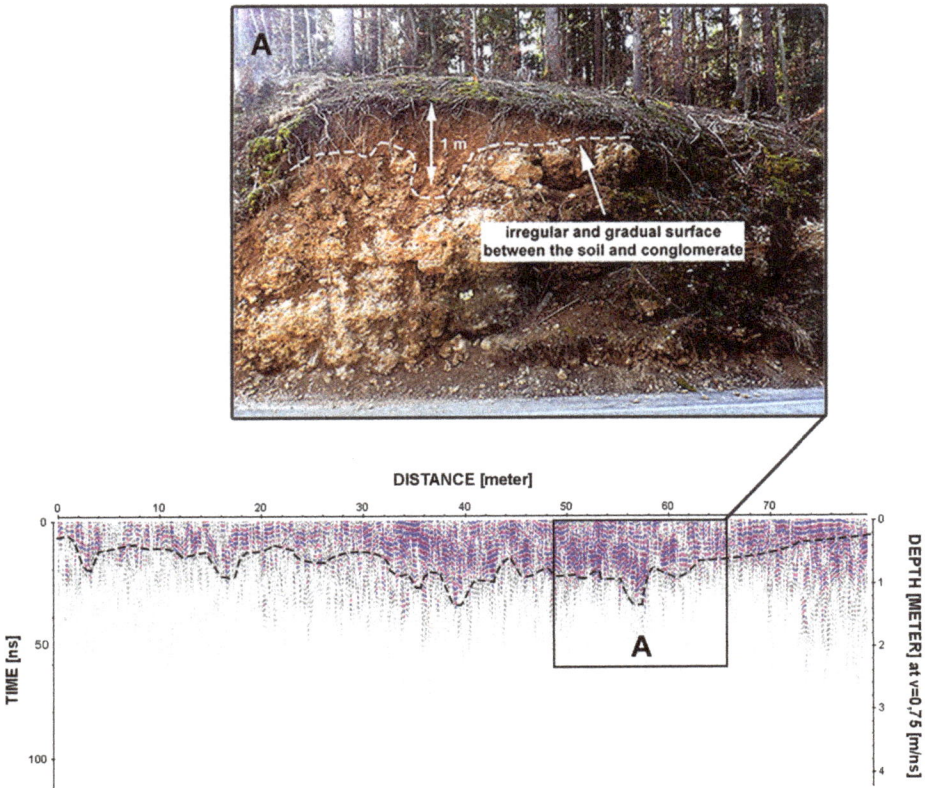

Figure 6. Profile 3: The contact between the soil cover and conglomerate bedrock on young conglomerate (Podbrezje) above a roadcut, using a 250 MHz antenna. Vertical exaggeration 1:6.

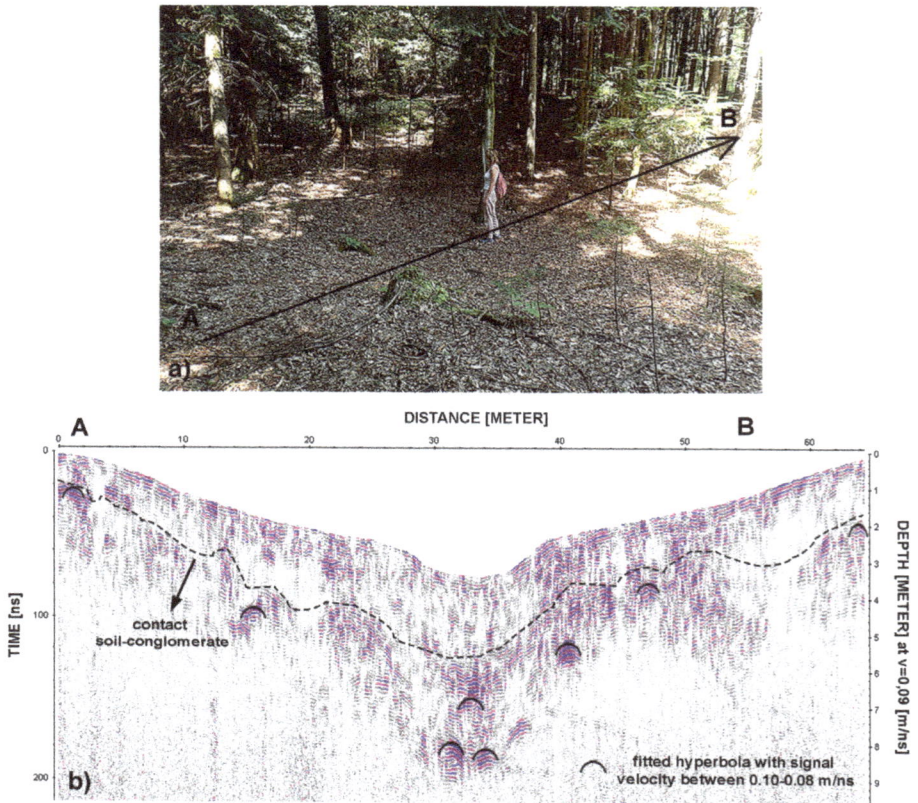

Figure 7. (**a**) The direction of Profile 4 acquired with a 250 MHz antenna over an uncultivated doline on young conglomerate (Podbrezje); (**b**) The soil/bedrock boundary is estimated by the appearance of hyperbolic diffractions caused by cavities and fractures within carbonate conglomerate. Vertical exaggeration 1:3.

A simplified model of anthropogenic reshaping of the doline was constructed using the detected Bt horizon on the GPR data from cultivated dolines (Figure 8). The primary shape of dolines is difficult to reconstruct. The primary depth of dolines can be approximately obtained from the GPR data, but that kind of determination is very time-consuming when a large number of dolines must be measured. GPR results show considerable morphological modifications of dolines due to anthropogenic intervention, as cultivated dolines are approximately 2 m shallower than uncultivated ones. This is why dolines on cultivated areas were excluded from further spatial analyses.

1. The primary shape of the doline

2. Doline reshaping due to anthropogenic tilling

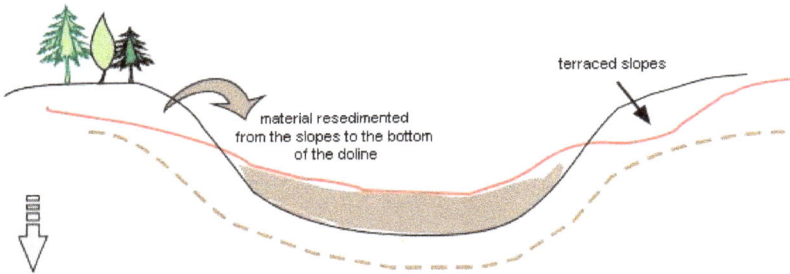

terraced slopes

material resedimented
from the slopes to the bottom
of the doline

3. The current shape of the doline

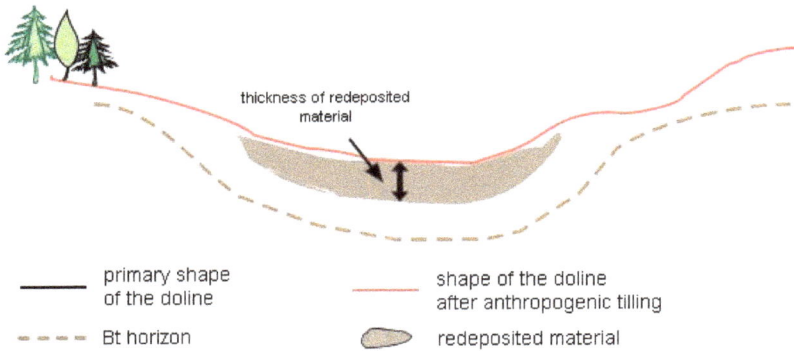

thickness of redeposited
material

——	primary shape of the doline	——	shape of the doline after anthropogenic tilling
- - - -	Bt horizon	⬭	redeposited material

Figure 8. A simplified model of agriculturally reshaped dolines. The Bt horizon in the current form of the doline as detected by the GPR survey indicates the redeposited material.

4.2. Test Sites for Spatial Analyses

Study sites on three different conglomerates (young, middle, and old) were selected on the basis of the maps done by Žlebnik [2], and Pavich and Vidic [22] on locations where the dating of both authors was accordant (Table 4) and where the surface is forested. They are at most 1.5 km apart and are separated by the valleys of the Sava and Lipnica rivers (Figure 9) which flow at the elevation of approximately 380 m. The oldest conglomerate (Poljšica) is at the altitude 470–490 m, similar to the middle conglomerate (Dobrava) which is located at 475–485 m, while the youngest conglomerate (Podbrezje) is at a lower altitude (450–460 m). On the basis of DEM examination and GPR data, only dolines on uncultivated areas were subject to spatial analyses.

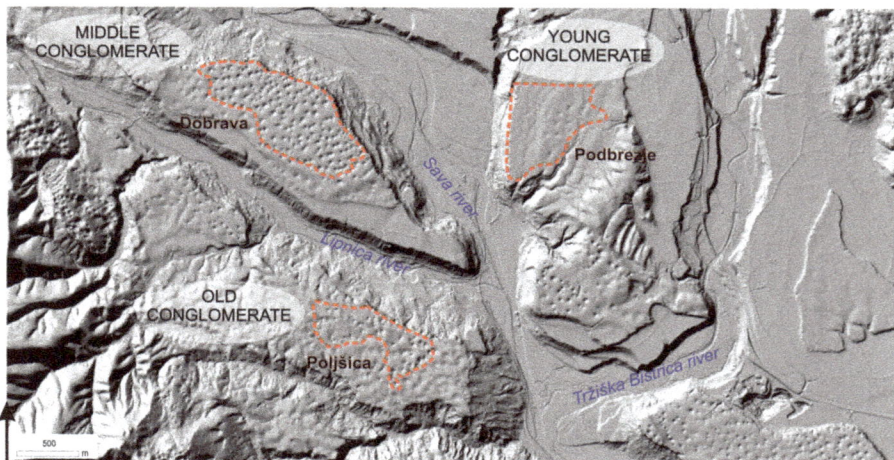

Figure 9. Test sites and relief characteristics as visible from 1 m resolution LiDAR data [37].

Table 4. Basic statistics of test sites.

	Podbrezje	Dobrava	Poljšica
Relative age of conglomerate	Young	Middle	Old
Area (m²)	382 108	775 562	309 349
Number of identified dolines	21	185	73

5. Results

5.1. The Circularity of the Doline Planar Shape

Doline planar shapes are highly circular and uniform on conglomerates of all ages. Mean values of the circularity index (Ic) for doline planar shapes (Figure 10) are 1.104–1.069, being close to the value 1.000 of the circle. The range of the majority of values is similar for all three sites (1.05–1.13), while the irregularity slightly increases with the age of conglomerate, as well as the dispersion of values. Circularity is not linked to the size of a doline planar shape, nor to doline depth (Figure 11). The independence of size and conglomerate age (and other mechanical and chemical properties of the rock) from generally high circularity of dolines indicates that circularity as one of the main characteristics of dolines is linked to external factors of karst surface reshaping, rather than to the temporal aspect of doline evolution. Spatial distribution of dolines classified by planar size does not indicate any particular patterns.

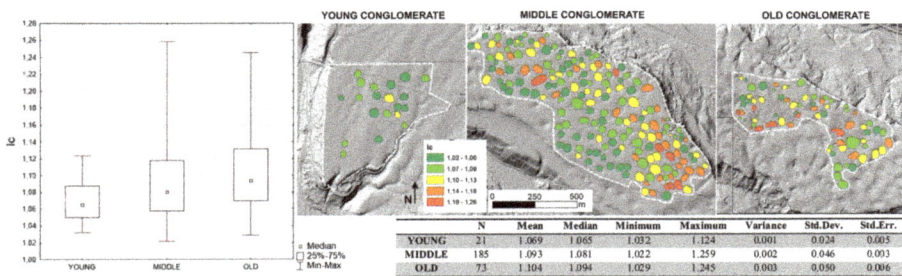

Figure 10. The circularity of dolines' planar shapes for conglomerates by age.

	N	Mean	Median	Minimum	Maximum	Variance	Std.Dev.	Std.Err.
YOUNG	21	1.069	1.065	1.032	1.124	0.001	0.024	0.005
MIDDLE	185	1.093	1.081	1.022	1.259	0.002	0.046	0.003
OLD	73	1.104	1.094	1.029	1.245	0.003	0.050	0.006

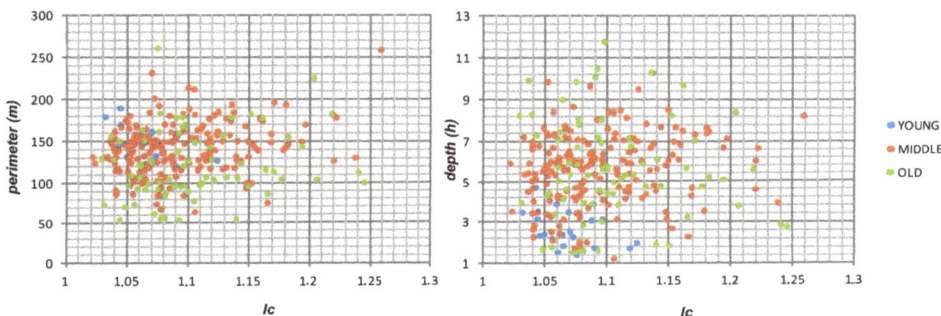

Figure 11. The correlations between circularity index (Ic) and perimeter and depth for conglomerates by age.

5.2. The Area of the Doline Planar Shape

Based on the calculated values (Figure 12), the planar size of the dolines on all conglomerates appears as a highly constant characteristic. The majority of values for all ages of conglomerate range from 800 to 1900 m^2, which is a very narrow class, while the dispersion of values increases considerably with the age of conglomerate. A slight decrease of doline planar size mean value follows the increasing age of conglomerate due to the appearance of dolines with very small planar size on the old conglomerate. It is of particular interest that the planar size of shallow dolines on young conglomerate (see also Section 5.3) does not deviate noticeably from the planar size of deep dolines on middle or old conglomerate, which means that a doline acquires its planar size in its early stage of development (Figure 13). The same was established by Ferrarese and Sauro [4] for dolines on Montello conglomerate and explained by diffuse porosity in conglomerate rock. The conceptual model applicable to the Dinaric solution dolines was used as a base to calculate that the expansion rate of a doline perimeter decreases with time as the reciprocal of a quadratic function, meaning that the increase of the planar shape size is fast in the beginning and slows down later [61]. A high variability of shapes (see Section 5.1), as well as a high variability of planar sizes were noticeable on old conglomerate, where big and very small dolines occur together. The size is the most uniform on young conglomerate, and the uniformity decreases with the increasing age of conglomerate. Spatial distribution of dolines by planar size does not show any specific patterns.

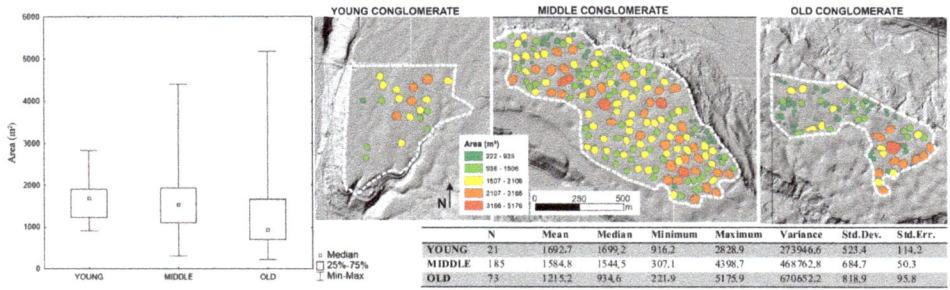

Figure 12. The area of a doline planar shape for conglomerates by age.

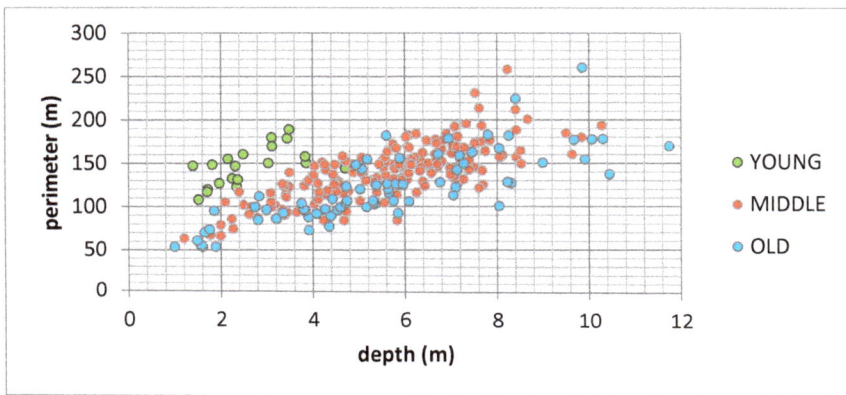

Figure 13. Depth to perimeter relationship for dolines on conglomerates by age.

5.3. Doline Depth

The considerable difference in doline depth range is noticeable between young conglomerate on one side, and middle and old conglomerate on the other (Figure 14). Dolines on young conglomerate are shallow (mean 2.7 m) and very uniform, deviating only 3.3 m in depth, while dolines on the other two sites (mean 5.6 m for middle and 5.5 m for old conglomerate) are on average much deeper and considerably more varied, deviating as much as 10.8 m in depth on old conglomerate. Generally, the depth of dolines increases with the increasing age of conglomerate, where the average on old conglomerate is lowered by the presence of very shallow depressions. The increase of doline depth with age is consistent with the usual time-dependent growth of natural phenomena, but shallow dolines on young conglomerate may also be the reflection of relatively slower karstification processes due to less favorable rock properties. Spatial distribution of dolines by depth does not indicate any particular patterns.

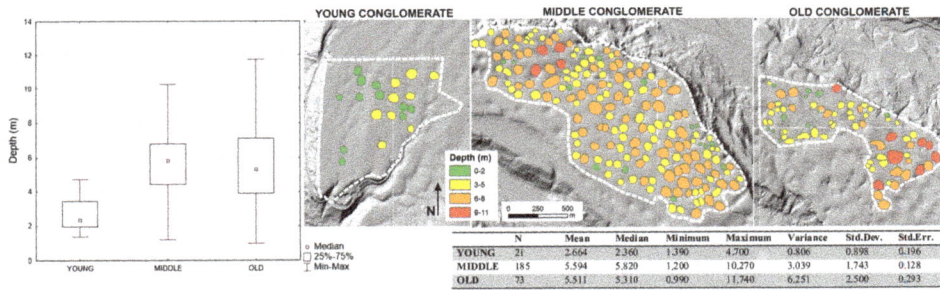

Figure 14. Doline depth for conglomerates by age.

5.4. Doline Distribution

Dolines cover 9% of young conglomerate, 38% of middle, and 29% of old conglomerate at the test sites. The density is the lowest (mean 0.152) but quite uniform on young conglomerate. On middle conglomerate, the high density (mean 0.397) is very uniform over the entire test site. Big dolines are especially densely distributed, a phenomenon which is particularly obvious on old conglomerate. Here, density is the least uniform as big dolines are located close to each other, while smaller dolines occupy empty areas where they are spread more uniformly and far apart (Figure 15). This phenomenon is especially evident on old and middle conglomerate. The distribution of dolines is not linked to linear tectonic structures. This was already observed on Montello conglomerate karst [4], where the same authors explained that it stemmed from the predominance of diffuse over fracture porosity. They concluded that the pattern of dolines on conglomerate reflects the influence of morphological elements rather than a fracture network.

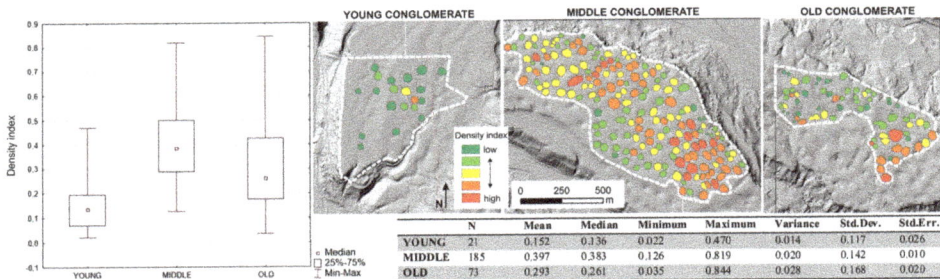

Figure 15. Doline distribution by the age of conglomerate index represented by density index.

5.5. Typization of Karst Surface Morphology as a Tool for Dating Conglomerates

Morphometrical and distributive analyses of dolines on unconsolidated carbonate gravel and on old, middle, and young carbonate conglomerates indicate the following general geomorphological characteristics:

- Unconsolidated gravel: the surface is flat with no surface features (Figure 16a).
- Young conglomerate: the surface is flat and characterized by scarce shallow surface features. Shallow linear depressions appear as nearly unrecognizable irregularities. Sporadically, large but shallow dolines develop. In some cases, their location seems to be linked to linear depressions (Figure 16b).
- Middle conglomerate: the surface is rather flat and entirely covered by mainly uniform, funnel-like deep dolines (Figure 16c).

173

- Old conglomerate: the surface is irregular; the depressions come in all sizes, depths, and shapes (Figure 16d) and do not entirely cover the surface. The largest and deepest dolines occur here (Type 1), as well as small and shallow (Type 2), and double ones (Type 3).

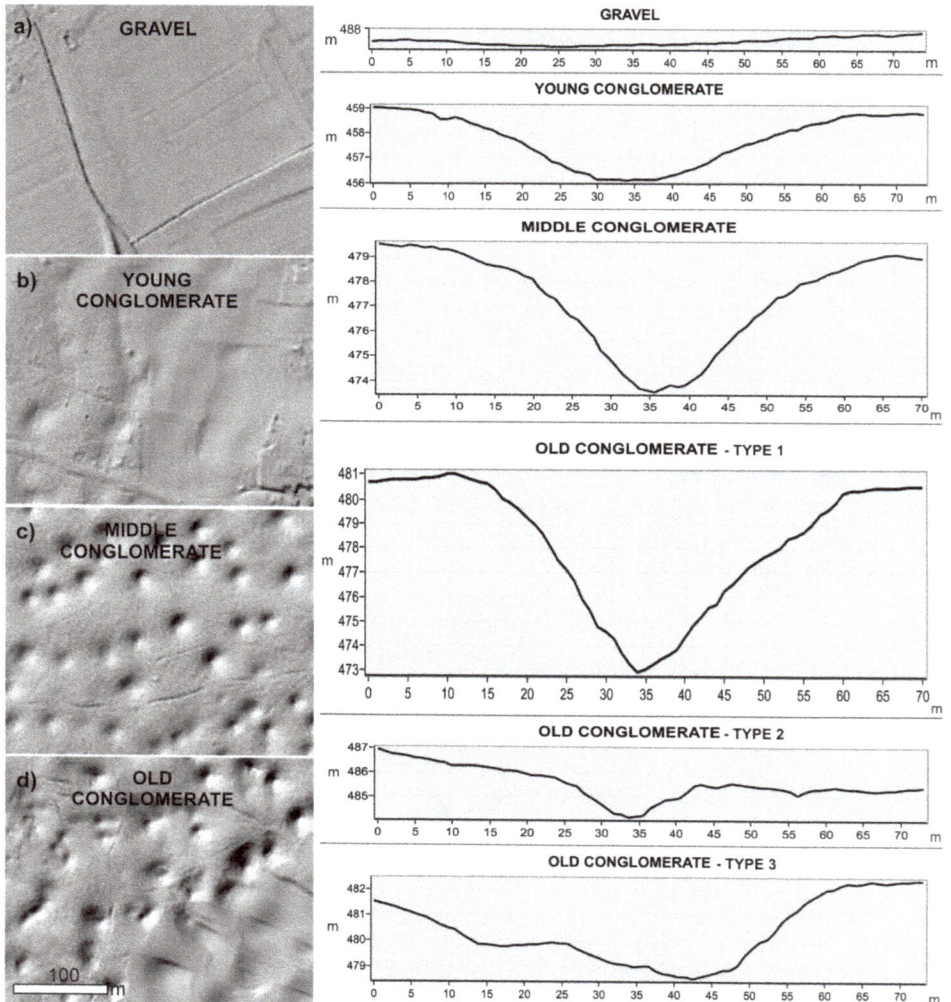

Figure 16. The typical surface morphology of gravel and karstified conglomerates by age with the profiles across characteristic dolines. (**a**) Gravel: flat surface with no surface features; (**b**) Young conglomerate: flat surface with scarce shallow surface features (dolines and linear depressions); (**c**) Middle conglomerate: rather flat surface with uniform dolines of high density; (**d**) Old conglomerate: irregular surface with depressions of high morphologic heterogeneity (profiles across the three types of dolines).

According to the acquired data, dolines on young conglomerate are the most uniform. They are characterized by large and the most circular planar shapes, while they are the shallowest compared to dolines on other conglomerates. Dolines on old conglomerate, however, are the most heterogeneous in each parameter, with the smallest average size and the least regular planar shape. The deepest dolines can be found on this conglomerate, but many shallow dolines also occur. The latter are different from those on young conglomerate, as they are small and more irregular (Table 5).

Table 5. Descriptive morphometrical and distributive characteristics of dolines for karstified conglomerates by age.

		Young Conglomerate	Middle Conglomerate	Old Conglomerate
Surface features		shallow linear depressions, dolines	dolines	dolines
Dolines	volumetric shape	bowl-like	funnel-like	Type 1: funnel-like Type 2: bowl-like
	planar shape	highly circular	highly circular	Type 1: highly circular Type 2: irregular
	planar size	uniform	uniform	Type 1: uniform Type 2: small
	depth	shallow	Type 1: deep Type 2: shallow	Type 1: deep Type 2: shallow
	slope	gentle	steep	Type 1: extremely steep Type 2: gentle
	frequency	rare	numerous	moderate
	uniformity	high	moderate	low
Distribution		sporadic	covering the entire surface	scattered

6. Discussion

The visual analysis of shaded relief derived from high-resolution LiDAR data itself already enables the recognition of relief differences on particular ages of carbonate conglomerates. However, morphometrical and distributive spatial analyses of dolines were performed to quantify the specific characteristics that are linked to the age of conglomerate. Noticeable differences in calculated parameters confirmed a satisfactory reliability of visual analysis in distinguishing karstified surfaces on carbonate conglomerates. Based on this, visual analysis was employed to extrapolate the data to the wider conglomerate area in the Ljubljana basin, which proved an easy and fast method for surface classification without time-consuming field examinations. Among all the analyzed parameters, the scarcity and the shallowness of dolines were the most distinctive characteristics of doline population developing on young conglomerates. As the distribution of dolines on middle and old conglomerates may sometimes appear similar, the degree of uniformity of morphometrical properties was the crucial factor for their distinction.

The reliability of geomorphological dating (Figure 17) was tested by comparing the geomorphological characteristics of the analyzed sites to all the areas dated by Žlebnik [2], and Pavich and Vidic [22]. The DEM from the high-resolution LiDAR data shows a high accordance of the surface karstification degree with relative dating based on the morphostratigrafic mapping and borehole data [2], as well as cosmogenic-nuclide burial dating and paleomagnetic methods [22]. The geomorphological dating method fails on very small areas as a certain amount of the surface should be exposed and karstified in order to generalize local morphometrical and distributive characteristics (marked blue in Figure 17).

Discrepancies with Žlebnik's [2] dating were found on two locations (A and B), and on another two with the dating done by Pavich and Vidic [22] (C and D) (Figure 17). A high morphological diversity of features and some extreme doline depths are more closely related to old conglomerate than with middle on Location A, which corresponds with the dating by Pavich and Vidic [22]. On Location B, the distribution of dolines is too dense to belong to young conglomerate, while the shallowness of dolines is due to human impact. After geomorphological dating, this site corresponds to middle conglomerate. Towards the south, the surface lacks any surface irregularities, so it was classified as a Würm gravel fill. Indicating shallow surface irregularities, Locations C and D correspond to young conglomerate. Both results of geomorphological dating at Locations C and D are in accordance with Žlebnik's dating [2].

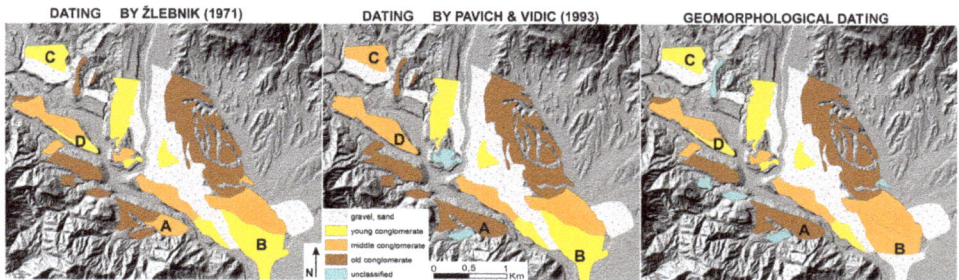

Figure 17. A comparison of geomorphological dating by the type of surface karstification with relative dating based on morphostratigrafic mapping and borehole data by Žlebnik [2] and absolute dating based on paleomagnetic and [10]Be analyses by Pavich and Vidic [22] in the wider area of the Ljubljana Basin. Discrepancies in geomorphological dating are marked with A–D.

7. Conclusions

Pleistocene conglomerates of different ages in the Ljubljana Basin were subject to different relative and absolute dating methods in the past. The different degrees of karstification, as well as doline shape and depth obtained on the terrain during geological mapping were used as additional indicators for determining the relative age of conglomerate terraces.

Nowadays, LiDAR data with high spatial resolution enables a better detection of dolines and a fast determination of different parameters needed for further morphometrical and distributive spatial analyses. The main purpose of the study was to classify all conglomerate terraces within the Ljubljana Basin on the basis of the geomorphological characteristics of dolines obtained on the three test sites including old, middle, and young conglomerates.

The surface on all three conglomerates is partially forested and also partially cultivated. The ground penetrating radar (GPR) method was used to select the appropriate dolines for further analyses. GPR results confirmed the observations from LiDAR images: cultivated dolines are shallower and have lower slope inclination due to redeposition of soil from the flanks into the doline. GPR data measured over cultivated dolines on old and middle conglomerate proves that cultivated dolines are approximately 2.5 m shallower in the bottom due to the human tilling, while some dolines are also slightly terraced. For that reason, cultivated dolines were not included in further morphometrical analyses.

Circularity index, planar size, depth, and density index of dolines were calculated at the three test sites for each relative age of conglomerates. Morphometrical and distributive analyses revealed that dolines have typical geomorphological characteristics on each of the conglomerates of a relative age. Results derived at all three test sites were further extrapolated to the wider area of the Ljubljana Basin and compared with relative dating [2] and absolute dating [22] to test the reliability of geomorphological dating. The results of spatial analyses generally confirmed former dating and proved that geomorphological dating can be a fast and reliable method for dating karstified conglomerates. Furthermore, the morphometrical and distributive data obtained from the dolines can be used as a basis for future research analyses aimed at understanding the karstification processes in conglomerates.

Acknowledgments: This study was conducted with the support of the research Program P1-0011 and the Ph.D. grant 1000-15-0510 financed by the Slovenian Research Agency. This work also benefited from networking activities carried out within the EU-funded COST Action TU1208 "Civil Engineering Applications of Ground Penetrating Radar". We would also like to thank Andrej Pipan for field assistance, France Šušteršič for comments

Remote Sens. **2017**, *9*, 1213

and advice during the preparation of this paper, and Tomaž Verbič for sharing his field experience related to the geophysical and pedological properties of conglomerates.

Author Contributions: Teja Čeru acquired and processed the GPR data and wrote Section 2, Section 3.2, Section 4.1, and Section 7 and contributed to Sections 1 and 5. Ela Šegina did the morphometrical and distributive spatial analyses and wrote Section 3.1, Section 4.2, Section 5, and Section 6 and contributed to Section 1. Andrej Gosar contributed to GPR measurements and data processing as well as to the preparation of the manuscript.

Conflicts of Interest: The authors declare no conflict of interest.

References

1. Šifrer, M. Kvartarni razvoj Dobrav na Gorenjskem (The Quarternary development of Dobrave in Upper Carniola (Gorenjska) Slovenia). *Geografski Zbornik* **1969**, *11*, 99–221.
2. Žlebnik, L. Pleistocen Kranjskega, Sorškega in Ljubljanskega polja. *Geologija* **1971**, *14*, 5–51.
3. Knez, M.; Šebela, S.; Gabrovšek, F. Geološke osnove ter jame (Geological settings and caves). In *Udin Boršt*; Kranjc, A., Ed.; Museo di Storia Naturale e Archeologia: Montebelluna, Italy, 2005; pp. 9–24.
4. Ferrarese, F.; Sauro, U. The Montello Hill: The "classical karst" of the conglomerate rocks. *Acta Carsol.* **2005**, *34*, 439–448. [CrossRef]
5. Telbisz, T.; Látos, T.; Deák, M.; Székely, B.; Koma, Z.; Standovár, T. The advantage of lidar digital terrain models in doline morphometry compared to topographic map based datasets—Aggtelek karst (Hungary) as an example. *Acta Carsol.* **2016**, *45*, 5–24. [CrossRef]
6. Pardo-Igúzquiza, E.; Valsero, J.J.D.; Dowd, P.A. Automatic detection and delineation of karst terrain depressions and its application in geomorphological mapping and morphometrical analysis. *Acta Carsol.* **2013**, *42*, 17–24. [CrossRef]
7. Carvalho, O.D.; Guimarães, R.; Montgomery, D.; Gillespie, A.; Gomes, R.T.; Martins, É.D.S.; Silva, N. Karst depression detection using ASTER, ALOS/PRISM and SRTM-Derived digital elevation models in the Bambuí Group, Brazil. *Remote Sens.* **2013**, *6*, 330–351. [CrossRef]
8. Kobal, M.; Bertoncelj, I.; Pirotti, F.; Kutnar, L. Lidar processing for defining sinkhole characteristics under dense forest cover: A case study in the Dinaric mountains. *Int. Arch. Photogramm. Remote Sens. Spat. Inf. Sci.* **2014**, *40*, 113–118. [CrossRef]
9. Jeanpert, J.; Genthon, P.; Maurizot, P.; Folio, J.-L.; Vendé-Leclerc, M.; Sérino, J.; Join, J.-L.; Iseppi, M. Morphology and distribution of dolines on ultramafic rocks from airborne LiDAR data: The case of southern Grande Terre in New Caledonia (SW Pacific). *Earth Surf. Process. Landf.* **2016**, *41*, 1854–1868. [CrossRef]
10. Černatic-Gregorič, A.; Zega, M. The impact of human activities on dolines (sinkholes): Typical geomorphologic features on Karst (Slovenia) and possibilities of their preservation. *Geogr. Pannonica* **2010**, *14*, 109–117. [CrossRef]
11. Breg Valjavec, M.; Zorn, M. Degraded karst relief: Waste-filled dolines. In *Advances in Environmental Research*, 1st ed.; Daniels, J.A., Ed.; Nova Science Publishers: New York, NY, USA, 2015; Volume 40, pp. 77–95.
12. Gabrovšek, F. Jame v konglomeratu: Primer Udin Boršta, Slovenija. (Caves in conglomerate: Case of Udin Boršt, Slovenia). *Acta Carsol.* **2005**, *34*, 507–519.
13. Gantar, J. Arneševa luknja. *Acta Carsol.* **1995**, *1*, 151–158.
14. Šter, D. Udin Boršt in njegov kras. *Proteus* **1995**, *57*, 237–244.
15. Lipar, M.; Ferk, M. Eogenetic caves in conglomerate: An example from Udin Boršt, Slovenia. *Int. J. Speleol.* **2011**, *40*, 53–64. [CrossRef]
16. Ferk, M.; Lipar, M. Eogenetic caves in Pleistocene carbonate conglomerate in Slovenia. *Acta Geogr. Slov.* **2012**, *52*, 7–33. [CrossRef]
17. Grad, K.; Ferjančič, L. *Basic Geological Map of Yugoslavia, Sheet Kranj, L33–65*; Federal Geological Survey of Beograd: Beograd, Serbia, 1976.
18. Šifrer, M. Porečje Kamniške Bistrice v pleistocenu. (The basin of Kamniška Bistrica during the pleistocene period). *Dela SAZU* **1961**, *10*, 211.
19. Meze, D. Porečje Kokre v pleistocenu. *Geografski Zbornik* **1974**, *14*, 98.
20. Kuščer, D. Kvartarni savski zasipi in neotektonika. *Geologija* **1990**, *33*, 299–313. [CrossRef]
21. Šercelj, A. *Peloidne Analize Pleistocenskih in Holocenskih Sedimentov Ljubljanskega Barja*; Razprave 9/9, 4 Raz.; SAZU: Ljubljana, Slovenia, 1966; pp. 431–472.

22. Pavich, M.J.; Vidic, N. Application of paleomagnetic and 10Be analyses to chronostratigraphy of Alpine glacio–fluvial terraces, Sava River Valley, Slovenia. In *Climate Change in Continental Isotope Records*; Swart, P., Ed.; Geophysical Monograph: Washington, DC, USA, 1993; Volume 78, pp. 263–275.

23. Vidic, N.J.; Lobnik, F. Rates of soil development of the chronosequence in the Ljubljana Basin, Slovenia. *Geoderma* **1997**, *76*, 35–64. [CrossRef]

24. Vidic, N.J. Soil-age relationships and correlations: Comparison of chronosequences in the Ljubljana Basin, Slovenia and USA. *Catena* **1998**, *34*, 113–129. [CrossRef]

25. Mihevc, A.; Bavec, M.; Häuselmann, P.; Fiebig, M. Dating of the Udin Boršt conglomerate terrace and implication for tectonic uplift in the northern part of the Ljubljana Basin (Slovenia). *Acta Carsol.* **2015**, *44*. [CrossRef]

26. Penck, A.; Brückner, E. *Die Alpen in Eiszeiten*; Tauchnitz: Leipzig, Germany, 1909; p. 1199. (In German)

27. Vidic, N.; Pavich, M.; Lobnik, F. Statistical analyses of soil properties on a quaternary terrace sequence in the upper Sava river valley, Slovenia, Yugoslavia. *Geoderma* **1991**, *51*, 189–211. [CrossRef]

28. Doctor, D.H.; Young, J.A. An evaluation of automated GIS tools for delineating karst sinkholes and closed depressions from 1-meter lidar-derived digital elevation data. In Proceedings of the 13th Multidisciplinary Conference on Sinkholes and the Engineering and Environmental Impacts of Karst, National Cave and Karst Research Institute, Carlsbad, New Mexico, 6–10 May 2013; pp. 449–458.

29. Obu, J.C.A.; Podobnikar, T. Algorithm for karst depression recognition using digital terrain models. *Geod. Vestnik* **2013**, *57*, 260–270. [CrossRef]

30. Šušteršič, F. The Pure Karst Model. *Cave Karst Sci.* **1996**, *23*, 25–32.

31. Triglav Čekada, M. Možnosti uporabe zračnega laserskega skeniranja (LiDAR) za geomorfološke študije. *Geogr. Vestnik* **2011**, *83*, 81–93.

32. Bondesan, A.; Meneghel, M.; Sauro, U. Morphometrical analysis of dolines. *Int. J. Speleol.* **1992**, *21*, 1–55. [CrossRef]

33. Šušteršič, F. Classic dolines of classical site. *Acta Carsol.* **1994**, *23*, 123–152.

34. Ćalić, J. Uvala—Contribution to the Study of Karst Depressions (with Selected Examples from Dinarides and Carpatho-Balkanides). Ph.D. Thesis, University of Nova Gorica, Slovenia, 2009.

35. Denizman, C. Morphometrical and spatial distribution parameters of karstic depressions, Lower Suwannee River Basin, Florida. *J. Cave Karst Stud.* **2003**, *65*, 29–35.

36. Šegina, E.; Benac, Č.; Rubinić, J.; Knez, M. Morphometrical analyses of dolines—The problem of delineation and calculation of basic parameters. *Acta Carsol.* **2017**, in press.

37. LiDAR (Public Information of Slovenia, the Surveying and Mapping Authority of the Republic of Slovenia, LiDAR, 2015) ARSO, Ministry of the Environment. Available online: http://gis.arso.gov.si/evode/profile.aspx?id=atlas_voda_Lidar@Arso (accessed on 21 November 2017).

38. Bric, V.; Berk, S.; Oven, K.; Triglav Čekada, M. *Aerofotografiranje in Aerolasersko Skeniranje Slovenije*; Slovensko združenje za geodezijo in geofiziko: Ljubljana, Slovenia, 2014; Volume 20, pp. 57–71.

39. Mongus, D.; Horvat, D. Napredno orodje za obdelavo lidarskih podatkov. *Geod. Vestnik* **2015**, *59*, 153–158.

40. Mongus, D.; Lukač, N.; Žalik, B. Ground and building extraction from LiDAR data based on differential morphological profiles and locally fitted surfaces. *J. Photogramm. Remote Sens.* **2014**, *93*, 145–156. [CrossRef]

41. Mongus, D.; Žalik, B. Computationally efficient method for the generation of a digital terrain model from Airborne LiDAR data using connected operators. *IEEE J. Sel. Top. Appl. Earth Obs. Remote Sens.* **2014**, *7*, 340–351. [CrossRef]

42. Batayneh, A.T.; Abueladas, A.A.; Moumani, K.A. Use of ground-penetrating radar for assessment of potential sinkhole conditions: An example from Ghor al Haditha area, Jordan. *Environ. Geol.* **2002**, *41*, 977–983. [CrossRef]

43. Delle Rose, M.; Leucci, G. Towards an integrated approach for characterization of sinkhole hazards in urban environments: The unstable coastal site of Casalabate, Lecce, Italy. *J. Geophys. Eng.* **2010**, *7*, 143–154. [CrossRef]

44. Gómez-Ortiz, D.; Martín-Crespo, T. Assessing the risk of subsidence of a sinkhole collapse using ground penetrating radar and electrical resistivity tomography. *Eng. Geol.* **2012**, *149–150*, 1–12. [CrossRef]

45. Rodriguez, V.; Gutiérrez, F.; Green, A.G.; Carbonel, D.; Horstmeyer, H.; Schmelzbach, C. Characterizing sagging and collapse sinkholes in a mantled karst by means of ground penetrating radar (GPR). *Environ. Eng. Geosci.* **2014**, *20*, 109–132. [CrossRef]

46. Pueyo-Anchuela, Ó.; López Julián, P.L.; Casas-Sainz, A.M.; Liesa, C.L.; Pocoví-Juan, A.; Ramajo Cordero, J.; Perez Benedicto, J.A. Three dimensional characterization of complex mantled karst structures. Decision making and engineering solutions applied to a road overlying evaporite rocks in the Ebro Basin (Spain). *Eng. Geol.* **2015**, *193*, 158–172. [CrossRef]

47. Margiotta, S.; Negri, S.; Parise, M.; Quarta, T.A.M. Karst geosites at risk of collapse: The sinkholes at Nociglia (Apulia, SE Italy). *Environ. Earth Sci.* **2016**, *75*, 1–10. [CrossRef]

48. Zarroca, M.; Comas, X.; Gutiérrez, F.; Carbonel, D.; Linares, R.; Roqué, C.; Mozafari, M.; Guerrero, J.; Pellicer, X.M. The application of GPR and ERI in combination with exposure logging and retrodeformation analysis to characterize sinkholes and reconstruct their impact on fluvial sedimentation. *Earth Surf. Process. Landf.* **2016**, *42*, 1049–1064. [CrossRef]

49. Jol, H.M. *Ground Penetrating Radar: Theory and Applications*, 1st ed.; Elsevier Science: Amsterdam, The Netherlands, 2009; p. 524.

50. Shih, S.F.; Doolittle, J.A. Using radar to investigate organic soil thickness in the Florida everglades. *Soil Sci. Soc. Am. J.* **1984**, *48*, 651–656. [CrossRef]

51. Huisman, J.A.; Hubbard, S.S.; Redman, J.D.; Annan, A.P. Measuring soil water content with ground penetrating radar: A Review. *Vadose Zone J.* **2003**, *2*, 476–491. [CrossRef]

52. Lunt, I.; Hubbard, S.; Rubin, Y. Soil moisture content estimation using ground-penetrating radar reflection data. *J. Hydrol.* **2005**, *307*, 254–269. [CrossRef]

53. Simeoni, M.A.; Galloway, P.D.; O'Neil, A.J.; Gilkes, R.J. A procedure for mapping the depth to the texture contrast horizon of duplex soils in south-western Australia using ground penetrating radar, GPS and kriging. *Aust. J. Soil Res. Aust.* **2009**, *47*, 613–621. [CrossRef]

54. Breiner, J.M.; Doolittle, J.A.; Horton, R.; Graham, R.C. Performance of ground-penetrating radar on granitic regoliths with different mineral composition. *Soil Sci.* **2011**, *176*, 435–440. [CrossRef]

55. Zhang, J.; Lin, H.; Doolittle, J. Soil layering and preferential flow impacts on seasonal changes of GPR signals in two contrasting soils. *Geoderma* **2014**, *213*, 560–569. [CrossRef]

56. McNeill, J.D. *Electrical Conductivity of Soils and Rock*; Technical Note TN-5; Geonics Limited: Mississauga, ON, Canada, 1980; 22p.

57. Olhoeft, G.R. Electrical properties from 10^{-3} to 10^{+9} Hz—Physics and chemistry. In *Physics and chemistry of porous media II: Ridgefield, CT, 1986*; Jayanth, R.B., Joel, K., Kenneth, W.W., Eds.; American Institute of Physics: Ridgefield, CT, USA, 1987; pp. 281–298.

58. Doolittle, J.A.; Collins, M.E. A comparison of EM induction and GPR methods in areas of karst. *Geoderma* **1998**, *85*, 83–102. [CrossRef]

59. Mala ProEx—Professional Explorer Control Unit. Operating Manual. Mala. Available online: http://www.guidelinegeo.com/wp-content/uploads/2016/07/MALA-ProEx-Control-Unit-Manual-v.2.0.pdf (accessed on 23 November 2017).

60. Reynolds, J.M. *An Introduction to Applied and Environmental Geophysics*, 2nd ed.; Wiley: New York, NY, USA, 2011; p. 712.

61. Šušteršič, F. A conceptual model of dinaric solution doline dynamics. *Cave Karst Sci.* **2017**, *44*, 66–75.

remote sensing

MDPI

Article

Application of Ground Penetrating Radar Supported by Mineralogical-Geochemical Methods for Mapping Unroofed Cave Sediments

Teja Čeru [1,*], Matej Dolenec [1] and Andrej Gosar [1,2]

1   Faculty of Natural Sciences and Engineering, University of Ljubljana, Aškerčeva 12, 1000 Ljubljana, Slovenia; matej.dolenec@geo.ntf.uni-lj.si (M.D.); andrej.gosar@gov.si (A.G.)
2   Seismology and Geology Office, Slovenian Environment Agency, Vojkova 1b, 1000 Ljubljana, Slovenia; andrej.gosar@gov.si
*   Correspondence: teja.ceru@ntf.uni-lj.si; Tel.: +386-40-752-084

Received: 9 March 2018; Accepted: 17 April 2018; Published: 20 April 2018

Abstract: Ground penetrating radar (GPR) using a special unshielded 50 MHz Rough Terrain Antenna (RTA) in combination with a shielded 250 MHz antenna was used to study the capability of this geophysical method for detecting cave sediments. Allochthonous cave sediments found in the study area of Lanski vrh (W Slovenia) are now exposed on the karst surface in the so-called "unroofed caves" due to a general lowering of the surface (denudation of carbonate rocks) and can provide valuable evidence of the karst development. In the first phase, GPR profiles were measured at three test locations, where cave sediments are clearly evident on the surface and appear with flowstone. It turned out that cave sediments are clearly visible on GPR radargrams as areas of strong signal attenuation. Based on this finding, GPR profiling was used in several other places where direct indicators of unroofed caves or other indicators for speleogenesis are not present due to strong surface reshaping. The influence of various field conditions, especially water content, on GPR measurements was also analysed by comparing radargrams measured in various field conditions. Further mineralogical-geochemical analyses were conducted to better understand the factors that influence the attenuation in the area of cave sediments. Samples of cave sediments and soils on carbonate rocks (rendzina) were taken for X-ray diffraction (XRD) and X-ray fluorescence (XRF) analyses to compare the mineral and geochemical compositions of both sediments. Results show that cave sediments contain higher amounts of clay minerals and iron/aluminium oxides/hydroxides which, in addition to the thickness of cave sediments, can play an important role in the depth of penetration. Differences in the mineral composition also lead to water retention in cave sediments even through dry periods which additionally contribute to increased attenuation with respect to surrounding soils. The GPR method has proven to be reliable for locating areas of cave sediments at the surface and to determine their spatial extent, which is very important in delineating the geometry of unroofed cave systems. GPR thus proved to be a very valuable method in supporting geological and geomorphological mapping for a more comprehensive recognition of unroofed cave systems. These are important for understanding karstification and speleogenetic processes that influenced the formation of former underground caves and can help us reconstruct the direction of former underground water flows.

Keywords: ground penetrating radar (GPR); X-ray diffraction (XRD); X-ray fluorescence (XRF); karst; cave sediments; unroofed caves

---

1. Introduction

Unroofed caves are surface karst features, the result of a general surface lowering originating from the dissolution of soluble rocks (denudation), mainly carbonates. They present the former underground features that are now exposed on the surface [1,2]. Unroofed caves as karst phenomena have been thoroughly studied mostly in Slovenia [1–7]. However, some international studies also describe these phenomena but mainly without speleological or geomorphological interpretations [8–11].

Many new findings were made in karstological science during motorway network constructions in Slovenia (see a review of all the research in the book edited by Knez and Slabe [12]). Many surface, epikarst and subsoil karst features were revealed during the construction works and over 350 new caves were opened in the Karst region, including unroofed caves [12]. Many karst features have been neglected, including unroofed caves in karst studies, therefore such research provides invaluable information and contributes to the fundamental knowledge of karstology. However, apart from karst investigations incorporated, for instance, in the earthworks preceding motorway construction, studying karst environment is mostly limited to observing features that are visible on the surface. Traditional destructive methods such as drilling and trenching are time-consuming, expensive and often unfeasible due to the rough and inaccessible karst terrain. Surface and underground features are closely linked and the pre-existence of underground karstification has conditioned the formation of surface features [5,6], which is why geophysical methods in general and especially ground penetrating radar (GPR), are very useful non-invasive and relatively fast methods to provide subsurface information.

The term unroofed cave also includes flowstones and other allochthonous fluvial cave sediments that appear on the surface, even when caves are not morphologically expressed due to strong surface reshaping [2]. Since rock sculpturing and larger outcrops of flowstones as typical proofs of speleogenesis are usually absent, unroofed caves are often preserved only by their fine-grained alluvium material, the so-called "cave sediments." In such cases, geophysical methods used in favourable conditions can provide information about the presence of cave sediments on the surface and underground features related to unroofed caves. Unroofed caves and related sediment-filled karst features and their characterization have recently been successfully studied with GPR on Krk Island in Croatia [13]. Subsurface information obtained using a GPR with a 50 MHz RTA (Rough Terrain Antenna) reveals that the valley-like depressions, irregular depressions completely or partially filled with sediment and some dolines are associated with a nearly 4 km-long unroofed cave (developed as a result of karst denudation) [13].

Clastic cave sediments that encompass autochthonous and allochthonous cave sediments [14] have been the subject of many studies because they are important in paleoclimatic and paleogeographic reconstructions and help us understand the relationship between sedimentary and speleogenetic processes [14–19]. Although the objective of this paper is not to study the origin of the cave sediments, it should be noted that cave sediments can be of autogenic (originating from the cave) or allogenic (surface material transported with the stream flow) origin and are not homogeneous throughout the cave [20]. However, their mineral composition and origin are not easy to study. When cave sediments are exposed on the surface due to denudation processes, they are under the influence of intensive pedological processes [21]. Their mineralogical composition begins to change and their mineralogy turns out to be much more complex and heterogeneous.

Although GPR is widely used to investigate different soils, fewer studies observe the impact of the mineralogical and geochemical compositions of the soil on GPR results. Mineralogical and geochemical changes in composition can strongly affect the GPR effectiveness in the sense of the penetration of electromagnetic (EM) waves into the subsurface [22,23]. Conditions in the field can be very heterogeneous and many factors such as water content, grain size, mineralogical composition, organic-matter content and others may influence the propagation of the GPR signal in sediments [24]. These conditions can never be fully known when conducting GPR surveying in the field but complementary mineralogical analyses can provide very useful information for interpreting GPR images. Sediment properties in different karst areas can vary strongly, which means that a geochemical-mineralogical characterization should be done for each individual research area, which is time-consuming and requires additional funds.

The main goal of this research was to study the potential of the GPR method in detecting unroofed caves in regions where they are not morphologically expressed on the surface but partly revealed only by cave sediments preserved to a smaller extent. Because the research took place over a longer time period in different seasons, we also observed the influence of the various field conditions on GPR results. At the test location, the sampling of cave sediments and soils on carbonate rocks (rendzina) was carried out for geochemical and mineralogical analyses. They were completed in order to determine which factors might significantly influence GPR signal attenuation and lower the EM wave velocity.

2. Study Area

The study area (Lanski vrh) is located in the south-eastern part of the Planinsko polje, within the classical karst of western Slovenia (Figure 1) and is tectonically a part of the Adriatic carbonate platform. Planinsko polje extends along the Idrija strike-slip fault zone in the NW-SE direction. The wider surroundings of Planinsko polje are built of Upper Triassic dolomite as well as Jurassic and Lower and Upper Cretaceous carbonate rocks [25]. The study area of Lanski vrh consists mainly of Lower Cretaceous bituminous limestones with fossils (*Requienia, Miliolidae*) interbedded with grained dolomites and Upper Cretaceous rudist limestones and limestones with chert. The wider area is densely covered with solution dolines [26], with some collapse dolines (cave roof collapses appearing on the surface) visible as larger depressions on the LiDAR high resolution shaded relief image (Figure 1). Apart from all the mentioned features, there are many explored caves and also remnants of unroofed caves, which often appear at the (Cretaceous) limestone-dolomite contacts.

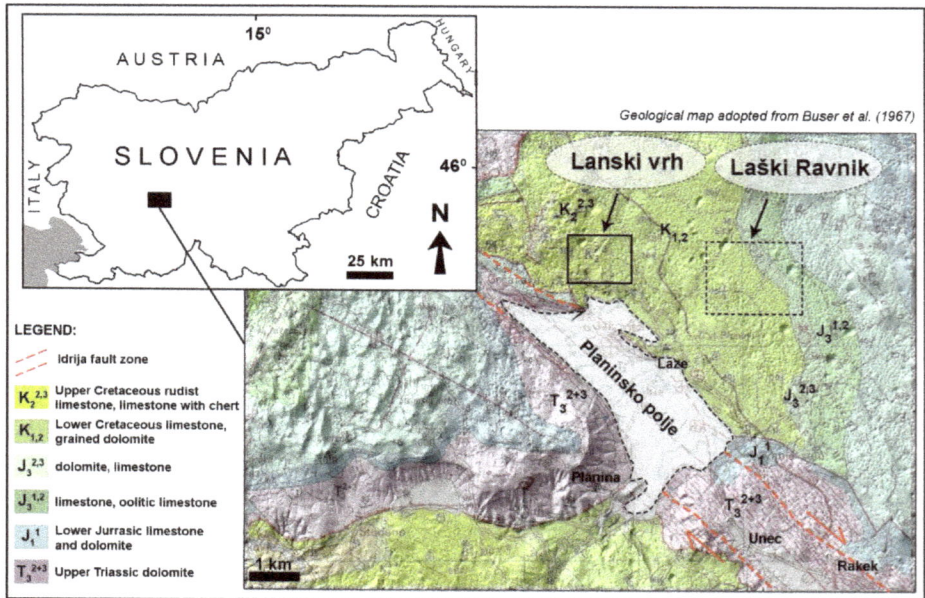

Figure 1. Location of the study area Lanski vrh and the basic geology [27] with the selected lithostratigraphic units of the narrow area on the LiDAR shaded relief image [28].

Cave sediments have already been systematically studied in the area of Laški Ravnik, which lies in the vicinity of our study area (Figure 1). Morphological and sedimentological indicators of unroofed caves, such as conglomerates, basal fills, flowstones, unroofed cave channels and openings of phreatic tubes were mapped in the area of Laški Ravnik [29]. The features now exposed on the surface were originally deep phreatic, oblique and relatively small with localized flowstone only [29].

Cave sediments in the study area (Lanski vrh) are mostly distinguishable by slightly yellowish or reddish colour compared to the brownish colour of the soil on carbonate rocks, the so-called "rendzina" (Figure 2a,b). In some places, we can find preserved flowstones (Figure 2c,d) that could be indicators for speleogenesis and septarian concretions, the latter being one of the most reliable indicators for speleogenesis. However, such indicators are generally not preserved in the area. Cave sediments can sometimes also be revealed by deepenings on forest roads, requiring the road surface to be frequently filled to keep it passable (Figure 3a). In these areas, water retention is evident, lasting even through dry periods in summer, so it is often necessary to dig gutters to drain the redundant water (Figure 3b).

The shape of the unroofed cave depends on the size, shape and type of the former underground cave, as well as on the incline of the surface which cut through the cave. A theoretical model of unroofed cave development is shown on the case of a simple horizontal cave passage (Figure 4). The underground passage developed in the first phase during karstification processes. In the second stage, the water table starts dropping, which results in the passage gradually becoming completely dry. During this stage, cave sediments are being deposited in the passage and speleothem growth begins. At some point, denudation reaches the passage and the cave starts to disintegrate, becoming a part of the surface topography. Therefore, cave sediments appear on the surface or are located right below the surface. In some places, the surface flowstone is well preserved but it mostly occurs locally in smaller pieces. Strong surface reshaping masks the primary form of the cave, which is why detecting cave sediments with GPR can help us follow and connect certain surface features into unroofed cave systems. The areas on which we focused in this study are presented in a simplified model (see the red ellipses in Figure 4).

Figure 2. Cave sediments and flowstone as typical indicators for speleogenesis: (**a**) Comparison of the colour difference between cave sediments and soils on carbonate rocks (rendzina); (**b**) Closer view of the cave sediments; (**c**) Cave sediments are accompanied with occurrences of flowstone in some places; (**d**) Outcrops of flowstone are often covered with moss, making it easier to recognize them compared to outcrops of limestones.

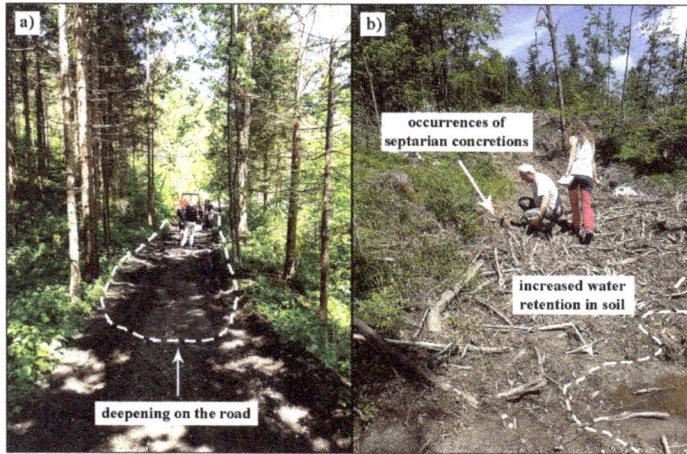

Figure 3. Indirect indicators for the possible presence of cave sediments: (**a**) Cave sediments are sometimes associated with surface deepening, so the road must be frequently filled to keep it passable; (**b**) Increased water retention in sediments even during dry periods is evident in certain places. In the area below the increased water retention septarian concretions were found as one of the most reliable indicators of speleogenesis.

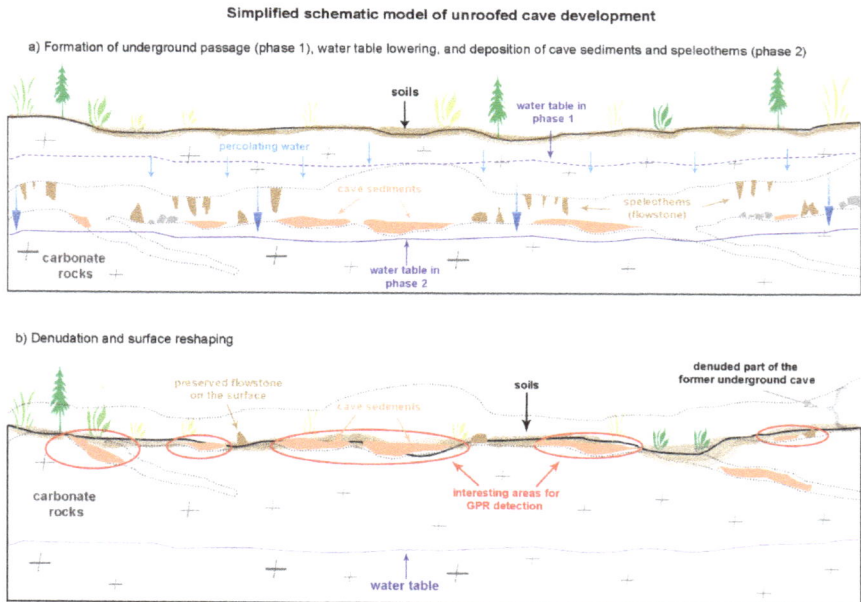

Figure 4. Schematic model of unroofed cave development: (**a**) Underground horizontal passage formed during karstification. When the water table started dropping, cave sediments were deposited and speleothems began to form. Dry passage filled with allochthonous and autochthonous sediments; (**b**) Denudation processes cut off the passage's upper parts, so cave sediments are revealed on the surface. Weathering and surface processes reshape the karst surface.

3. Materials and Methods

First, we selected the locations of testing profiles where the presence of cave sediments was proven by direct indicators of speleogenesis such as flowstones, septarian concretions and the typical colour of cave sediments. We chose three such locations to test how the cave sediments are reflected on radargrams and to evaluate the GPR method for accurate characterizations of such areas. Profiles 1, 2 and 3 at testing locations (Figure 5) were measured where undeniable indicators proved the presence of cave sediments. At one test location (Profile 1), we obtained samples for further mineralogical-geochemical analyses to identify the possible differences in mineral composition between cave sediments and soils on carbonate rocks that can influence the GPR image.

After measuring the testing profiles, where cave sediments are clearly evident on GPR images as strongly attenuated areas, profiling was performed on other locations (Profiles 4, 5 and 6), where direct indicators for unroofed caves are absent or inadequate and appear only sporadically. We selected these locations to detect the spatial extent of cave sediments where there is no direct evidence for their presence, even though outcrops of flowstones and/or cave sediments are visible in the vicinity (Profiles 4 and 5). Occasionally, indirect indicators such as slight deepenings and water retention in sediments even during dry periods can indicate their presence. During the profiling, we marked all the possible signs for the presence of cave sediments on radargrams for an easier integration of GPR data and geomorphological mapping. Some profiles that are not presented in this study are given in orange lines in Figure 5.

Since the extensive research also included other research issues which took place over a longer period of time, we conducted the GPR surveys in various field conditions in different seasons (see Table 1). In this paper, only radargrams of Profiles 1, 2a and 2b are presented, measured in various field condition. Most of the profiles were measured in dry periods to avoid additional attenuation in wet field conditions.

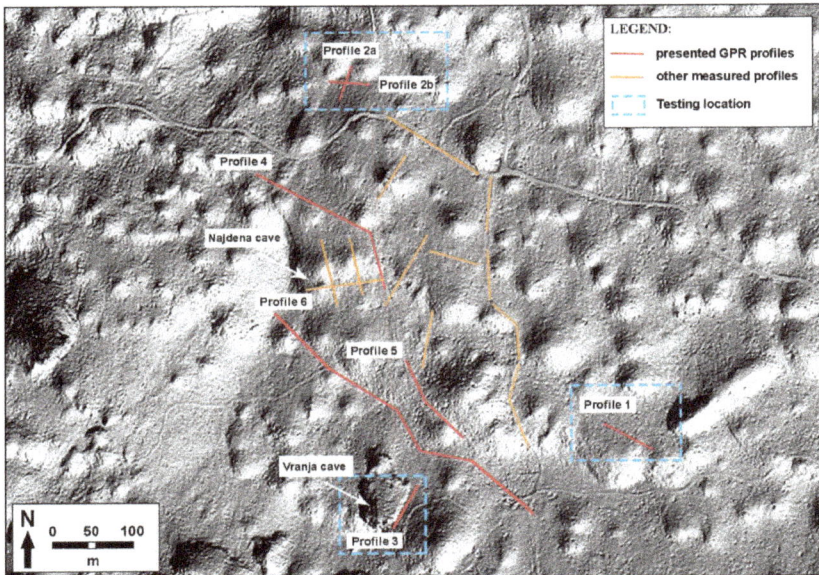

Figure 5. Locations of ground penetrating radar (GPR) profiles at the investigated area on the LiDAR shaded relief image. Testing locations (blue rectangles) were chosen where cave sediments and flowstone occur on the surface.

Table 1. Presented profiles with basic characteristics.

Profile	Antenna Frequency	General Direction	Period of Measuring	Field Conditions	Indicators for the Presence of Cave Sediments	
Profile 1	50 and 250	SE-NW	July 2016 March 2017	very dry dry	colour of sediments, flowstone	
Profile 2a	50 and 250	SSW-NNE	March 2017 January 2018	dry moderately wet	colour of sediments, small pieces of flowstone	Testing Profiles
Profile 2b	50 and 250	E-W	March 2017 January 2018	dry moderately wet	colour of sediments, small pieces of flowstone	
Profile 3	250	SSW-NNE	July 2016	very dry	colour of sediments, flowstone	
Profile 4	50 and 250	SSE-NNW, ESE-WNW	January 2018	moderately wet	deepening on the road, water retention, septarian concretions in cave sediments	
Profile 5	250	NNW-SSE	July 2016	very dry	deepening on the road, water retention	
Profile 6	50	SE-NW	July 2016	very dry	flowstone at one location	

In the second stage of the research, at the location of Profile 1 (Figure 5), we drilled three boreholes with auger drilling equipment to collect the samples for the X-ray diffraction (XRD) and X-ray fluorescence (XRF) analyses. Borehole V1 and V2 were located in the area of cave sediments, while borehole V3 (at a distance of 10 m from borehole V2) was located in the soil on carbonate rocks (rendzina). Manual drilling into cave sediments was possible to the depth of 85 cm in V1 and 95 cm in V2. The depth of 85 cm in V3 represents the contact between soil and the carbonate bedrock. Samples were collected every 20–35 cm. The colour of samples by depth is shown in Table 2.

Table 2. The soil colour of the samples.

	Sample	Depth (cm)	Colour (Munsel Soil Chart)
cave sediments V1	1V1	0–20	7.5YR 5/6–7.5YR 5/8
	2V1	20–50	5YR 3/4
	3V1	50–85	5YR 3/4
cave sediments V2	1V2	0–20	7.5YR 4/4
	2V2	20–40	5YR 4/4
	3V2	40–70	5YR 4/4
	4V2	70–95	5YR 4/4
soil on carbonate rocks (rendzina) V3	1V3	0–20	10YR 5/3
	2V3	20–50	10YR 4/3
	3V3	50–85	10YR 4/4–10YR 4/6

3.1. Ground Penetrating Radar (GPR)

The number of various applications of the GPR method has increased in the past 20 years along with the number of studies examining various issues in karst environments. This indicates a wide range of potential uses for the GPR method in the last ten years [30–34]. GPR as a high-resolution geophysical technique has also been widely used in soil surveys to estimate taxonomic composition, to detect diagnostic soil horizons and evaluate their depth, to assess variations in soil properties, to determine soil water content and to determine subsurface flow pathways and water table [35–42]. The most commonly used antennas for soil investigations regarding the soil condition and subsurface features are those with centre frequencies between 100 and 500 MHz or lower-frequency antennas if the soils cause high attenuation [43]. Apart from the many mentioned applications of the GPR method, its use in karst environments has been one of the most successful applications [44].

In our study, we used the ProEx (MALÅ Geoscience, Sweden) GPR with a 50 MHz unshielded Rough Terrain Antenna (RTA) and a shielded 250 MHz antenna to observe the information obtained by measuring with two different frequencies and systems (shielded and unshielded).

Profile measurements were carried out along macadam roads or forest paths due to the even and cleared terrain, which allows better ground contact for the antenna and thus provides higher-quality data. The survey was performed during various field conditions but we obtained the major part of the measurements in dry periods to minimize the influence of moisture on signal attenuation in sediments, which can be inherently strongly conductive depending on their mineralogical composition. When studying regolith bodies (the term regolith as suggested by Šušteršič [26] covers the great variety of heterolithic sedimentary bodies in karst environments when their origin is not known), which are known as highly conductive, sediments properties should be known for a consistent and certain interpretation of the GPR data. However, karst is one of the most complex and unpredictable environments, where features are very heterogeneous and the soil/bedrock interface is uniform, which represents a challenge for geophysical exploration [45]. Therefore, any additional information about the subsurface constitute an important contribution in the interpretation of geophysical data.

All profiles were processed and analysed using the Reflex-Win software of Sandmeier Geophysical Research. The following processing steps were employed: editing, subtract-mean (dewow), time zero correction, background removal, gain (manually-defined exponential corrections) and bandpass filtering. Lastly, we used subtracting average to eliminate periodic ringing noise and applied topographic correction where necessary. The raw data and the two radargrams that include some processing steps are presented on the case of Profile 1 (Figure 6). Various gain functions were applied to profiles in order to achieve a good visualization of the GPR images. Energy decay and automatic gain control (AGC) with different time windows were applied first but were not sufficient. Only the manually-defined exponential corrections, where the operator can emphasize the signal in the areas of assumed structures gave the most effective results.

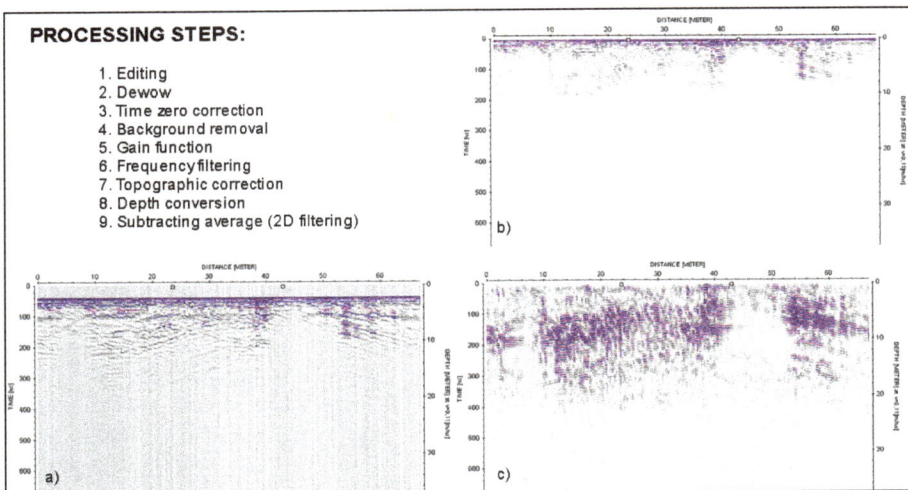

Figure 6. Processing steps (1–9) applied in profiles. Raw data and some processing steps are presented for Profile 1: (**a**) Raw data; (**b**) After dewow and time zero correction; (**c**) After background removal, manually-defined exponential gain correction, frequency and 2D filtering.

3.2. X-ray Diffractometry (XRD)

To characterize the crystalline minerals of the samples, we applied X-ray powder diffraction using a Philips PW3710 X-ray diffractometer equipped with CuKa radiation, a proportional counter and a secondary graphite monochromator operated at a voltage of 40 kV with a filament current 30 mA. Scanning speed was 3.4° 2θ per minute in the range from 2° to 70° 2θ.

Clay minerals were identified from the clay fractions of bulk samples obtained from an oriented glass slide using a combination of a kitchen-grade blender, an ultrasonic probe and ultracentrifugation [46]. The following diagnostic treatments were carried out for all the samples: ethylene glycol solvation at 70 °C overnight, glycerol solvation at 70 °C and heating to 550 °C for 2 h each. Diffraction patterns were identified with X'Perth HighScore Plus 4.6a software using the PAN-ICSD database and Rietveld refinement that applied the Pseudo-Voigt function for quantitative mineral phase analysis.

3.3. X-ray Fluorescence (XRF)

The elemental composition of the samples was measured by the XRF method with a handheld ThermoFisher NITON XL3t-GOLDD 90S-He analyser using the factory setting Mining mode. We determined the presence of Si, Al, Fe, Mg, Ca, K, S, Cu, Mn, Mo, Nb, Pb, Rb, Sr, Zn, Zr, Ti and Ba. During the measurements helium gas was supplied at the analysed site for better light elements (Mg to S) energy detection. At the beginning and at the end of sample measurement, accuracy and precision were assessed against pre-calibration with 24 international standards (NIST and USGS) and calibration with NIST-1d and NIST-88b. To better constrain the relationship between elemental and mineralogical variability among the samples, statistical analyses (PCA—principal component analysis) were performed using STATISTICA 8.0 and Grapher 10.0 software.

4. Results

4.1. GPR Results at Testing Locations

At the three test locations (Figure 5), the typical yellowish or reddish colour of cave sediments and outcrops of flowstones prove that we are researching the remnants of a former underground cave. Some profiles are presented by comparing the results obtained with different frequencies (Profiles 1, 2a, 2b, 4). Areas with higher wave attenuation related to larger regolith bodies including cave sediments are clearly visible on both radargrams recorded with the 50 MHz antenna as well as with the 250 MHz. Therefore, measurements of Profiles 3, 5 and 6 are presented only by a one-frequency GPR image (50 or 250 MHz).

At the testing location of Profile 1, we obtained measurements where cave sediments and flowstone occur on the forest path (Figure 7a). Attenuation resulting in the absence of reflections is clearly visible both on 50 MHz and 250 MHz images (Figure 7b,c). Penetration of the signal is completely restricted in the area of cave sediments, compared to its surroundings where the depth of penetration for the 50 MHz and 250 MHz antennas reached approximately 15–20 m and 4–8 m, respectively. In the area of cave sediments, we drilled boreholes V1 and V2 to the depth of 85 cm and 95 cm, respectively (Figure 7d). The depth of the boreholes does not represent the total thickness of cave sediments because manual drilling with an auger was impossible to greater depths. The signal in the area of borehole V3 (Figure 7e), where the contact between soil and the bedrock occurs at the depth of 85 cm, is not as attenuated as in the area of cave sediments (see the comparison of signal amplitudes in Figure 7c).

In the area of Profiles 2a and 2b (Figure 8a), cave sediments were revealed on the surface by a fallen tree where the roots brought up underground material. In addition, we can find pieces of flowstone in the vicinity. Two profiles were measured in two directions perpendicular to each other to detect the spreading of the cave sediments. Radargram of Profile 2a clearly shows the direction and extent of cave sediments, where signal is strongly attenuated (Figure 8b,c). The attenuated area in the radargram of Profile 2b is limited to the outcrops of cave sediments on the surface which indicates that the direction of this profile was oriented transversally to the spatial extent of the sediments (Figure 8c).

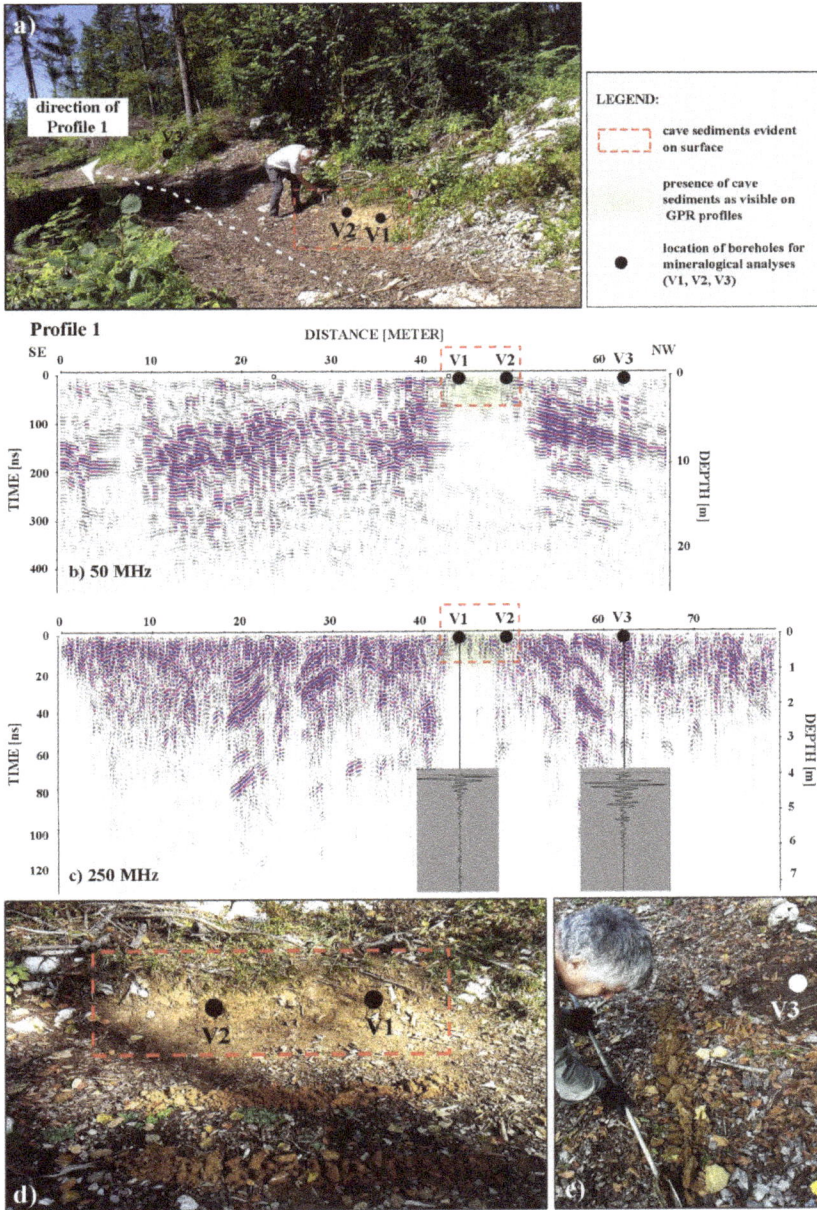

Figure 7. (**a**) Direction of Profile 1 acquired over an area of cave sediments; (**b**) A processed radargram with highly attenuated area presenting cave sediments with 50 MHz antenna and (**c**) with 250 MHz antenna (vertical exaggeration 1:4); (**d**) Locations of V1 and V2 boreholes in cave sediments; (**e**) Sampling with a manual drilling machine in soils on carbonate rocks (V3).

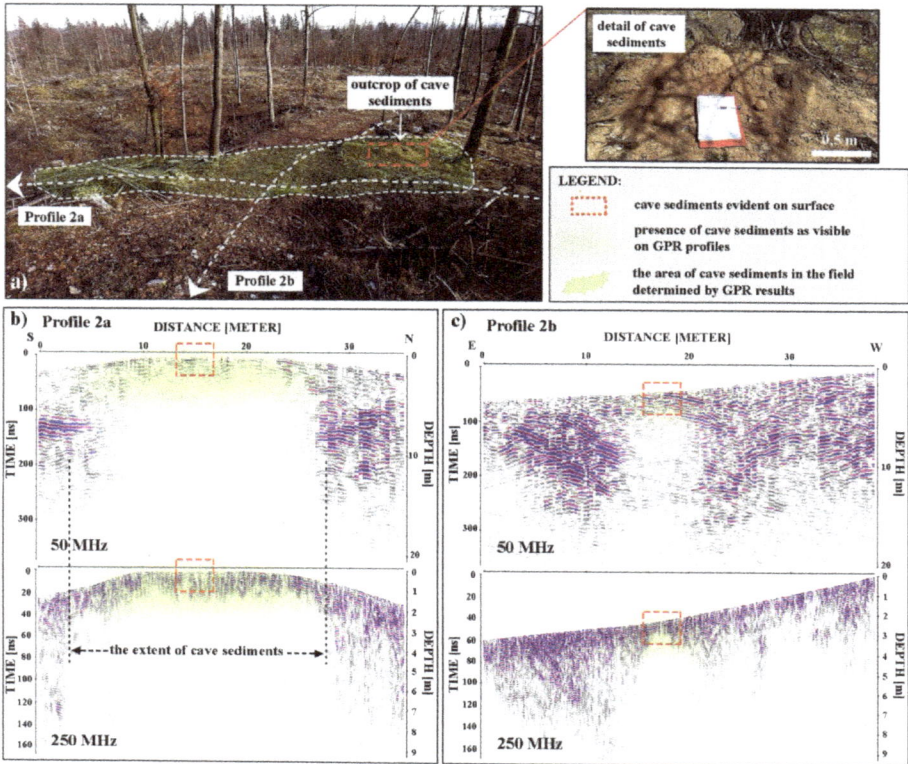

Figure 8. (**a**) Direction of perpendicular Profiles 2a and 2b recorded with a 50 and 250 MHz antenna (vertical exaggeration of the 250 MHz GPR images is 1:2) in the area of the cave sediments outcrop to determine their spatial extent; (**b**) Extent of cave sediments was determined in Profile 2a; (**c**) Boundaries of cave sediments detected by GPR are limited to the area of the outcrop visible on the surface.

Larger attenuated areas restricted to 22 m in length in the areas of Profiles 1, 2a and 2b, as visible on the radargrams, suggest that these areas are related to cave sediments, where pieces of flowstone also indicate their presence. Geomorphologically, these areas are relatively flat with an elevated surrounding relief. They could represent the very last surface manifestation of the segments of the unroofed cave system and cave channels, like the ones visible in the nearby area of Laški Ravnik (see Figure 3 in [29]). An additional sign for identifying unroofed caves are the geomorphological features of the surrounding terrain. Depending on the cave type and the manner in which denudation cut the cave, unroofed caves can appear as elongated dolines, series of dolines, shafts or linear to irregular depressions completely filled with sediments [2,6,7]. Apart from that, surface processes can strongly reshape and modify the segments of the former cave, so they are sometimes completely masked, meaning that their speleogenetic origin cannot be demonstrated.

In the area of Profiles 1, 2a and 2b, an elongated depression (Figure 9a) and series of dolines with intermediate ridges (Figure 9b) support the existence of segments of unroofed cave in these areas.

Figure 9. Surface topography in the vicinity of detected cave sediments: (**a**) An elongated depression at the location of Profile 1; (**b**) The surface morphology in the area of Profiles 2a and 2b, where depressions pass from one to another with intervening small ridges.

The last test location is situated in the vicinity of the collapse doline which is also the entrance to the Vranja cave. Profile 3 was measured in the area of a flowstone outcrop and the processed profile revealed a large attenuated area which extends over the larger area beyond the measured line (Figure 10). The penetration of the signal is completely restricted (distance 6–50 m) compared to its surroundings where the depth of penetration for the 250 MHz antenna reached to the approximate depth of 6–7 m, which can be seen at the beginning of the profile.

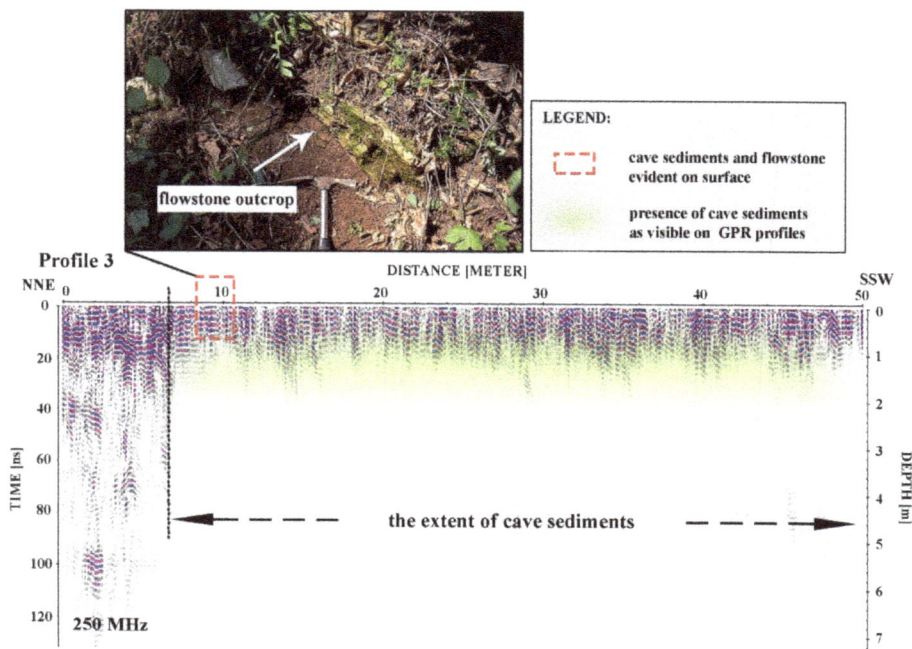

Figure 10. Radargram acquired with a 250 MHz antenna, where the spatial extent of cave sediments was revealed along Profile 3.

4.2. A Comparison of GPR Images Conducted in Different Time Periods

Studying the effect of seasonal soil water dynamics on GPR signal changes is usually conducted to reveal subsurface flow dynamics and patterns [35]. In this study, profiles were measured in different site conditions to determine how a difference in water content influences the GPR data and to verify if measuring is appropriate in all time periods regardless of the weather conditions before profiling.

A comparison of radargrams conducted in various field conditions is illustrated in Figure 11. We selected only two profiles to show the results because comparisons of other profiles are quite similar. As we can see from the radargrams of Profiles 1 and 2b, anomalous zones (highly attenuated areas associated with cave sediments) are clearly visible regardless of the field conditions during the measuring. Comparing the radargrams of Profile 1, we can see some smaller differences in the continuity of the reflections. In very dry conditions (radargram measured in July 2016 in Figure 11a), greater continuity of a slightly dipping reflector (distance 30–40 m at the depth of 1–3 m) can be seen in comparison to the radargram conducted in a slightly wetter but still generally dry time period (radargram measured in March 2017 in Figure 11a). On the GPR images of Profile 2b there is essentially no difference between the results of the surveys conducted in a relatively dry time period (radargram measured in March 2017 in Figure 11b) and in moderately wet field conditions (radargram measured in January in Figure 11b). We can conclude that measuring in different weather conditions does not significantly affect the quality of GPR images for the purpose of this study where detecting cave sediments was the main goal.

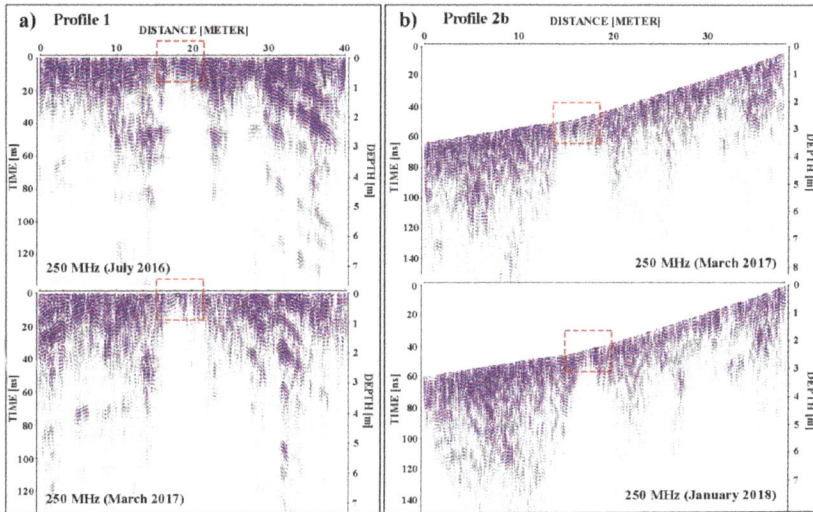

Figure 11. The GPR images obtained in various field conditions using the 250 MHz antenna and the area of cave sediments (dashed red lines): (**a**) Profile 1 acquired in July 2016 (very dry field conditions) and in March 2017 (dry field conditions); (**b**) Profile 2b acquired in March 2017 (dry field conditions) and in January 2018 (moderately wet field conditions).

4.3. GPR Results at Other Locations

After profiling at the test locations, where cave sediments are clearly evident on the profile, we measured other profiles (Profiles 4, 5 and 6 presented in this study) where direct indicators for the presence of cave sediments are absent or appear only locally. Profiles were also positioned in areas where traces of water retention were observed in the field (Profiles 4 and 5).

In Profile 4, three anomalous areas are clearly seen on the radargram (Figure 12b,c). First, the concave downward anomaly in area A is seen at the beginning of the profile (distance 15–70 m in the 50 MHz image and 22–65 m in the 250 MHz image). The location of the anomaly detected by GPR corresponds with the position of the two successive elongated depressions (Figure 12a). Considering the GPR results and the geomorphological characteristics of the closer area, we concluded that this area was a depression that was filled and compacted with material during the construction of the road. Furthermore, the filled depression below the road and the two elongated depressions most probably belong to the segments of an unroofed cave, especially due to the fact that flowstone occurs in two places along the depressions on the line from the road to the entrance of Najdena cave. The concave downward reflection in the profile is completely contrary to our expectations for a filled circular depression. The study completed by Goodman and Piro [47], in which they modelled a deeply buried circular trench, revealed that the shapes recorded on raw radargrams may have a completely opposite structure in the ground and can be mistakenly interpreted as a reflection pattern due to a buried cylindrical object.

The anomalous areas B and C in Profile 4 are characterized by a strong attenuation of signal on both radargrams (Figure 12b,c). The signal amplitude is totally attenuated in area C compared to its surroundings (see signal amplitudes in Figure 12c).

After re-examining the terrain where the anomalies were detected, we observed four potential areas of water retention recognizable as slight deepenings on the road (green rectangles in Figure 12). In the immediate vicinity of the detected anomaly C, septarian concretions also prove the presence of the cave sediments. However, only the areas of anomalies B and C detected by GPR correspond to the observed deepenings in the field. Furthermore, the locations of these two areas are also consistent with the surrounding topography, where highly attenuated areas B and C related to cave sediments correspond to depressions (Figure 12d,e). The recognized deepening on the road at the distance of 60 m is most probably connected with anomaly A.

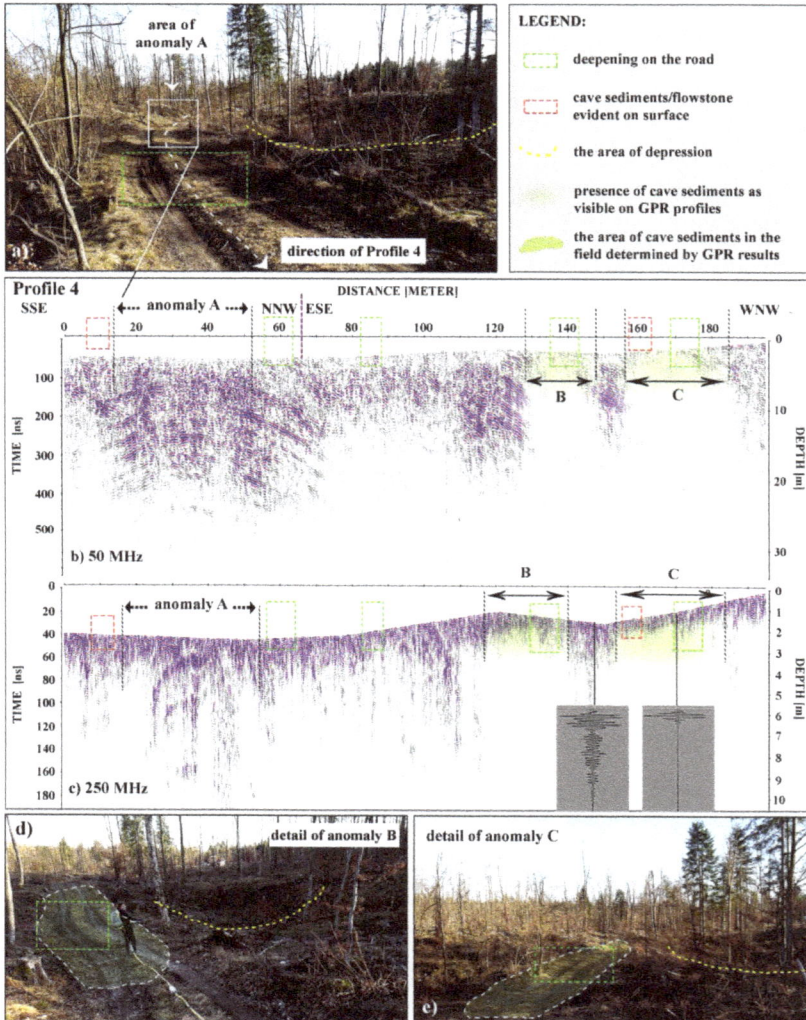

Figure 12. (**a**) The starting point and direction of Profile 4 with the surface depression which appears in the area of anomaly A detected by GPR. The surface deepening is also visible on the road; (**b**) The radargram conducted with the 50 MHz antenna with the detected anomaly A and attenuated areas B and C interpreted as cave sediments; (**c**) The radargram conducted with the 250 MHz antenna (note the vertical exaggeration 1:5) with a comparison of signal amplitudes, where the signal is strongly attenuated by the depth in the area of cave sediments even after the processing steps; (**d**) Surface morphology in the area of anomaly B, where the depression and the deepening on the road occur—the attenuated areas were interpreted as cave sediments; (**e**) The extent of cave sediments on the field as detected by GPR and the accompanying occurrences of the road deepening.

Profile 5 was positioned on a large area of a deepening on the road, where the road needs to be frequently filled with coarse-grained material and gutters must be dug to drain the redundant water (Figure 13a). In the vicinity, large outcrops of flowstone appear on the surface on a steep slope below the road (Figure 13b,c). According to the results of the profiles at testing locations, we expected a similar GPR image with a strongly attenuated area but the radargram recorded with the 250 MHz antenna (the same results were also obtained with the 50 MHz antenna) does not show any clear evidence of cave sediments. The signal is slightly more attenuated in the area between 0–35 m compared to the other parts of the radargram (Figure 13d) but not as strongly as in Profiles 1, 2, 3 and 4. This could be explained by the fact that a lot of material has been deposited in this area through the frequent reconstructions of the road. The strong reflection visible at the distance of 18 m down to the depth of 9 m could be the result of reflections from a larger block of filling material. However, due to the unclear results obtained by GPR, we interpreted this area as having a possible presence of cave sediments (Figure 13d).

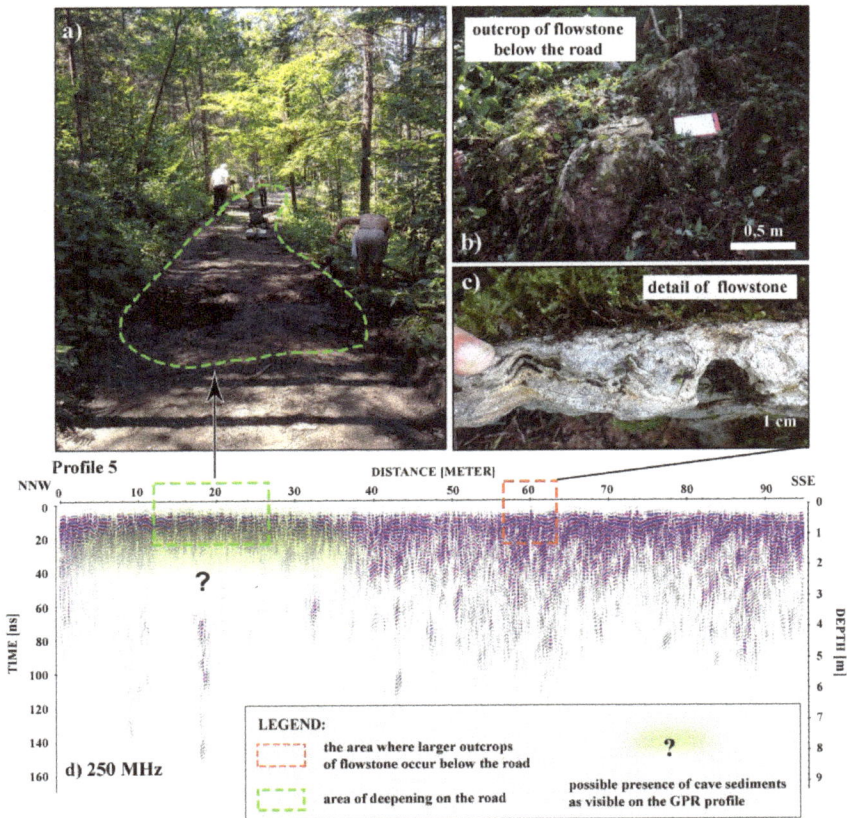

Figure 13. (a) Roadwork, where slightly deepened areas need to be filled with coarse-grained material to keep the road in function; (b) Large outcrops of flowstone occur below the road; (c) A detail of flowstone structure; (d) In the radargram of Profile 5, the area of the possible presence of cave sediments is not as obvious as in other profiles.

Profile 6 was positioned along the road to observe a possible connection with the detected attenuated area in Profile 3 and to discover other possible interconnections between different depressions. As it is visible in Figure 14, four large areas of high attenuation are evident. Superficial indicators for cave sediments are only evident at one location (at the distance of 235–265 m). However, the attenuated area at 155–175 m is almost certainly related to cave sediments, because its position corresponds to the cave sediments detected in Profile 3. The other two locations (at the distance of 105–125 m and 400–455 m) most probably belong to cave sediments as well but we should not exclude the possibility of other sedimentary regolith bodies. These areas were therefore classified as having a possible presence of cave sediments, even though there are no superficial indicators for their presence.

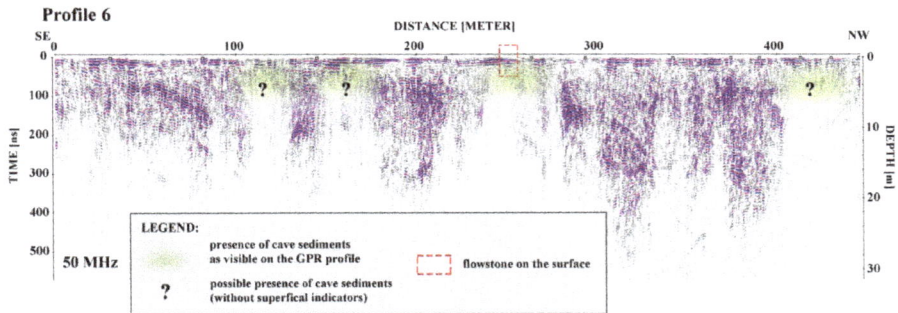

Figure 14. The radargram of Profile 6 with four detected attenuated areas. Flowstone only occurs on the surface at one location, so other locations were interpreted as "possible locations of cave sediments" (see legend). The attenuated area at the distance of 155–170 m corresponds to the direction of the cave sediments extent detected in Profile 3.

4.4. XRD and XRF Results

The quantitative XRD analysis shows a rather similar mineralogical composition of cave sediments and soils on carbonate rocks (rendzina), where vermiculite (Vrm), chlorite group minerals (Chl), muscovite/illite (Ms/Ill), kaolinite (Kln) and quartz (Qz) were detected in all samples (Figure 15). Hematite (1–2%) is present in some samples of cave sediments and rutile was detected in one surface sample from V1. Na-plagioclase feldspar (7–10%) are present only in soils on carbonate rocks.

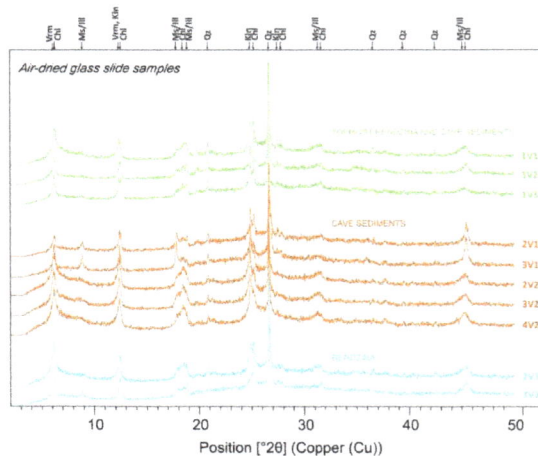

Figure 15. X-ray diffraction (XRD) patterns of air-dried oriented samples with the determined minerals (Vrm-vermiculite, Chl-chlorite group minerals, Ms/Ill-muscovite/illite, Kln-kaolinite, Qz-quartz).

Principal component analysis (Figure 16) explained 64.63% of the variance in the data by the first two components (PCA1 and PCA2). It is a procedure for finding hypothetical components that account for as much of the variance in multidimensional data as possible [48]. Samples from location V3 are associated with the high values of Si, Sr, Mg and Ca related to the well-crystallized carbonates, Na-plagioclase feldspar, quartz and chlorite group minerals. These samples also contain lesser amounts of clay minerals (kaolinite, illite and vermiculite). A high proportion of Al, Fe, Mn, Ti, S, Rb, Ba, As, Zn, Zr, Nb, Cu, Pb and Mo belongs to samples of cave sediments (V1 and V2). Most of them are related with clays (Kln, Vrm), as well as Fe/Al/Ti oxides/hydroxides. Fe/Al/Ti oxides/hydroxides were not detected by XRD due to broad reflections and poor crystallinity of such minerals.

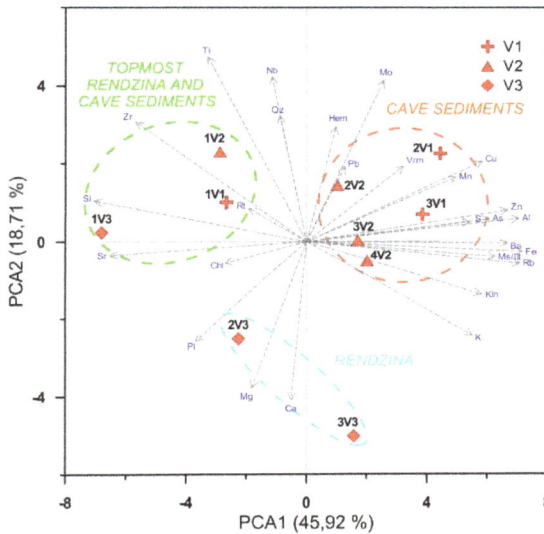

Figure 16. PCA (Principal Component Analysis) diagram for the overall XRF and XRD datasets.

4.5. Integration of Results

The synthesis of results from different methods used in this study is shown in Figure 17. For a comprehensive determination, many techniques and methods should be used to minimize the ambiguities in interpretation. Field work, which includes geomorphological mapping and locating all surface indicators of speleogenesis on a LiDAR map, is very important. Based on terrain observations, GPR profiling can be properly planed and locations for the mineralogical and geochemical analyses can be selected. For a proper reconstruction of unroofed cave systems, each location of cave sediments must be specifically investigated due to heterogeneous mineral composition of the cave sediments. The lateral comparison of signal attenuation between the cave sediments and soils in surrounding proved to be very good indicator for the cave sediments presence. However, signal attenuation can occur also in areas of filled shafts or other regolith bodies. Due to the very complex and heterogeneous characteristics of karst underground system, which is full of voids, different features of various sizes and shapes, automatic recognition of cave sediments cannot be provided. Besides that, cave sediments can have very heterogeneous mineral composition and their extent can be of various size and shape. In order to provide possible automation of the proposed procedure (e.g., [49]), additional mineralogical-geochemical analyses should be done in the future and the electrical resistivity tomography (ERT) as a complementary method can be conducted in the further research. However, due to the locally rough karst terrain and the time-consuming nature of ERT, we decided to apply only the GPR method for the purpose of this study.

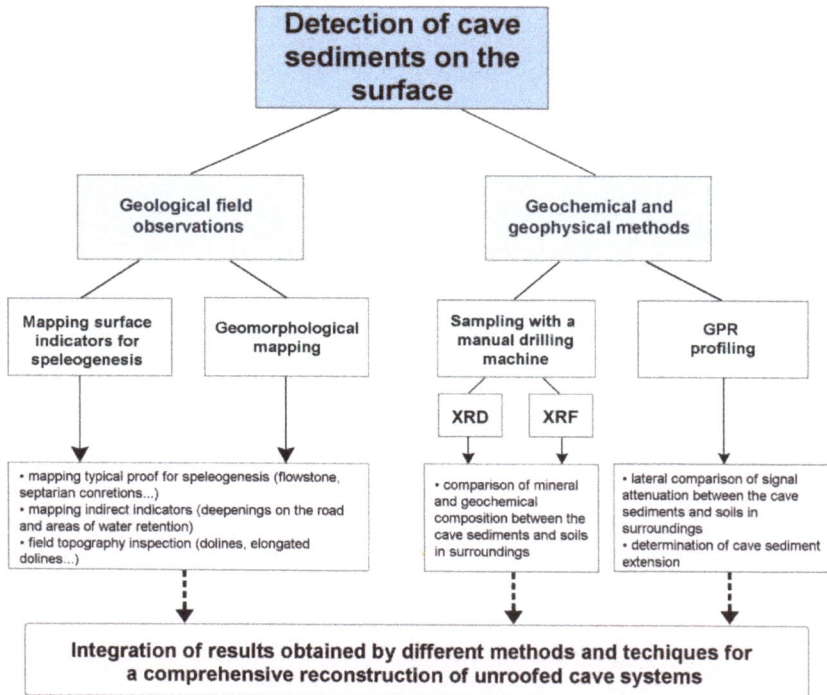

Figure 17. Diagram shows the used methods with the main results, as needed to delineate unroofed cave systems into a whole.

5. Discussion

Cave sediments at all testing locations (Profiles 1, 2 and 3), where outcrops of flowstone indicate their presence, are clearly shown as highly attenuated areas on radargrams (Figures 7, 8 and 10). Signal amplitude is completely attenuated and the depth of penetration consequently restricted in these areas (Figures 7c and 12c). As field examinations and geological mapping are mostly insufficient to follow the extent of cave sediments in the whole area of interest, due to limited occurrences of indicators on the surface, GPR proved to be a very useful tool to determine their spatial extent. The spatial extent of cave sediments and the direction of the unroofed cave system were determined in two testing locations (see Profiles 2a, 2b and 3 in Figures 8b and 10).

The relatively complex terrain of the study area, full of dolines, complicates the task of recognizing the various features as a cohesive whole model, compared to unroofed cave systems observed in Classical Karst, where segments of unroofed cave systems are easier to identify (e.g., [1,2,5,7]). To study the location, size and shape of unroofed caves, high-resolution LiDAR data proved to be very useful [2]. Where LiDAR data are not available, aerial photos can also be an indispensable tool in planning the proper location of GPR profiles [13]. Nowadays, remotely gathered data (e.g., LiDAR) in conjunction with the ground penetrating radar method and field observations enable easier and comprehensive studying of various geomorphological issues [50–58]. Previously unrecognized geomorphological features can now be revealed by LiDAR data in combination with different geophysical methods. It is important to note that all GPR results should be correlated and integrated with field topography observations. The extent of cave sediments derived from GPR profiling was further placed on the LiDAR image to facilitate the final interpretation (Figure 18). In addition, the terrain was meticulously re-examined after GPR measurements to determine any possible connections of the detected anomalies with the surrounding topography (Figures 9 and 12a,d,e).

Cave sediments were also clearly evident in other GPR profiles, where cave sediments or flowstone occur only occasionally in the vicinity of the GPR lines. All detected anomalies visible as highly attenuated areas in the presented profiles are related to the larger sedimentary regolith bodies. In the areas where superficial indicators complement their speleogenetic origin, these thicker sediment occurrences can be defined as cave sediments (Figures 12 and 13).

However, it should be noted that it is not always possible to distinguish between various larger regolith bodies filled with sediments on the basis of their mineralogy. Due to their vertical shape, only karst cutters can collect sufficient insoluble residue, all other accumulation of regolith bodies must be admixed materials of diverse origins or be completely allogenic in origin [20]. With GPR we can detect the increased thickness of sediments, while the sediments can be of various origins. In karst environments, the origin of various surface features cannot always be determined but anomalous areas detectable by GPR contribute to the information which is otherwise often invisible from the surface.

During the extensive research, measurements were carried out in different field conditions to observe how cave sediments reflect on the radargrams depending on different water content. Loamy and clayey materials strongly attenuate radar signals, so various horizons or layers of soil materials cannot be distinguished when their electrical properties are too similar. The penetration depths can be severely restricted in clayey materials in areas of karst [37,59]. However, such attenuated areas can be good indicators for the presence of thicker sediments, which is important evidence in detecting unroofed caves [13].

The depth range of GPR surveys is dependent mostly on field conditions including hydrogeological characteristics, climate and weather conditions (moisture content, infiltration, subsurface water flow, etc.) and many other factors. The impact of the above-mentioned factors on the characteristics of a certain terrain is very complex, especially in karst environments. Only a small number of published papers studies the influence of seasonal changes in soil moisture on GPR data [60–64]. Slowik [62] demonstrated that the depths of penetration were highest at low groundwater levels but the depth range containing useful information was the same as at high water levels. Furthermore, Slowik [65] conducted GPR on fluvial, lacustrine and anthropogenic landforms to determine the influence of hydrogeological conditions, silt content and measurements parameter settings on depth range and resolution. He determined that the highest depths of penetration were reached at low groundwater levels but some sedimentary structures were better imaged at high groundwater levels. Zhang et al. [35] showed that GPR reflections between different horizon interfaces became clearer as soil became wetter at one test location, while GPR reflections in the soil-bedrock interface and the weathered-unweathered rock interface became intermittent as soil became wetter at another test location. From the results of different studies, we can conclude that water content related to seasonal changes can have different effects on the GPR image. The results of the mentioned studies demonstrate the value of repeated GPR surveys in different field conditions to select the optimal time for GPR surveys for the specific terrain conditions and for a better insight in the interpretation of radargrams. However, the results of our study show that different field conditions did not substantially affect the quality of GPR data when detecting cave sediments which are highly conducted materials by themselves, irrespective of the water content during measuring.

The mineralogy of cave sediments and soils on carbonate rocks is similar, there are only small variations in the abundance of each mineral. Mineral composition of cave sediments includes more clay minerals (kaolinite, illite and vermiculite) and iron/aluminium oxides/hydroxides. Hematite causing the red colour was detected in some samples of cave sediments, while Na-plagioclase feldspar only occurred in soils on carbonate rocks. However, the mentioned differences in mineralogy, especially the proportion and the type of clay minerals and a higher content of Fe, Mn and Al elements, most likely contribute to increased attenuation in cave sediments compared to soils on carbonate rocks. However, mineral composition of soils and an abundance of different minerals play an important role in the performance of GPR. For example, Breiner et al. [22] observed that soils on a bedrock in granitic terrain, which contains more mafic minerals (5% hornblende, 20% biotite), are more attenuating compared to soils on a bedrock where mafic mineral content is lower (<1% hornblende and 20% biotite). They concluded that especially increased biotite content severely restricts the performance of GPR. Van Dam et al. [66] discovered that iron oxides lower electromagnetic wave velocity. Further laboratory analyses proved that the amount of iron oxide in a material correlates with volumetric water content and, therefore with dielectric properties. The correlation is caused by the larger specific surface and capillary-retention capacity of iron oxides like goethite and can thus have a profound influence on the GPR signal [66].

Based on the results of this study we can conclude that in addition to the thickness of the cave sediments, small differences in the amounts of a particular type of clay minerals can affect attenuation in the soils. Even very small quantities of hematite and other iron oxides and hydroxides greatly increase the overall soil conductivity. Apart from that, all mentioned factors indirectly affect increased water retention in areas of cave sediments compared to their surroundings. Variations in the abundance of clay minerals and a consequently higher water content severely restrict GPR signal propagation.

Distinction between cave sediments and other regolith bodies (sediments in dolines, filled shafts, cutters) in karst can be very difficult, especially since cave sediments can be very heterogeneous and similar to the soils as the product of bedrock weathering. For that reason, attenuated areas detected by GPR were labelled as having a possible presence of cave sediments, where superficial indicators for their presence are absent (Figure 18). GPR has limitations when the size of regolith bodies and their mineral composition are too similar, therefore caution is needed when interpreting the results. On the basis of GPR data in combination with geomorphological mapping and LiDAR data, certain karst features can be interconnected. However, additional profiles (outside forest paths) should be conducted in order to gain information needed for a proper determination of the geometrical extent of unroofed cave systems.

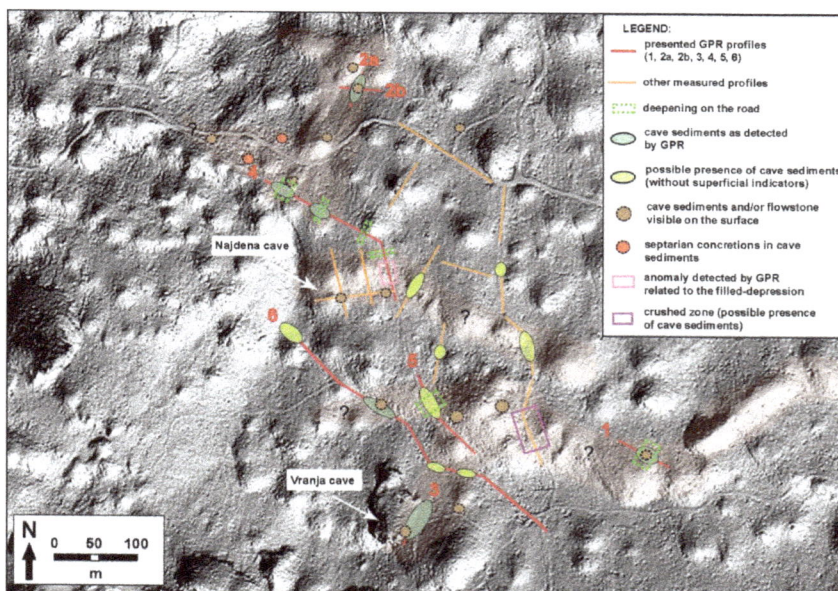

Figure 18. Profiles with interpreted anomalies detected by GPR and the occurrences of the superficial indicators of cave sediments.

6. Conclusions

Systematic mapping of unroofed cave features such as cave sediments and flowstones can yield a better insight into the spatial distribution of cave systems. It can contribute to the knowledge of former underground cave systems, their history and most importantly, to a better understanding of the karst development in the sense of an interconnection between the underground and the surface. Any new information helps us determine the origin of surface features including the most common form, a doline, the formation of which is still not clearly understood. Outcrops of cave sediments and other regolith bodies are usually limited in spatial extent, covered or completely reshaped due to surface processes. Most of the indicators for unroofed caves disappear over time. In areas where direct indicators for unroofed caves, such as cave sediments and flowstones, are preserved only locally, GPR has proven to be a useful method to determine the spatial extent of cave sediments. Results obtained by geological and geomorphological mapping in combination with GPR can thus help us recognize individual cave systems and reconstruct the direction of the former underground streams.

The results of XRD and XRF analyses show that mineralogical-geochemical characteristics of cave sediments are not obviously different from the characteristics of soils on carbonate rocks. However, we found that minor changes in mineralogical composition and an abundance of clay minerals can strongly affect the GPR image. The quantity and type of clay minerals are important factors affecting the dielectric properties and attenuation in the soil. As observed, minor differences can lead to significant changes in signal penetration.

Furthermore, greater amounts of clay minerals and the presence of small amounts of iron and aluminium oxides as well as hydroxides control the water quantity and its retention time in the soil, which are the key factors for increased attenuation in the areas of cave sediments.

Further detailed XRD analyses with changes such as a longer scanning time, Mg and K-saturation as well as glycerol solvation are in progress to distinguish all types of clay and mix-layered clay minerals. Some additional explorations of different locations for GPR profiling and for further mineralogical-geochemical analyses are planned since the mineral composition of cave sediments is quite heterogeneous. The suggested surveys will contribute to a better understanding of the impact of the mineralogical-geochemical composition on the GPR performance in sediments, where mineral composition and consequently a certain proportion of water are of great importance for material conductivity properties.

Acknowledgments: This study was conducted with the support of the research Program P1-0011 and the Ph.D. grant 1000-15-0510 financed by the Slovenian Research Agency. This work also benefited from networking activities carried out within the EU-funded COST Action TU1208 "Civil Engineering Applications of Ground Penetrating Radar." The authors are grateful to members of the Society for Cave Exploration from Ljubljana for their contribution during field measuring. We would also like to thank Jaka Flis for field assistance and France Šušteršič for comments and advice during the preparation of this paper and for sharing his field experience in karstology of the study area. We also highly appreciate the copy edits done by Kaja Bucik Vavpetič.

Author Contributions: Teja Čeru acquired and processed the GPR data and wrote Sections 1, 2, 3.3, 4.1–4.3, 4.5, 5 and 6. Matej Dolenec did the mineralogical and geochemical analyses and wrote Sections 3.2, 3.3 and 4.4. Andrej Gosar contributed to GPR measurements and data processing as well as to the preparation of the manuscript.

Conflicts of Interest: The authors declare no conflicts of interest.

References

1. Knez, M.; Slabe, T. Unroofed caves and recognising them in karst relief (discovered during motorway construction at Kozina, South Slovenia). *Acta Carsol.* **1999**, *28*, 103–112. [CrossRef]

2. Mihevc, A. *Uporaba Lidarskih Posnetkov v Geomorfologiji Krasa na Primeru Brezstropih Jam. Zbornik Posvetovanja Raziskave s Področja Geodezije in Geofizike 2015*; Slovensko združenje za geodezijo in geofiziko: Ljubljana, Slovenia, 2015; pp. 141–149.

3. Mihevc, A. Brezstropa jama pri Povirju. *Naše Jame* **1996**, *38*, 65–75.

4. Mihevc, A. *Speleogeneza Divaškega Krasa*; ZRC: Ljubljana, Slovenia, 2001; p. 180.

5. Mihevc, A.; Slabe, T.; Šebela, S. Denuded caves: An inherited element in the karst morphology; the case from Kras. *Acta Carsol.* **1998**, *27*, 165–174.

6. Šušteršič, F. Interaction between a cave system and the lowering karst surface; case study: Laški ravnik. *Acta Carsol.* **1998**, *27*, 115–138. [CrossRef]

7. Knez, M.; Slabe, T. Unroofed caves are an important feature of karst surfaces: Examples from the classical karst. *Z. Geomorphol.* **2002**, *46*, 181–191.

8. Mais, K. Roofless caves, a polygenetic status of cave development with special references to cave regions in the Eastern Calcareous Alps in Salzburg and Central Alps, Austria. *Acta Carsol.* **1999**, *28*, 145–158. [CrossRef]

9. Osborne, R.A.L. Karst geology of Wellington Caves: A review. *Helictite* **2001**, *37*, 3–12.

10. Klimchouk, A. Cave un-roofing as a large-scale geomorphic process. *Speleogenesis Evolution Karst Aquifers* **2006**, *4*, 1–11.

11. Ufrecht, W. Evaluating landscape development and karstification of the Central Schwäbische Alb (Southwest Germany) by fossil record of karst fillings. *Z. Geomorphol.* **2008**, *52*, 417–436. [CrossRef]

12. Knez, M.; Slabe, T. *Cave Exploration in Slovenia: Discovering Over 350 New Caves during Motorway Construction on Classical Karst, (Cave and Karst Systems of the World)*; Springer: Cham, Switzerland, 2016; p. 324.

13. Čeru, T.; Šegina, E.; Knez, M.; Benac, Č.; Gosar, A. Detecting and characterizing unroofed caves by ground penetrating radar. *Geomorphology* **2018**, *303*, 524–539. [CrossRef]

14. White, W.B. Cave sediments and paleoclimate. *J. Cave Karst Stud.* **2007**, *69*, 76–93.

15. Zupan Hajna, N.; Pruner, P.; Mihevc, A.; Schnabl, P.; Bosák, P. Cave sediments from the Postojnska-Planinska cave system (Slovenia): Evidence of multi-phase evolution in epiphreatic zone. *Acta Carsol.* **2008**, *37*, 63–86. [CrossRef]

16. Sasowsky, I.D. Clastic sediments in caves—Imperfect recorders of processes in karst. *Acta Carsol.* **2007**, *36*, 143–149. [CrossRef]

17. Arriolabengoa, M.; Iriarte, E.; Aranburu, A.; Yusta, I.; Arrizabalaga, A. Provenance study of endokarst fine sediments through mineralogical and geochemical data (Lezetxiki II cave, northern Iberia). *Quat. Int.* **2015**, *364*, 231–243. [CrossRef]

18. Martini, I. Cave clastic sediments and implications for speleogenesis: New insights from the Mugnano cave (Montagnola Senese, Northern Apennines, Italy). *Geomorphology* **2011**, *134*, 452–460. [CrossRef]

19. Bónová, K.; Bella, P.; Bóna, J.; Spišiak, J.; Kováčik, M.; Kováčik, M.; Petro, L. Heavy minerals in sediments from the Mošnica Cave: Implications for the pre-Quaternary evolution of the middle-mountain allogenic karst in the Nízke Tatry Mts., Slovakia. *Acta Carsol.* **2014**, *43*, 297–317. [CrossRef]

20. Šušteršič, F.; Rejšek, K.; Mišič, M.; Eichler, F. The role of loamy sediment (terra rossa) in the context of steady state karst surface lowering. *Geomorphology* **2009**, *106*, 35–45. [CrossRef]

21. Mihevc, A.; Zupan Hajna, N. Clastic sediments from dolines and caves found during the construction of the motorway near Divača, on the Classical Karst. *Acta Carsol.* **1996**, *25*, 169–191.

22. Breiner, J.M.; Doolittle, J.A.; Horton, R.M.; Graham, R.C. Performance of ground-penetrating radar on granitic regoliths with different mineral composition. *Soil Sci.* **2011**, *176*, 435–440. [CrossRef]

23. Van Dam, R.M.; Hendrickx, J.M.H.; Cassidy, N.J.; North, R.E.; Dogan, M.; Borchers, B. Effects of magnetite on high-frequency ground-penetrating radar. *Geophysics* **2013**, *78*, H1–H11. [CrossRef]

24. Van Dam, R.L. Causes of ground-penetrating radar reflections in sediment. Ph.D. Thesis, Vrije Universiteit, Amsterdam, The Netherlands, 2001; p. 101.

25. Čar, J. Geološka zgradba požiralnega obrobja Planinskega polja (Geological setting of the Planina polje ponor area). *Acta Carsol.* **1981**, *10*, 75–105.

26. Šušteršič, F. A conceptual model of dinaric solution doline dynamics. *Cave Karst Sci.* **2017**, *44*, 66–75.

27. Buser, S.; Grad, K.; Pleničar, M. *Basic Geological Map of Yugoslavia, Sheet Postojna, L33–77*; Federal Geological Survey of Beograd: Beograd, Srbija, 1967.

28. LiDAR (Public Information of Slovenia, the Surveying and Mapping Authority of the Republic of Slovenia, LiDAR, 2015) ARSO, Ministry of the Environment. Available online: http://gis.arso.gov.si/evode/profile.aspx?id=atlas_voda_Lidar@Arso (accessed on 5 March 2018).

29. Šušteršič, F. Cave Sediments and Denuded Caverns in the Laški Ravnik, Classical Karst of Slovenia. In *Studies of Cave Sediments: Physical and Chemical Records of Paleoclimate*; Sasowsky, I.D., Mylroie, J., Eds.; Springer: New York, NY, USA, 2004; pp. 123–134.

30. Silva, O.L.; Bezerra, F.H.R.; Maia, R.P.; Cazarin, C.L. Karst landforms revealed at various scales using LiDAR and UAV in semi-arid Brazil: Consideration on karstification processes and methodological constraints. *Geomorphology* **2017**, *295*, 611–630. [CrossRef]

31. Kruse, S.; Grasmueck, M.; Weiss, M.; Viggiano, D. Sinkhole structure imaging in covered karst terrain. *Geophys. Res. Lett.* **2006**, *33*, L16405. [CrossRef]

32. Pueyo-Anchuela, Ó.; López Julián, P.L.; Casas-Sainz, A.M.; Liesa, C.L.; Pocoví-Juan, A.; Ramajo Cordero, J.; Perez Benedicto, J.A. Three dimensional characterization of complex mantled karst structures. Decision making and engineering solutions applied to a road overlying evaporite rocks in the Ebro Basin (Spain). *Eng. Geol.* **2015**, *193*, 158–172. [CrossRef]

33. Guidry, S.A.; Grasmueck, M.; Carpenter, D.G.; Gombos, A.M. Jr.; Bachtel, S.L.; Viggiano, D.A. Karst and early fracture networks in carbonates, Turks and Caicos Islands, British West Indies. *J. Sediment. Res.* **2007**, *77*, 508–524. [CrossRef]

34. Fernandes, A.L.; Medeiros, W.E.; Bezerra, F.H.R.; Oliveira, J.G.; Cazarin, C.L. GPR investigation of karst guided by comparison with outcrop and unmanned aerial vehicle imagery. *J. Appl. Geophys.* **2015**, *112*, 268–278. [CrossRef]

35. Zhang, J.; Lin, H.; Doolittle, J. Soil layering and preferential flow impacts on seasonal changes of GPR signals in two contrasting soils. *Geoderma* **2014**, *213*, 560–569. [CrossRef]

36. Simeoni, M.A.; Galloway, P.D.; O'Neil, A.J.; Gilkes, R.J. A procedure for mapping the depth to the texture contrast horizon of duplex soils in south-western Australia using ground penetrating radar, GPRS and kriging. *Aust. J. Soil Res.* **2009**, *47*, 613–621. [CrossRef]

37. Doolittle, J.A.; Collins, M.E. A comparison of EM induction and GPR methods in areas of karst. *Geoderma* **1998**, *85*, 83–102. [CrossRef]

38. Stroh, J.C.; Archer, S.; Doolittle, J.A.; Wilding, L. Detection of edaphic discontinuities with ground-penetrating radar and electromagnetic induction. *Landsc. Ecol.* **2001**, *16*, 377–390. [CrossRef]

39. Čeru, T.; Šegina, E.; Gosar, A. Geomorphological Dating of Pleistocene Conglomerates in Central Slovenia Based on Spatial Analyses of Dolines Using LiDAR and Ground Penetrating Radar. *Remote Sens.* **2017**, *9*, 1213. [CrossRef]

40. Gish, T.J.; Walthall, C.L.; Daughtry, C.S.; Kung, K.J. Using soil moisture and spatial yield patterns to identify subsurface flow pathways. *J. Environ. Qual.* **2005**, *34*, 274–286. [CrossRef] [PubMed]

41. Huisman, J.A.; Snepvangers, J.J.J.C.; Bouten, W.; Heuvelink, G.B.M. Mapping spatial variation in surface soil water content: Comparison of ground-penetrating radar and time domain reflectometry. *J. Hydrol.* **2002**, *269*, 194–207. [CrossRef]

42. Kowalsky, M.B.; Finsterle, S.; Rubin, Y. Estimating flow parameter distributions using ground-penetrating radar and hydrological measurements during transient flow in the vadose zone. *Adv. Water Res.* **2004**, *27*, 583–599. [CrossRef]

43. Jol, H.M. *Ground Penetrating Radar: Theory and Applications*, 1st ed.; Elsevier Science: Amsterdam, The Netherlands, 2009; p. 524.

44. Ehsani, M.R.; Daniels, J.J.; Allred, B.J. *Handbook of Agricultural Geophysics*, 1st ed.; CRC Press: Boca Raton, FL, USA, 2008; p. 432.

45. Waltham, T.; Bell, F.; Culshaw, M. *Sinkoles and Subsidence. Karst and Cavernous Rocks in Engineering and Construction*; Springer: Berlin, Germany, 2010; pp. 181–204.

46. Moore, D.M.; Reynolds, R.C. *X-Ray Diffraction and the Identification and Analysis of Clay Minerals*, 2nd ed.; Oxford University Press: Oxford, UK; New York, NY, USA, 1997; p. 378.

47. Goodman, D.; Piro, S. *GPR Remote Sensing in Archaeology, Geotechnologies and the Environment*; Springer: Heidelberg, Germany, 2013.

48. Krebs, C.J. *Ecological Methodology*; Addison Wesly Longman: New York, NY, USA, 1998.

49. Li, W.; Cui, X.; Guo, L.; Chen, J.; Chen, X.; Cao, X. Tree Root Automatic Recognition in Ground Penetrating Radar Profiles Based on Randomized Hough Transform. *Remote Sens.* **2016**, *8*, 430. [CrossRef]

50. Bernhardson, M.; Alexanderson, H. Early Holocene dune field development in Dalarna, central Sweden: A geomorphological and geophysical case study. *Earth Surf. Process. Landf.* **2017**, *42*, 1847–1859. [CrossRef]

51. Sinclair, S.N.; Licciardi, J.M.; Campbell, S.W.; Madore, B.M. Character and origin of De Geer moraines in the Seacoast region of New Hampshire, USA. *J. Quaternary Sci.* **2018**, *33*, 225–237. [CrossRef]

52. Nooren, K.; Hoek, W.Z.; Winkels, T.; Huizinga, A.; Van der Plicht, H. The Usumacinta–Grijalva beach-ridge plain in southern Mexico: A high-resolution archive of river discharge and precipitation. *Earth Surf. Dynam.* **2017**, *5*, 529–556. [CrossRef]

53. Oliver, T.S.N.; Tamura, T.; Hudson, J.P.; Woodroffe, C.D. Integrating millennial and interdecadal shoreline changes: Morpho-sedimentary investigation of two prograded barriers in southeastern Australia. *Geomorphpology* **2017**, *288*, 129–147. [CrossRef]

54. Kasprzak, M.; Sobczyk, A. Searching for the void: Improving cave detection accuracy by multi-faceted geophysical survey reconciled with LiDAR DTM. *Z. Geomorphol.* **2017**, *61*, 45–59. [CrossRef]

55. Helfricht, K.; Kuhn, M.; Keuschnig, M.; Heilig, A. Lidar snow cover studies on glaciers in the Ötztal Alps (Austria): Comparison with snow depths calculated from GPR measurements. *Cryosphere* **2014**, *8*, 41–57. [CrossRef]

56. Colucci, R.R.; Forte, E.; Boccali, C.; Dossi, M.; Lanza, L.; Pipan, M.; Guglielmin, M. Evaluation of internal structure, volume and mass of glacial bodies by integrated lidar and ground penetrating radar surveys: The case study of Canin eastern glacieret (Julian Alps, Italy). *Surv. Geophys.* **2015**, *36*, 231–252. [CrossRef]

57. Malehmir, A.; Andersson, M.; Mehta, S.; Brodic, B.; Munier, R.; Place, J.; Maries, G.; Smith, C.; Kamm, J.; Bastani, M.; et al. Post-glacial reactivation of the Bollnäs fault, central Sweden—A multidisciplinary geophysical investigation. *Solid Earth* **2016**, *7*, 509–527. [CrossRef]

58. Bubeck, A.; Wilkinson, M.; Roberts, G.P.; Cowie, P.A.; McCaffrey, K.J.W.; Phillips, R.; Sammonds, P. The tectonic geomorphology of bedrock scarps on active normal faults in the italian apennines mapped using combined ground penetrating radar and terrestrial laser scanning. *Geomorphology* **2015**, *237*, 38–51. [CrossRef]

59. Leucci, G.; Margiotta, S.; Negri, S. Geophysical and geological investigations in a karstic environment (Salice Salentino, Lecce, Italy). *J. Environ. Eng. Geophys.* **2004**, *9*, 25–34. [CrossRef]

60. Boll, J.; van Rijn, R.P.G.; Weiler, K.W.; Ewen, J.A.; Daliparthy, J.; Herbert, S.J.; Steenhuis, T.S. Using ground-penetrating radar to detect layers in a sandy field soil. *Geoderma* **1996**, *70*, 117–132. [CrossRef]

61. Lunt, I.A.; Hubbard, S.S.; Rubin, Y. Soil moisture content estimation using ground-penetrating radar reflection data. *J. Hydrol.* **2005**, *307*, 254–269. [CrossRef]

62. Słowik, M. Influence of measurement conditions on depth range and resolution of GPR images: The example of lowland valley alluvial fill (the Obra River, Poland). *J. Appl. Geophys.* **2012**, *85*, 1–14. [CrossRef]

63. Tatum, D.; Francke, J. Radar suitability in aeolian sand dunes—A global review. In Proceedings of the 14th International Conference on Ground Penetrating Radar, Shanghai, China, 4–8 June 2012; Volume III, pp. 1–706. [CrossRef]

64. Truss, S.; Grasmueck, M.; Vega, S.; Viggiano, D.A. Imaging rainfall drainage within the Miami oolitic limestone using high-resolution time-lapse ground-penetrating radar. *Water Resour. Res.* **2007**, *43*, 1–15. [CrossRef]

65. Słowik, M. Analysis of fluvial, lacustrine and anthropogenic landforms by means of ground-penetrating radar (GPR): Field experiment. *Near Surf. Geophys.* **2014**, *12*, 777–791. [CrossRef]

66. Van Dam, R.L.; Schlager, W.; Dekkers, M.J.; Huisman, J.A. Iron oxides as a cause of GPR reflections. *Geophysics* **2002**, *67*, 536–545. [CrossRef]

MDPI

St. Alban-Anlage 66

4052 Basel

Switzerland

Tel. +41 61 683 77 34

Fax +41 61 302 89 18

www.mdpi.com

Remote Sensing Editorial Office

E-mail: remotesensing@mdpi.com

www.mdpi.com/journal/remotesensing

www.ingramcontent.com/pod-product-compliance
Lightning Source LLC
Chambersburg PA
CBHW051845210326
41597CB00033B/5783